**RECORD COPY**
Do not remove from office

La Consentida

# *La* CONSENTIDA

*Settlement, Subsistence, and Social Organization in an Early Formative Mesoamerican Community*

GUY DAVID HEPP

UNIVERSITY PRESS OF COLORADO
*Louisville*

© 2019 by University Press of Colorado

Published by University Press of Colorado
245 Century Circle, Suite 202
Louisville, Colorado 80027

All rights reserved
Printed in the United States of America

 The University Press of Colorado is a proud member of the Association of University Presses.

The University Press of Colorado is a cooperative publishing enterprise supported, in part, by Adams State University, Colorado State University, Fort Lewis College, Metropolitan State University of Denver, Regis University, University of Colorado, University of Northern Colorado, Utah State University, and Western State Colorado University.

∞ This paper meets the requirements of the ANSI/NISO Z39.48–1992 (Permanence of Paper).

ISBN: 978-1-60732-852-0 (cloth)
ISBN: 978-1-60732-853-7 (ebook)
DOI: https://doi.org/10.5876/9781607328537

Library of Congress Cataloging-in-Publication Data

Names: Hepp, Guy David, author.
Title: La Consentida : settlement, subsistence, and social organization in an Early Formative Mesoamerican community / Guy David Hepp.
Description: Louisville : University Press of Colorado, [2019] | Includes bibliographical references and index.
Identifiers: LCCN 2018056062| ISBN 9781607328520 (cloth) | ISBN 9781607328537 (ebook)
Subjects: LCSH: Indians of Mexico—Mexico—Oaxaca (State)—Antiquities. | Excavations (Archaeology)—Mexico—Oaxaca (State) | La Consentida Archaeological Project | Indians of Mexico—Mexico—Oaxaca (State)—Economic conditions. | Indians of Mexico—Mexico—Oaxaca (State)—Social life and customs. | Indians of Mexico—Agriculture—Mexico—Oaxaca (State) | Indigenous peoples—Mexico—Oaxaca (State) | Oaxaca (Mexico : State)—Antiquities.
Classification: LCC F1219.1.O11 H45 2019 | DDC 972/.7401—dc23
LC record available at https://lccn.loc.gov/2018056062

Cover photograph by the author.

To my family
and
a los costeños,
a los de hoy y a los de ayer,
a los nativos y a los recién llegados

# Contents

| | |
|---|---|
| *Acknowledgments* | ix |
| 1 La Consentida as an Early Formative Mesoamerican Village | 3 |
| 2 Debating the Early Formative Period | 17 |
| 3 Methods and Mapping for the La Consentida Archaeological Project | 52 |
| 4 La Consentida's Occupational History | 68 |
| 5 Settling Down: The Shift to Sedentism | 85 |
| 6 Diet and Changing Culinary Tastes | 104 |
| 7 Social Organization: Diverse Identities at La Consentida | 135 |
| 8 No Village Is an Island: Interregional Interaction and Exchange | 171 |
| 9 La Consentida: A Community in Transformation | 194 |
| *Appendix 1: Description of Excavated Deposits* | 205 |
| *Appendix 2: The Tlacuache Ceramic Assemblage* | 237 |
| *Notes* | 275 |
| *Works Cited* | 279 |
| *Index* | 317 |

# Acknowledgments

Archaeology is not really the work of one person, but instead is a collaborative effort of field researchers, local people, permitting agencies, funding sources, colleagues and friends, family, and mentors. In undertaking the research presented in this book, I have benefitted from countless brainstorming sessions, studies by and conversations with other scholars, and unwavering support from friends and family. It is impossible to name everyone who has influenced my work, but I will use this opportunity to mention a few. First, I want to thank the funding agencies that supported this project. I am grateful to the National Science Foundation, the Fulbright–García Robles Scholarship committee, the University of Colorado Graduate School and the Department of Anthropology, the CU Latin American Studies Center, the Center to Advance Research and Teaching in the Social Sciences, the Colorado Archaeological Society, the Montrose Community Foundation, the Florida State University Department of Anthropology, and California State University, San Bernardino. Without financial backing and other support from these groups, I could not have completed this book. I also thank the Instituto Nacional de Antropología e Historia (INAH) for permitting this research. I would also like to thank the Society for American Archaeology for recognizing the dissertation on which this book is based with their 2016 Dissertation Award.

The La Consentida Archaeological Project (LCAP) has benefitted from the labors of a diverse group of scientists and field assistants. LCAP volunteer and staff archaeologists included Kyle Urquhart, Susan Chandler-Reed, Alan Reed, Martín

Cuitzeo Domínguez Núñez, Adam Andrus, and Jamie Forde. These assistants oversaw labor teams, organized artifacts, completed paperwork, made scale drawings, took photographs, and helped excavate important features. Sarah "Stacy" Barber of the University of Central Florida generously provided ground-penetrating radar and mapping equipment at key points in the 2009 and 2012 field seasons. Carlo Lucido and Jeff Brzezinski helped with mapping, and Carlo returned for some crucial profile drawing. José "Pepe" Aguilar provided bioarchaeological expertise and companionship as the 2012 field season drew to a close and I lived alone in San José del Progreso for months with only broken pottery and tarantulas for company. All fieldwork was made possible with the help of local workers. Over several seasons of research, I worked with many residents of San José del Progreso and the modern town of La Consentida. The local knowledge, funny stories, delicious and strange foods, and music they shared with me are among my many reasons for returning to Oaxaca season after season. Although there is not room to mention everyone who worked on the project, particularly good friends among them include Felix Herrera, Giovanni Pinto, Angel "Chucho" Reyes Soriano, Rey David Rosario, Anselmo Ramos, Ramiro Pinto, Jesús Pacheco, and Leticia Gómez. Without the generous use of the land and bridge owned by the Soltelo brothers (Vicente and Gerónimo), our team would have been swimming across a crocodile-infested canal to reach the site every day.

My conversations with other Mesoamericanists about the results of research at La Consentida have been helpful in refining my thinking about the site. Although these scholars haven't always agreed with my results or with my interpretations of them, conversations with Marc Winter, Cira Martínez López, Joseph Mountjoy, Rosemary Joyce, Jon Lohse, Julia Guernsey, Payson Sheets, Jeffrey Blomster, and John Clark (to name but a few) have influenced the chapters that follow and my thinking about Mesoamerica in general. I am also grateful for the mentorship of my PhD advisor, Arthur Joyce. His advice and encouragement over about the last fifteen years have been driving forces, and I'm glad he introduced me to La Consentida.

The LCAP has included collaboration with a local community museum (Museo Comunitario Yucusaa / Tututepec) and resulted in the production of an educational display at the museum. I've been invited to give several public talks and radio interviews about the project, both on the coast and in and around Oaxaca City. I am indebted to Robin Cleaver, Barbara Cleaver, Paul Nunn Cleaver, Roberto Lepe, Gina Machorro, Sheila Clarke, and Adriana Giraldo for helping make these presentations possible. My time at the Cuilapan de Guerrero INAH research facility near Oaxaca City was important for the laboratory studies that helped form this study. I'm grateful to many of the permanent and visiting denizens of Cuilapan, including Marc Winter, Cira Martínez López, José Cervantes Pérez, Félix Rivera Pacheco,

Héctor López Calvo, Gonzalo Sánchez Santiago, and Ismael Vicente Cruz for their insight, suggestions, friendship, and for inviting a mere gringo to play on their *fútbol* team. Biologist Silvia Pérez Hernández's participation in the 2014 faunal study was a key addition to investigating La Consentida's diet. Back in Colorado, Paul Sandberg and Jim Millette helped me tackle the data with suggestions for statistical analyses.

My first visit to La Consentida was in summer 2008, when Art Joyce led a crew of researchers on something of a vision quest into the coastal jungle of the Parque Nacional de Chacahua. I was already almost sure I wanted to do my dissertation work at the site. After hours of searching for La Consentida, we finally stumbled (hot, sweating, and exhausted) upon the main platform. In subsequent years, La Consentida has often seemed to shift location, even eluding rediscovery when GPS points should lead me right to it. Perhaps the site likes to hide until those who seek her pay an ample blood offering to the fire ants, mosquitos, killer bees, and ticks. Such journeys are allegorical of this project as a whole. At times, organizing research that would do justice to La Consentida has felt like an overwhelming task. Some days the mosquitos seemed to exact more than their fair toll. It is to the support and participation of people I've mentioned here (and others) that I owe much of the credit for seeing this through.

Finally, and at the risk of falling into cliché, I want to thank the ancient La Consentida community itself. My curiosity about the lives of these early Mesoamerican people, which I believe were lived at a pivotal moment in the history of the Americas, has never diminished despite the challenges of the project. Their art, beliefs, and modes of making a living in a difficult climate and without modern amenities have fascinated and bolstered me when the project seemed most daunting. It is in this spirit of interest in and passion about the lives of Oaxaca's people (both ancient and modern) that I dedicate this book to the memory of two friends I made and lost over the course of the research I present here, Paul Cleaver and Angel "Chucho" Reyes Soriano. Although he was a transplant, Paul loved Oaxaca as much as anyone I have ever met. He had an appetite for knowledge, and a drive to establish a Puerto Escondido intelligentsia, that I will never forget. He seemed to me one of the last real Renaissance men and appeared to have at least some knowledge about every topic I could think of to discuss with him. The stories he told at his bed and breakfast, El Hotel Tabachín, were so fantastical that I was never sure which ones to believe and would often jot down notes for later fact checking. Did he really meet Che Guevara? Was he really friends with novelists and poets of the Beat Generation? We did not know each other long enough, but I'm glad to call him a friend. A totally different sort of character, Chucho was like a coiled spring of potential kinetic energy unleashed upon the fieldwork of this project. With him, my job as a project director was to channel this energy and force of personality to the benefit of the research. I didn't realize at first that we were also forging a friendship. His smile, his guileless way of making friends, and his gentleness with all manner of animals (including the deadly ones) remain in the minds of those who knew him. In early fall 2015, aged only in his early twenties, Chucho fell victim to an absurd brand of violence that mars the otherwise enduring peace and beauty of Pacific coastal Mexico. Pablo and Chucho, you won't easily be forgotten.

# La Consentida

# 1

## La Consentida as an Early Formative Mesoamerican Village

The Mesoamerican Early Formative period (approx. 2000–1000 cal BC) was a time of social transformation.[1] In the preceding Archaic period (approx. 7000–2000 cal BC), mobile hunter-gatherers had moved seasonally across the landscape and experimented with a few domesticates, such as squash, maize, beans, and root crops. By the end of the Formative, Mesoamericans lived in permanent towns and cities, relied on agriculture, and were ruled by powerful royal dynasties. The Late Archaic and Early Formative periods set the stage for these dramatic changes, but the exact timing of and the possible connections between transitions to sedentism, agriculture, and social complexity are debated in Mesoamerican archaeology (Blake and Clark 1999; Clark 2004a; Clark et al. 2007; Killion 2013; Lesure and Blake 2002; Love 2007; Webster 2011). Traditionally, sedentism has been seen as beginning with the Early Formative and as hastening ethnolinguistic divergence among previously mobile and fluid cultural groups (Flannery and Marcus 2003; Hopkins 1984; Winter et al. 1984). Arguably, the extreme terrain in parts of Mesoamerica would tend to accentuate cultural and language differences among increasingly sedentary communities. Some recent scholarship suggests, however, that certain populations remained semi-mobile during the Early Formative, despite interactions with their agrarian neighbors (e.g., P. Arnold 1999; Rosenswig 2011) and that contact and exchange among Formative period communities was more complex than implied by models of sedentary isolationism (Blomster 2004; Pool et al. 2010). Although some researchers (Coe 1981; Coe and Flannery 1967; Sanders and Webster 1978) have argued that

DOI: 10.5876/9781607328537.c001

the economic basis for sedentism was maize agriculture (supplemented with other crops such as squash and beans), others (e.g., P. Arnold 2009; Blake et al. 1992; Clark et al. 2007; Smalley and Blake 2003; VanDerwarker 2006) propose that maize in coastal zones was a feasting food that, along with other, limited-use horticultural products, supplemented a broad diet consisting mostly of wild resources collected in estuarine or floodplain settings. The origin of Mesoamerican social complexity is another topic of disagreement. The timing of initial complexity differed among regions, with areas such as Mazatán apparently experiencing hierarchical hereditary inequality by about 1600 cal BC (Clark 1991, 1997; Hill and Clark 2001). In comparison, the Gulf Coast region likely did not see such formalized hierarchies until later in the Early Formative, and regions of the Soconusco outside of Mazatán did not do so until the Middle Formative (1000–400 BCE) (Love 2002; Pool 2007). Traditional definitions of social complexity focus on such hereditary hierarchies (Feinman and Marcus 1998; Spencer and Redmond 2004), while a smaller group of researchers has considered the ways in which complex heterarchical distinctions influence social landscapes (e.g., Crumley 1995, 2004; McIntosh 1999; Pauketat and Emerson 2007; Vega-Centeno Sara-Lafosse 2007:169). My purpose in enumerating these discussions is not to choose sides in all cases, but rather to demonstrate that the archaeology of the Early Formative period is very much an ongoing discussion.

Early Formative sites are found in diverse ecological settings across Mesoamerica (figure 1.1). Many coastal sites occur near estuaries, especially in the Soconusco region (Blake and Clark 1999; Clark 2004a; Lesure 2009, 2011a). Despite decades of research in various environmental and geographic settings, large areas (such as much of Mexico's Pacific coast) remain enigmas regarding Early Formative history. This circumstance has resulted in explanatory models for Early Formative social transitions that are based on research in only a few regions. Worldwide, the establishment of villages[2] (and the dietary and social implications of that process) presents major archaeological research problems, but no consensus exists as to its causes (e.g., Banning 2003; Bar-Yosef and Belfer-Cohen 1989; Binford 1968; Boyd 2006; Byrd 1994; Choe and Bale 2002; Flannery and Marcus 2012; Joyce and Henderson 2001; McClung de Tapia and Zurita-Noguera 2000; Weisdorf 2005). In the hopes of addressing these issues, I have asked with the research summarized in this book what relationships existed between settlement, subsistence, and social organization at La Consentida, an Early Formative period site in coastal Oaxaca, Mexico.

The majority of the investigations at La Consentida (under the aegis of the La Consentida Archaeological Project, or LCAP) have taken place during seven field and laboratory seasons totaling over twenty months of research (Hepp 2011a, 2011b, 2014, 2015; Hepp and Joyce 2013; Hepp and Reiger 2014; Hepp et al. 2017). Based on seven Early Formative AMS radiocarbon dates (table 1.1), which provide

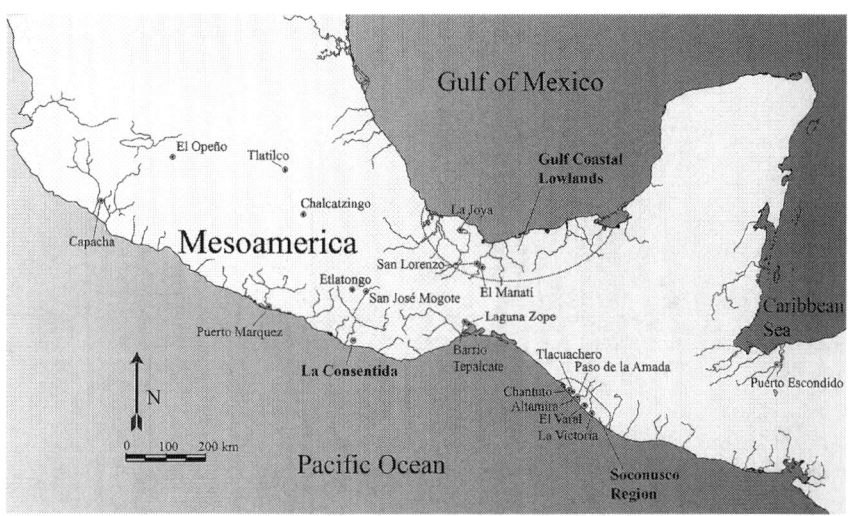

**FIGURE 1.1.** Map of key sites mentioned in the text

a calibrated date range of 1950–1525 cal BC, when reported with 2σ probability and 1885–1565 cal BC, when reported with 1σ probability,[3] La Consentida represents the earliest well-dated Formative period site discovered thus far in Oaxaca (Hepp 2011a, 2015, in press; A. Joyce 2010:71–72).[4] The contexts from which these radiocarbon samples were collected are stratigraphically controlled and are unequivocally associated with ceramics, mounded earthen architecture, and formal mortuary contexts that I interpret as early cemeteries. Dated deposits include well-preserved hearths sealed between layers of platform fill, burned food adhering to a jar fragment from a midden, and bone collagen from a human burial (see table 1.1, figure 1.2, chapter 3). With the exception of an eighth sample, which was likely contaminated, the radiocarbon dates are quite consistent across the site. More specific details about the dated contexts (and all strata excavated during the LCAP) can be found in chapter 4 and appendix 1. These radiocarbon dates are older than those for other Early Formative Oaxacan deposits of the Tierras Largas (1400–1200 BCE, or 1650–1500 cal BC) and Lagunita (1500–1100 BCE, or 1750–1350 cal BC) phases (table 1.2).[5] Some (e.g., Flannery and Marcus 1994:375) have proposed that the Espiridión phase predates Tierras Largas, though it has produced no radiocarbon dates and is now in question as truly distinct from Tierras Largas.[6] Radiocarbon dates also establish La Consentida as contemporary with the Barra phase (1900–1700 cal BC) of the Soconusco. Comparison of these phases demonstrates that La Consentida has yielded some of Mesoamerica's

**TABLE 1.1.** AMS radiocarbon dates from La Consentida (calibrated with IntCal 13 curve by OxCal 4.3.2). Reported with both 1σ and 2σ probability and rounded to five-year increments.

| AMS radiocarbon date (uncalibrated) | 2σ calibration | 1σ calibration | Material / Lab number | Context |
|---|---|---|---|---|
| 3480 ± 60 (A. Joyce 2005:17) | 1950–1640 cal BC | 1885–1740 cal BC (p = .64)<br>1715–1695 cal BC (p = .04) | Wood carbon (Beta-131037) | Floor or occupation layer |
| 3480 ± 40 | 1900–1690 cal BC | 1880–1835 cal BC (p = .24)<br>1830–1745 cal BC (p = .45) | Carbon-rich sediment (AA92453) | LC09 A-F4 hearth in Platform 1 |
| 3445 ± 35 | 1885–1665 cal BC | 1870–1845 cal BC (p = .13)<br>1810–1800 cal BC (p = .02)<br>1780–1690 cal BC (p = .54) | Wood carbon (AA101267) | LC12 A-F19 occupation layer |
| 3435 ± 45 | 1885–1635 cal BC | 1875–1845 cal BC (p = .11)<br>1815–1800 cal BC (p = .03)<br>1780–1680 cal BC (p = .54) | Carbon-rich sediment (AA101269) | LC12 E-F10 probable hearth in midden |
| 3420 ± 35 | 1880–1840 cal BC (p = .08)<br>1825–1795 cal BC (p = .04)<br>1785–1625 cal BC (p = .84) | 1766–1660 cal BC | Carbonized food (AA104836) | Burned food adhering to pottery from LC12 H-F4-s2 midden |
| 3360 ± 45 | 1755–1525 cal BC | 1740–1710 cal BC (p = .10)<br>1700–1610 cal BC (p = .58) | Carbon-rich sediment (AA92454) | LC09 B-F15 hearth in Platform 1 |
| 3335 ± 20 | 1690–1600 cal BC (p = .76)<br>1585–1530 cal BC (p = .20) | 1665–1610 cal BC (p = .65)<br>1575–1565 cal BC (p = .03) | Collagen from human femur (PRI-5423A/B [H6]) | Direct date on human remains from burial B12-I14 using XAD purification[1] |
| 2435 ± 35 | 755–680 cal BC (p = 0.22)<br>670–610 cal BC (p = 0.11)<br>595–405 cal BC (p = 0.63) | 730–690 cal BC (p = .14)<br>660–650 cal BC (p = .02)<br>545–410 cal BC (p = .52) | Carbon-rich sediment (AA101268) | Above domestic structure floor[2] |

[1] Refer to chapter 3 for a discussion of the methods used to process this human bone date.

[2] This date is considered suspect, based on its shallow deposition and proximity to modern plant roots. The sample may have been contaminated or may represent a burning event subsequent to site abandonment. It also occurs at a plateau on the calibration curve.

earliest known ceramics, mounded earthen architecture, and cemetery contexts (table 1.2). The site thus provides a unique opportunity to address debates in Early Formative period studies. As I discuss in chapters 8 and 9, the early dates complicate current models for the adoption of ceramics in Mesoamerica (e.g., Clark and Blake 1994), and suggest that there may have been two contemporaneous pottery traditions in the region by as early as 1900 cal BC.

## SITE AND REGIONAL BACKGROUND

The lower Río Verde Valley is located on the western Oaxaca coast (figure 1.3). Although sediment cores indicate maize cultivation and anthropogenic land clearance going back into the late Archaic, archaeological research since the 1980s has suggested that the region was sparsely populated until the Middle Formative period (Goman et al. 2005, 2013; A. Joyce 1991a; Joyce and Goman 2012). Contrary to recently published reports, however, it is not true that there was "virtually no occupation during the Archaic or Early Formative period" (see Rosenswig 2015:135). The region is best known ethnohistorically for the site of Tututepec, the capital of a Postclassic period (800–1521 CE) Mixtec empire (Joyce et al. 2004; Levine 2007, 2011; Spores 1993). Before the arrival of the Mixtecs, the area saw several periods of centralization and destabilization with a settlement and political hierarchy centered at Río Viejo, the seat of short-lived Terminal Formative (150 BCE–200 CE) and Late Classic period (500–800 CE) polities (Barber and Joyce 2007; A. Joyce 1991a, 2005, 2006, 2010, 2013a).

Prior to recent research at La Consentida, little was known about Early and Middle Formative sites in the lower Río Verde Valley. La Consentida was initially discovered by archaeologists during a regional reconnaissance in 1986 (A. Joyce 1991a:85, 116–17). The site is located about 6.5 km from the modern Pacific coastline and falls within the boundaries of the Chacahua National Park.[7] La Consentida is named after a small town located between the park and the local stretch of Mexico's Highway 200. During the Early Formative period, the site was probably positioned within about 4 kilometers of an open bay (Goman et al. 2005, 2013; A. Joyce and Goman 2012). Based on artifacts and earthen architecture visible at the surface, La Consentida covers at least 4.5 ha and is dominated by an earthen structure (Platform 1) measuring approximately 300 × 100 × 5 m.[8]

Preliminary work at La Consentida in 1988 (A. Joyce 1991a:406, 2005; Winter 1989) formed part of the Río Verde Formative Project and included surface collections, sediment sampling, and excavation of a single test unit. A charcoal sample from this excavation, which was performed atop the western edge of Platform 1, produced an AMS radiocarbon date of 3480 ± 60 (Beta-131037; wood charcoal; δ13C

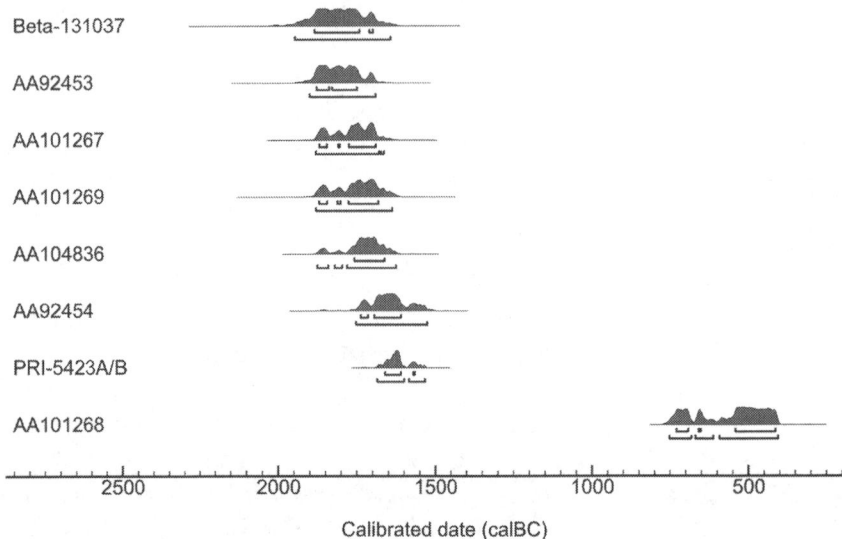

**FIGURE 1.2.** AMS radiocarbon dates from La Consentida (calibrated with IntCal 13 curve by OxCal 4.3.2). Reported with both 1σ and 2σ probability and rounded to five-year increments

= −24.4‰) or 1950–1640 cal BC (table 1.1; A. Joyce 2005; Winter 1989). This early date surprised the research team (who had expected to find Late or Middle Formative period deposits) and sparked interest in further investigations at the site. The 1988 pilot research recovered eroded medium brown ware sherds indicating a vessel assemblage of bottles, bowls, jars, and possibly platters and braziers. Also recovered were informally produced gray obsidian percussion fragments. These fragments seem to be largely debitage or randomly fractured waste material rather than purposeful flakes or formal tools. They evince a lithic industry focused on making sharp cutting edges regardless of tool shape, rather than producing material-conserving blades. In 2000, Arthur Joyce and colleagues (2009a:347, 2009b:522–25) carried out a surface survey and GPS mapping project at the site. The results of this mapping project are being revised with new total station mapping data, as discussed in chapter 3.

## PROJECT SCOPE AND OUTLINE OF THE BOOK

To date, the LCAP has focused on identifying relationships between transitions in sedentism, subsistence, and social organization at an Early Formative period site. Chapter 2 frames this research within the context of key debates regarding

TABLE 1.2. Comparison chronology of Mesoamerican Early Formative period phases

| Phase name | Site/region | Uncalibrated date | Calibrated date |
|---|---|---|---|
| Tlacuache | Oaxaca coast (La Consentida) | 1600–1350 BCE | 1950–1500 cal BC |
| Barraa (Clark 1994:544) | Soconusco | 1600–1450 BCE | 1900–1700 cal BC |
| Purrón (Clark and Gosser 1995; MacNeish et al. 1970:21) | Tehuacán Valley | 1600–1400 BCE | 1900–1680 cal BC |
| Espiridión (Flannery and Marcus 1994:375) | Valley of Oaxaca | 1600–1400 BCE est. | 1900–1650 cal BC est. |
| Ojochi (Cyphers and Zurita-Noguera 2012:146) | Gulf Coast | 1550–1400 BCE est. | 1750–1550 cal BC |
| Poxb (Brush 1965:194; Clark and Cheetham 2002:314) | Guerrero coast (Puerto Marquez) | 1500 BCE est. | 1750 cal BC est. |
| Early Ajalpan (estimate based on Clark and Gosser 1995) | Tehuacán Valley | 1400–1100 BCE est. | 1680–1350 cal BC est. |
| Lagunita (Reyes González and Winter 2010:151; Zeitlin 1978) | Isthmus | 1500–1100 BCE | 1750–1350 cal BC est. |
| Locona (Clark and Cheetham 2002:295) | Soconusco | 1450–1300 BCE | 1700–1550 cal BC |
| Tierras Largas (Drennan 2003a:363; Marcus and Flannery 1996:75) | Valley of Oaxaca | 1400–1200 BCE | 1650–1500 cal BC est. |

*continued on next page*

TABLE 1.2—*continued*

| Phase name | Site/region | Uncalibrated date | Calibrated date |
|---|---|---|---|
| Bajío (Cyphers and Zurita-Noguera 2012:146) | Gulf Coast | 1400–1300 BCE est. | 1600–1500 cal BC |
| Cruz A (Blomster 2004; Spores 1984:18–19; Winter 1992) | Mixteca Alta | 1400–1150 BCE | 1600–1400 cal BC est. |
| Ocotillo (Joyce and Henderson 2001) | Honduran coast | 1400–1100 BCE | 1600–1350 cal BC est. |
| Ocós (Lesure 2011b:13) | Soconusco | 1300–1150 BCE est. | 1550–1400 cal BC |
| Tulipan (P. Arnold 2003:31) | Gulf coastal highlands | 1300–1150 BCE | 1550–1400 cal BC est. |
| Chicharras (Hirth et al. 2013) | Gulf Coast (San Lorenzo) | 1250–1150 BCE | 1500–1400 cal BC |
| Capacha (I. Kelly 1980; Mountjoy 2006) | Colima | 1200–900 BCE | 1450–1150 cal BC est. |
| El Opeño (Mountjoy 2006; Oliveros Morales 1974; Oliveros Morales and de los Ríos 1993) | Michoacán | 1200–900 BCE est. | 1450–1150 cal BC est. |
| San José (Marcus and Flannery 1996:75) | Valley of Oaxaca | 1200–900 BCE | 1450–1150 cal BC est. |

*continued on next page*

TABLE 1.2—continued

| Phase name | Site/region | Uncalibrated date | Calibrated date |
|---|---|---|---|
| Cherla (Lesure 2011b:13) | Soconusco | 1150–1050 BCE est. | 1400–1300 cal BC est. |
| Cotorra 1-A (Bachand 2013:14) | Chiapa de Corzo | 1150–1050 BCE | 1400–1300 cal BC est. |
| San Lorenzo A (Coe and Diehl 1980a) | Gulf Coast (San Lorenzo) | 1150–950 BCE est. | 1400–1200 cal BC |
| Cruz B (Blomster 2004; Spores 1984:18–19; Winter 1992) | Mixteca Alta | 1150–850 BCE | 1400–1100 est. |
| Coyame (P. Arnold 2003:31) | Gulf coastal highlands | 1150–850 BCE | 1400–1100 est. |
| Golfo (Reyes González and Winter 2010:151; Zeitlin 1978) | Isthmus | 1100–800 | 1350–1050 cal BC est. |
| Cotorra 1-B (Bachand 2013:14) | Chiapa de Corzo | 1050–1000 BCE | 1300–1250 cal BC est. |
| Cuadros (G. Lowe 2007) | Soconusco | 1050–950 BCE est. | 1300–1200 cal BC |

[a] The beginnings of the date ranges for the Tlacuache and Barra phases differ due to discrepancies in calibration techniques. Specifically, Tlacuache phase dates have been calibrated with the updated IntCal 13 curve by OxCal 4.3.2 (Reimer et al. 2013). The longer error ranges of most Barra dates, relative to Tlacuache dates, also complicate comparison (see table 1.1; Clark 1994:app. 3). I estimate that the Barra and Tlacuache phases had roughly contemporaneous beginnings, though the latter remains a longer phase pending possible division into subphases.

[b] Brush (1965:194) reported a radiocarbon date from Puerto Marquez of 2940 ± 130 BCE, and an initial Pox Pottery date of 2440 ± 140 BCE. Clark and Cheetham (2002:314) revised this date to approximately 1500 BCE on the basis of stylistic similarities with Tierras Largas.

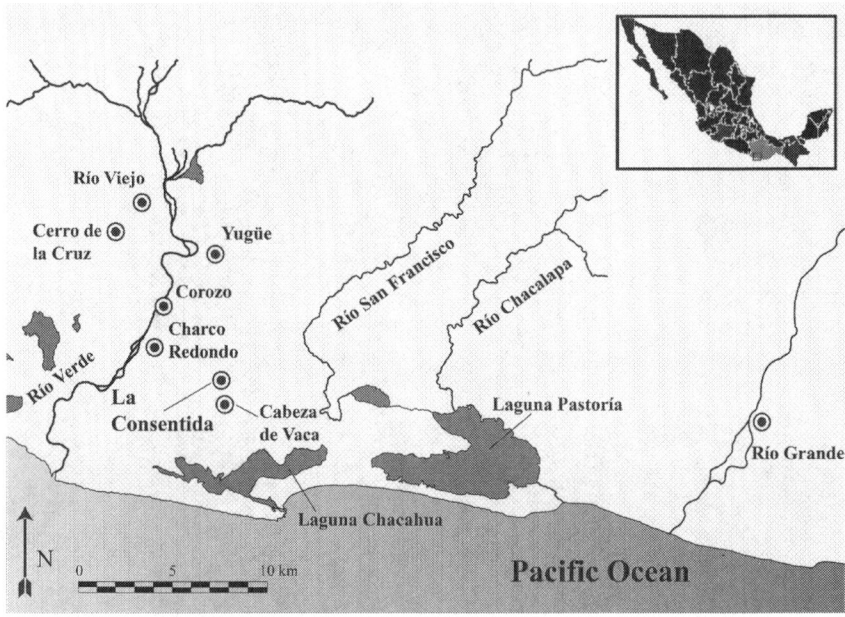

**FIGURE 1.3.** Map of key archaeological sites in Oaxaca's lower Río Verde Valley

these socioeconomic changes as they relate to the archaeology of Early Formative period Mesoamerica. Chapter 2 also discusses some material correlates for identifying sedentism, agriculture, and social complexity in the archaeological record. The LCAP has included surface survey, mapping, ground-penetrating radar, large-scale excavations, and laboratory study. The mapping phase updated preexisting information and revealed the dimensions and locations of Platform 1 and several earthen substructures atop it. Refer to chapter 3 for a discussion of research methods, terminology, and mapping results. Chapter 3 also presents several kinds of maps to help readers visualize the site's dimensions and spatial organization. At later sites in the region, platforms similar to La Consentida's Substructures 1–7 often supported domestic architecture and/or public buildings (Barber 2005:140–41, 235; A. Joyce 1991a:292). On the basis of this comparison, horizontal excavations atop these mounds have been one focus of the LCAP. Excavations also sought refuse middens, largely as a way to extend the regional ceramic sequence and to locate floral and faunal remains to aid dietary reconstruction. Chapter 4 presents a discussion of the occupational history at La Consentida. This information is meant to complement the specific descriptions of excavated deposits found in appendix 1. I pay particular attention in both sections to strata relevant for understanding La

Consentida's occupational history. Wherever possible, I discuss excavated contexts chronologically, using radiometric, stratigraphic, and ceramic data as supporting evidence for their relative dates of deposition. Where chronological relationships are less clear, I organize context descriptions stratigraphically and by operation area.

For interpretations of excavation and laboratory data specific to each component of the project's research questions, refer to chapters 5, 6, and 7. Chapter 5 addresses evidence for domestic mobility and sedentism. Chapter 6 presents evidence for La Consentida's subsistence economy. Chapter 7 offers evidence for social organization at the site. These discussions focus on architectural stratigraphy indicating shifting patterns of communal labor, iconography, evidence for personal adornment, and mortuary and ritual deposits. Iconography relevant to discussions of social organization includes figurines suggesting practices of bodily adornment and the expression of diverse social identities. Figurine analysis is an important step in interpreting social organization and identity in ancient Mesoamerica (Blomster 2009; Cyphers Guillén 1993; Faust and Halperin 2009; Hepp and Joyce 2013; Hepp and Rieger 2014; Lesure 1997a, 1999a; Marcus 1998, 2009). Ceramic figurines recovered from diverse contexts at La Consentida, including human burials, indicate an emphasis on the human form and especially on the depiction of women. Ceramic musical instruments from the site are among the earliest known in Mesoamerica and appear to predate similar instruments of the Tierras Largas phase (Hepp et al. 2014; Ramírez Urrea 1993:143). See chapters 7 and 8 for results of figurine and musical instrument analysis. Chapter 8 discusses evidence for interregional interaction and trade, including patterns identified through the study of ceramic vessel forms and decorative styles and obsidian X-ray fluorescence (XRF) sourcing data. Various lines of evidence suggest interaction with diverse regions including the Valley of Oaxaca, Central Mexico, the Gulf Coast, and possibly West Mexico.

In chapter 9, I summarize the evidence from each of the main components of the LCAP research agenda, consider how these social phenomena were interrelated, and present the final interpretations of the project to date. I conclude that La Consentida presents good evidence for a transition toward sedentism during site occupation, which appears to have lasted for about two and a half centuries during the Early Formative period. The community's diet was diverse but likely included more maize than did that of contemporaneous peoples of the Soconusco (Blake et al. 1992; Chisholm and Blake 2006) and Gulf Coast (Killion 2013). Dental pathologies, ground stone tools, and ceramic vessel styles suggest a possible shift from an emphasis on maize in beverage form to the processing of maize flour with stone manos and metates. In general, however, the Early Formative occupants of La Consentida did not eat the heavily maize-reliant diet of coastal Oaxaca's later pre-Hispanic peoples (Joyce et al. 2017). In terms of social organization, the La Consentida community appears to have been heterarchically complex, with perhaps the first glimmers of the ascribed hierarchies of the kind better documented in later Mesoamerican contexts.

As mentioned above, appendix 1 presents descriptions of excavated deposits useful for understanding the occupational history of La Consentida as well as the contexts from which carbon dates and specific finds were recovered. Appendix 2 contains detailed information about the ceramic vessel forms at La Consentida, as well as within-site patterns of pottery discard. These ceramics represent a previously unknown assemblage in the lower Río Verde Valley and thus require description as a new complex and phase in the regional ceramic sequence. Tlacuache-phase pottery (named in honor of a modern village located near the site) includes various types of jars, conical and semispherical bowls, bottles, and a few tecomates (neckless ceramic jars). In its current form, the Tlacuache phase is long (approximately 250 to 450 years, depending on if and how the dates are calibrated), presents a diverse ceramic assemblage, and may reflect some chronological variation. On the basis of future analysis, the phase may eventually be subdivided.

## PROJECT SIGNIFICANCE

La Consentida was occupied during some of the most revolutionary social transformations in the history of the Americas. Archaeologists working in many areas of the world debate the causal mechanisms behind sedentism, agriculture, the demise of egalitarianism, and the establishment of social complexity (e.g., Banning 2003; Boyd 2006; Choe and Bale 2002; Joyce and Henderson 2001; McClung de Tapia and Zurita-Noguera 2000). Studies of Formative period Mesoamerica are especially rife with debates over the timing of and causal relationships between these transitions. Positions in these debates correlate strongly with regional research foci, suggesting that a diverse material record inspires diverse interpretations (P. Arnold 1999, 2009; Blake et al. 1992; Blake and Clark 1999; Clark and Cheetham 2002; Flannery and Marcus 2003; Marcus and Flannery 1996). These different explanatory models also reflect major theoretical positions of their day, such as the ecological functionalism of the 1960s and 1970s, and the practice-based approaches of the 1990s and 2000s. La Consentida is uniquely suited to inform these debates for several reasons. First, the site's probable location near an open bay, in contrast to the estuarine environments of most coastal Early Formative sites, makes its ecological setting somewhat unique (Goman et al. 2005, 2013; Mueller et al. 2013). Second, because La Consentida was apparently abandoned by the late Early or early Middle Formative period, excavations at the site have exposed broad areas of early deposits rather than narrow windows through Classic or Postclassic period overburden. Third, the site's earthen architecture suggests communal labor efforts and perhaps the origins of social hierarchy associated with organizing work parties.

On a regional level, the LCAP represents a unique opportunity to expand understandings of ecological conditions and social organization at one of the earliest known villages on the Pacific coast of Mexico. Because La Consentida was apparently occupied *before* the development of local estuaries and the expansion of the Río Verde floodplain, it can provide information about settlement, subsistence, and social organization in the valley before it was intensively occupied during the late Middle Formative (A. Joyce 2005, 2010:180–95). The regional ceramic chronology (table 1.3; A. Joyce 2010:16) has never before included information for the Early Formative or early Middle Formative periods. With the newly identified Tlacuache phase (see appendix 2), the LCAP has expanded this regional chronology and promotes greater chronological depth of regional ceramic analysis and interregional comparison than has been possible previously. Similarly, ceramic iconography from the site permits a more deeply diachronic study of changing styles of decorated pottery, figurines, and musical instruments than has been possible before (e.g., Hepp 2007; Hepp et al. 2014; Hepp and Joyce 2013).

More broadly, research at La Consentida relates to general anthropological concerns, such as how Mesoamerican peoples negotiated the novel social and ecological conditions of increasingly sedentary and agrarian communities (Banning 2003; R. Joyce and Henderson 2001; R. Kelly 1992; R. Pearson 2006). Early Formative villages were occupied during major social transitions involved in the establishment of Mesoamerica (which may be defined, depending on the context, as a cultural and geographical entity or as a network of interaction and shared practices), but relatively few sites comprise the period's known material record (Blake and Clark 1999; Blomster 1998; Clark 1991, 1994; R. Joyce 2004b; Kirchhoff 1943; Lesure 2004). Identifying archaeological evidence of Early Formative period settlement practices, subsistence, and social organization, and refining explanatory models for their transformation, will be one of the most productive ways to address key debates in Mesoamerican studies in the future (see chapter 2). La Consentida, located in a region practically unknown in Early Formative archaeology, offers an opportunity for comparison with better-known areas to assess the applicability of current models for social transformation beyond the specific regions in which they have been developed. It is my hope that work at La Consentida can also help to redefine the role of Mexico's Pacific coast in the Early Formative roots of Mesoamerican culture.

In this book, I argue that transitions in settlement, subsistence, and social organization at La Consentida were intimately linked. Accelerator Mass Spectrometry

**TABLE 1.3.** Lower Río Verde regional ceramic sequence with uncalibrated radiocarbon dates (see A. Joyce 1991, 2010)

| Phase | Period | Date |
|---|---|---|
| Yucudzaa | Late Postclassic | 1100–1522 CE |
| Yugüe | Early Postclassic | 800–1100 CE |
| Yuta Tiyoo | Late Classic | 500–800 CE |
| Coyuche | Early Classic | 250–500 CE |
| Chacahua | Late Terminal Formative | 100–250 CE |
| Miniyua | Early Terminal Formative | 150 BCE–100 CE |
| Minizundo | Late Formative | 400–150 BCE |
| Charco | Late Middle Formative | 700–400 BCE |
| ? | Late Early–Middle Formative | 1350–700 BCE |
| Tlacuache | Early Formative | 1600–1350 BCE |

(AMS) radiocarbon dates from multiple, secure contexts demonstrate that the site's ceramics, mounded earthen architecture, and formalized mortuary contexts are among Mesoamerica's first. As I will discuss in the following chapters, Mesoamerica is too diverse for research at a single site to "lay to rest" ongoing debates about the Early Formative. Evidence from La Consentida does have the potential to impact those debates, however, as the site represents a unique example of the socioeconomic transformations that took place in an Early Formative village. Before discussing the evidence from the site in detail, I will first use chapter 2 to examine the different theoretical approaches and explanatory models applied to Early Formative period archaeology.

# 2

## Debating the Early Formative Period

As discussed in chapter 1, the primary goal of the research presented here has been to investigate relationships between settlement, subsistence, and social organization at La Consentida. As part of that undertaking, it is appropriate to review the state of current research and debate on these topics. The refinement of models for Early Formative period social organization, settlement, and subsistence is crucial to the future of Mesoamerican archaeology because proposed Archaic to Early Formative period social changes were not only precursors of later developments, but also exemplify broader transformations that occurred in many parts of the world during the early and middle Holocene epoch (Banning 2003; Bar-Yosef and Belfer-Cohen 1989; Binford 1983; M. Cohen 1985; Ford 1969; Hayden 1990, 1995, 2009). These changes are the basis for fundamental anthropological questions regarding the history of social and ecological transformations in the New World and globally. In this chapter, I summarize debates most relevant to the research questions of the LCAP. I also explain my own positions regarding these issues and discuss material correlates appropriate for the study of Early Formative social dynamics at La Consentida.

Archaeological study has indicated that Early Formative period Mesoamericans established the region's first villages, increased their dependency on domesticates in comparison to their Archaic period forebears, organized communal labor projects, and developed status distinctions that led ultimately to hierarchical hereditary inequalities (Blake and Clark 1999; Clark 2004a; A. Joyce 2010:64–83; R. Joyce 2004a). Much debate regarding the archaeology of the Early Formative has focused on possible ties between sedentism, agriculture, and social complexity (Clark 2004a:44–45; Love 2007; Marcus and Flannery 1996). For example, archaeological evidence suggests that the first hereditary, hierarchical social inequalities occurred at various times in different regions, with areas such as Mazatán likely developing such inequalities by 1600 cal BC (Clark 1991; Hill and Clark 2001), while neighboring regions not doing so until the Middle Formative (Lesure 2011a; Love 2002). Disagreements regarding Early Formative subsistence focus not only on relative degrees of maize reliance but also on the significance of wild resources and other crops. Legumes and tubers, for instance, have been proposed as significant Early Formative cultigens (Chandler-Ezell et al. 2006; Clark et al. 2007; Davis 1975; Isendahl 2011:464; G. Lowe 1967, 1975, 1977; Pohl et al. 1996). The dietary importance of nonmaize crops is difficult to discern, however, especially in areas with poor organic preservation.

The social changes of the Early Formative have been studied in various regions including the Soconusco, the Gulf Coast, the Basin of Mexico, and the highland valleys of Tehuacán and Oaxaca (P. Arnold 1999, 2000, 2009; Blake and Clark 1999; Clark 1991, 1994, 2004a; Clark and Blake 1994; Flannery 1968b; Flannery and Marcus 2003; Lesure 2009, 2011a; Lesure and Blake 2002; MacNeish and Eubanks 2000; Marcus and Flannery 1996; Sanders et al. 1979; Tolstoy 1975). The very beginning of this period, sometimes termed the initial Early Formative, is poorly understood in most areas (see table 1.2). Much of the information for this time of transition from Archaic period lifeways comes from the Soconusco region of Pacific coastal Chiapas and Guatemala (e.g., Blake and Clark 1999; Clark 1991, 1994, 2004a; Clark and Blake 1994; Lesure 2009, 2011a; Lesure and Blake 2002; Love 2007). Because it was a time of key social transformations, evidence from the initial Early Formative is crucial for resolving debates in Mesoamerican archaeology.

Scholars have proposed competing explanations for why such dramatic and geographically widespread social changes occurred during the Early Formative. Diverse ecological, demographic, economic, and political models for these changes, and for similar changes elsewhere in the world, have been the subjects of thoughtful review (J. Arnold 1993; Clark and Blake 1994; Flannery and Marcus 2003; Hayden 1995; Marcus 2008; Price and Feinman 1995; Tolstoy 1989a). Some of these models may be appropriate for certain regions and not others, a circumstance that highlights the historically and ecologically contingent nature of these cultural changes. For example, people in the neighboring Mazatán and Río Naranjo subregions of the Soconusco seem to have developed social complexity (measured according to variables such as settlement hierarchies, public architecture, and exotic grave offerings) at different times (Clark 1991, 1994:126; Lesure and Blake 2002; Love 2002, 2007). Also, some sites (such as Paso de la Amada in the Mazatán region) have produced evidence for key social transitions, such as mobilizing labor for large-scale architecture, despite apparently uniform wealth among households (Lesure and Blake 2002; Love 2007). I argue in the following sections that some models explain Early Formative social transitions better than others because they consider a broader set of causal factors or more fully incorporate the results of study in adjacent regions. For example, recent research into the importance of maize in Early Formative diets (e.g., Chisholm and Blake 2006; Clark et al. 2007; Pohl et al. 1996; Pope et al. 2001; Smalley and Blake 2003) benefits from a more diverse array of phytolithic, palynological, isotopic, and lithic data than did previous discussions of the importance of farming to the establishment of sedentism (e.g., Coe and Flannery 1967; Flannery 1968b; MacNeish 1969, 1972).

## DEBATING EARLY FORMATIVE PERIOD SETTLEMENT PRACTICES

The traditional definition of the Mesoamerican culture area, as summarized by John Clark and David Cheetham (2002:280), holds that the sedentary, agricultural, and socially stratified societies of the Late Formative through Postclassic periods grew out of the first farming villages of the Early Formative. Evidence for the first permanent villages has been identified in diverse regions of Mesoamerica, ranging from highland valleys (Marcus and Flannery 1996; Whalen 1983) to marshlands or estuaries (Berry and McAnany 2007; Voorhies 1989) and coastal plains (Blake et al. 1992; Clark and Blake 1994; Clark and Cheetham 2002; Rosenswig 2006, 2007). Such data, which include settlement patterns and population estimates, suggest that some of Mesoamerica's first villages coincided temporally with evidence for agriculture, including ceramic technologies and nonportable ground stone (Blanton et al. 1979; Flannery 1968b, 2009a; Kowalewski 1990:42; Marcus and Flannery 1996).

In the lower Coatzacoalcos River drainage of the Gulf Coast, monumental architecture and a dense settlement pattern indicate that the population was sedentary and dominated by San Lorenzo as a regional capital (and by several secondary political centers) by as early as the San Lorenzo A phase (1400–1200 cal BC) (P. Arnold 2009; Santley and Arnold 1996:244–45; Symonds 2000:64–80; Symonds et al. 2002). Sedentism at nearby sites may have begun even earlier, with Estero Rabón reaching 60–80 ha in size by 1600–1300 cal BC (P. Arnold 2009:402; Borstein 2002). Stacey Symonds (2000:64–66) argued that Early Formative settlement in the lower Coatzacoalcos region became dense during the Bajío phase (1600–1500 cal BC) and may relate to the use of local rivers for trade, transportation, and subsistence. It was also around this time that significant labor investments were made to modify the landscape at San Lorenzo (Symonds et al. 2002). Evidence for sedentism in other regions includes durable architecture at San José Mogote by the Tierras Largas phase (1650–1500 cal BC) (Drennan 2003b:47; Marcus and Flannery 1996:109–10). At Paso de La Amada, the community produced a monumental ballcourt by around 1600 cal BC (Hill and Clark 2001). Such communally constructed features imply the investment of human labor in specific areas of the landscape. Large domestic structures, such as Paso de la Amada's Mound 6 (which was apparently associated with the adjacent early ballcourt), may also suggest sedentism (Hill and Clark 2001; Lesure 1997a, 1999b; Lesure and Blake 2002). As I discuss below, however, and as demonstrated by sites such as Poverty Point in Louisiana, the presence of monumental constructions does not constitute *a priori* evidence of sedentism (Gibson 2000).

Despite suggestions of sedentism in certain regions, some scholars (e.g., P. Arnold 1999; Lesure 2009) have offered new interpretations of Early Formative settlement patterns by arguing that the shift toward sedentism was gradual, particularly in coastal regions such as the Soconusco, the Gulf Coast, and the Caribbean coast of Honduras. For example, Arnold (1999:160) found that the site of La Joya, located in the Tuxtla Mountains adjacent to the Gulf Coast, was a sedentary agricultural community during the Late Formative, but lacked the mounded earthen architecture, durable domestic structures, formal storage facilities, nonportable artifacts, and consistency in the placement of features that would identify it as sedentary during the Early Formative. Ground stone from Early Formative La Joya suggests the multipurpose use of portable tools rather than heavy stone metates or mills for flour processing, which would arguably evince sedentism. Philip Arnold (1999:160) also suggested that Tulipan phase (1550–1400 cal BC) tecomates at La Joya may have been all-purpose vessels suited to a semimobile lifestyle, by virtue of their restricted mouths and portability when enclosed in a net. Arnold (1999, 2009) thus concluded that Early Formative settlement practices were diverse, even within particular regions.

On the basis of finds at La Joya, Arnold extrapolated his model to other regions. He argued, for example, that a high proportion of tecomates among Barra phase ceramics in Mazatán suggests that Early Formative communities on the Pacific coast also remained semimobile (P. Arnold 1999:161). Such an interpretation allows for the possibility that large, subsistence-related artifacts such as tecomates and nonportable ground stone might have been cached at sites intermittently revisited by semisedentary peoples. Clark and Cheetham (2002:311) rejected this assertion on the basis that Paso de la Amada and other Mazatán sites exhibited public architecture such as plazas and a clay-surfaced ballcourt by as early as 1650 cal BC. Other Soconusco scholars have not been so quick to dismiss alternative arguments about Early Formative residential mobility and multipurpose use of ceramics, however. Richard Lesure (2009:259, 261; see also Blake et al. 1992; Voorhies 1989:116) identified estuarine sites with ceramic assemblages dominated by tecomates as probable seasonally occupied resource extraction points that supplied more permanent inland settlements. Such practices of seasonal resource exploitation in coastal wetlands may date back to the Chantuto (5000–1900 cal BC) occupations of the late Archaic (Voorhies 2004; Voorhies and Kennett 2011). I agree with Arnold (1999) and Lesure (2009) that reinterpretations of artifacts such as tecomates should prompt us to question direct correlations between specific types of material culture and sedentism (see also Clark and Cheetham 2002; Skibo and Feinman 1999). I find such reconsiderations of material culture informative for research at La Consentida, where ceramic vessels were used by a community that may have initially been semisedentary (see chapter 5).

In order to interpret an ancient community's settlement practices, it is important to consider what material correlates might indicate degrees of mobility and sedentism. Several criteria have been recognized cross-culturally as significant for assessing ancient mobility. Among these are the portability and versatility of tools used frequently, such as for daily food processing. Small, light tools such as mortars and pestles are more easily transported than are large items such as stone metates. Multipurpose tools, such as small ground stone mauls that might also be employed as grinders, offer ease of transport and flexibility of purpose at the possible expense of task-specific efficiency (Clark et al. 2007; McDonald 1991:85; Torrence 1983). Evidence of sedentism may thus include heavy, single-purpose ground stone tools such as manos and metates, which also evince flour processing from grains such as maize. These relatively nonportable tools suggest that the communities using them may have been tied to specific places on the landscape through subsistence practices (P. Arnold 1999:159–60; Rosenswig 2006). By contrast, portable and multipurpose ground stone tools (in the absence of heavy, single-purpose tools) have been considered evidence of nonsedentary occupation at Early Formative La Joya in the Gulf coastal highlands, and at Archaic and Paleo-Indian sites throughout Mesoamerica (P. Arnold 1999:159–60, 2009:404).

John Clark and colleagues (2007) studied Archaic to Early Formative period transitions in ground stone and boiling stone technology in the Soconusco. Although Archaic sites have been notoriously difficult to locate across Mesoamerica (see Borejsza et al. 2014), Clark and colleagues (2007:25) recognized fire-cracked rock (FCR) and certain types of ground stone tools as among the few artifact types held in common between Archaic and Early Formative sites. These authors (2007:28) identified an inverse correlation between increasing numbers of plain tecomates and decreasing quantities of FCR at Soconusco sites between 1900 and 1550 cal BC. This pattern corresponds with an increase in manos relative to a decrease in FCR between 1900 and 1150 cal BC (Clark et al. 2007:31). What these data indicate, according to Clark and colleagues, is that the use of hot stones dropped into water for cooking (an Archaic period practice) decreased apace with increasing use of ceramics and maize processing, which grew in significance throughout the Early Formative and became especially important by the Middle Formative. This pattern suggests that dietary practices changed along with the shift in technology, as maize flour and liquids boiled in ceramic vessels grew in significance. As I will discuss in chapter 6, ground stone and skeletal data suggest a similar trend at La Consentida.

The patterns discussed above suggest that maize reliance was not a necessary economic condition for the origins of sedentism in Mesoamerica. What these data *do not* provide is an answer to why maize reliance did not take hold in the Soconusco until the Middle Formative, well after populations both there and on the Gulf Coast were apparently sedentary (Blake et al. 1992; Cyphers et al. 2013; VanDerwarker 2006). Although the small, portable ground stone mortars and pestles of the Archaic gradually gave way to less portable manos and metates of the Formative period, the arrival of ceramic technology occurred abruptly in the Soconusco. For Clark and colleagues (2007:25; contra P. Arnold 1999), Barra phase pottery marks the beginnings of dedicated sedentism. Varying interpretations of the first ceramic vessels in the Tuxtla region versus in the Soconusco suggest an underlying debate about the nature of ceramics themselves, where Arnold (1999) notes that pottery may have been used by semimobile peoples, while Clark and colleagues (2007) see ceramics as an indicator of sedentism. As I will demonstrate in subsequent chapters (and with reference to evidence from ground stone tools, for example), data from La Consentida lead me to agree that ceramics were sometimes used by semisedentary groups in Mesoamerica.

Another key component of models for Early Formative settlement is the study of public architecture. Robert Santley and Philip Arnold (1996:228) noted that there was no evidence for earthen platform construction in the Tuxtla region during the Early Formative. This pattern might seem to corroborate Arnold's (1999, 2009) finding that the people of the Tuxtla Mountains remained semimobile during this time, but such a conclusion would indicate a facile correlation between monumental or public architecture and sedentism. Regardless, recent research by Cyphers and Zurita-Noguera (2012) may lead to revisions of the history of public architecture in the region, as earthen mounds in the Gulf Coastal wetlands are now dated to the Bajío phase (1600–1500 cal BC). Despite the presence of early, mounded earthen architecture on the Gulf Coast, it is clear that Early Formative villages were often small and ephemeral in comparison to later settlements. In coastal Honduras, for example, Rosemary Joyce and John Henderson (2007:643) found evidence that domestic structures remained relatively unsubstantial into the Ocotillo phase (1600–1350 cal BC). Furthermore, the use of public and/or monumental architecture as indisputable evidence of sedentism may not be appropriate. The megalithic landscapes of ancient Europe, the towering stone pillars of Turkey's Göbekli Tepe, and Archaic period earthworks of North America all suggest that semimobile peoples can amass large labor parties and invest resources in a place as part of producing a sacred landscape (R. Bradley 1993, 1998, 2000, 2005; Gibson 2000; Pauketat and Alt 2003; Peters and Schmidt 2004; Sherratt 1990; Tilley 1994, 2007).

Another criterion for assessing a community's degree of domestic mobility is its approach to storage. Storage of surplus goods is not unique to sedentary populations, but "formal" (e.g., slab or clay-lined) storage features have been recognized as an index of sedentism among the Maya of the northern Yucatán (Smyth 1989:90) and in the American Southwest (Kent 1992). In highland Oaxaca, Early Formative storage features included large, bell-shaped pits requiring significant labor for their construction (Winter 2009:27–29). Such formal storage facilities, argued Michael Smyth (1989:90, 92), suggest a community's own predictions about its future mobility. It was, in part, the lack of such continuity in site organization that led Arnold (1999) to argue that La Joya was an encampment for semimobile peoples well into the Formative period. Diachronically shifting patterns in the placement of domestic features, in some instances leaving "hot spots" of overlapping structures and storage features, suggested to Arnold (1999:160, 2009) that Tulipan phase La Joya was a site of repeated reoccupation by seasonally mobile groups. Research elsewhere in the Americas has indicated that mobile peoples interact with their surroundings as socially constructed spaces rather than merely for resource collection. Mark Mitchell (2008) found, for example, that burned rock middens in southeastern Colorado possessed historical biographies for the hunter-gatherers who used them. He argued that the interplay of human social action and the landscape that shapes and *is shaped* by those relationships are determining factors in practices of mobility and land use (Mitchell 2008:60). The adoption of sedentism is thus not merely an economic choice, but also a social one. I agree that the mere presence of repeatedly used food-processing or storage features does not constitute sufficient evidence of sedentism. Also, the use of storage features to assess domestic mobility may be hampered by regional variations in storage practices, perhaps making it unsupportable to rely on them alone as sufficient evidence for sedentism. As I will discuss below, I find that the only way to adequately address the issue of domestic mobility is by basing one's conclusions on as many variables as possible. Even when that is accomplished, the material record of domestic mobility may be unclear (see chapter 5).

Debates about Early Formative sedentism such as those outlined above demonstrate that researchers can support differing conclusions using the same data. Some types of features—domestic buildings, formal storage facilities, mounded earthen architecture, and the like—suggest that a community invested a good deal of labor in a specific place rather than (or in *addition to*) maintaining multiple seasonal occupation sites. When such features remained consistent in their placement through time, argue some authors (e.g., P. Arnold 1999; Rosenswig 2006:336), the evidence suggests that a community remained permanently in one place rather than periodically reoccupying a site. Similarly, certain types of artifacts, such as nonportable metates, suggest that a community's food processing took place in a specific spot and likely included making flour from domesticates such as maize (Clark et al. 2007; McDonald 1991:85; Torrence 1983). The argument that heavy grinding stones indicate sedentism is complicated, however, by evidence for the seasonal reoccupation of resource extraction areas in regions such as the Soconusco (e.g., Voorhies 1989). Once transported to a location, a metate could be cached by a semimobile group planning to revisit a site the following season. Such potentially confounding factors underscore the need to marshal as many lines of evidence as possible in the study of complex social and ecological factors involved in domestic mobility.

On the basis of my review of the literature, I find that there is no single line of evidence that provides an absolute indication or unambiguous material signature of sedentism. Rather than existing as a strict dichotomy, nomadism and sedentism are extremes within a spectrum of domestic mobility. This conclusion is supported by evidence that many ethnographically recognized peoples have practiced complicated forms of mobility that shift according to seasonal, environmental, and political circumstances (e.g., R. Kelly 1992; Kent 1992; Marshall 2006). Only when the preponderance of data (drawn from such diverse lines of evidence as domestic architecture, communal labor projects, mortuary practice, and subsistence) suggests that a community placed greater emphasis on either mobility or sedentism is it appropriate for us to label them as such (see P. Arnold 1999; Clark and Cheetham 2002; Rosenswig 2006).

One important step in investigating evidence for sedentism at La Consentida is the analysis of domestic contexts. Although mounds at Early Formative coastal sites are sometimes the result of shellfish or salt processing (e.g., Lesure 2009:185), comparison with data from later sites in the lower Río Verde region suggests that Substructures 1–7 at La Consentida likely supported domestic or public architecture (Barber 2005:140–41, 235; A. Joyce 1991a:292). The analysis of features atop these mounds is thus one way to infer whether La Consentida was continually or intermittently occupied, particularly if buildings or their supporting earthen architecture were consistent in their placement through time and required significant labor for construction (P. Arnold 1999:160). Hypothetically, though, earthen mounds of any size could have been used to stabilize even the most ephemeral and temporary of structures built by a semimobile group. Analysis of refuse associated with these features may thus further strengthen inferences of mobility, particularly if they contain such evidence as seasonally available resources collected throughout the year. Durable architecture such as structures at San José Mogote (Drennan 2003b:47; Marcus and Flannery 1996:109–10) and the ballcourt at Paso de La Amada (Hill and Clark 2001) could also lend credence to the interpretation of sedentism because it suggests significant amounts of labor invested in a single area. Although European and Near Eastern Neolithic data (e.g., Bradley 1993, 1998, 2000, 2005; Peters and Schmidt 2004; Sherratt 1990; Tilley 1994, 2007) indicate that public and/or monumental constructions are not sufficient evidence of sedentism, they nonetheless represent one line of evidence useful for assessing degrees of mobility. The beliefs behind the production of monuments by nonsedentary peoples may be part of a process that ultimately led to sedentism, rather than ex post facto evidence of sedentism.

"Durability" of structures is a relative term but can be established through comparison of building practices and labor investments with those of known sedentary communities (Abrams 1989). As Brian Boyd (2006; see also Joyce and Goman 2012) suggested, a cemetery may indicate that a community was tied to a site as a part of its symbolic landscape. Indications that symbolic associations with the deceased reaffirmed a community's ties to a place may include long-term reuse of a cemetery, anthropomorphic figurines (perhaps indicating ancestor veneration) as offerings with burials, or the interment of the deceased in other contexts such as below household foundations or floors. The spectrum of variation among mobile and sedentary peoples may be similar in its production of a complicated material record to that between horticulturalists and agriculturalists. It is to the discussion of Early Formative subsistence, itself intimately related to domestic mobility, that I now turn my attention.

## EARLY FORMATIVE SUBSISTENCE

As discussed above, agriculture has traditionally been considered part of a suite of factors (along with, and as *tied to*, sedentism) that promoted social complexity in Mesoamerica (e.g., Coe and Flannery 1967; Flannery 1972a, 1973; Flannery and Marcus 2003; MacNeish 1992; Marcus and Flannery 1996; Sanders 1968; Sanders and Webster 1978). At what point, though, did people experimenting with domesticates truly become farmers? Was this change really contemporaneous with the transition to sedentism? In order to address debates regarding the relationship between subsistence and mobility, it is necessary to differentiate between "horticulture" and "agriculture" (P. Arnold 2009; Clark et al. 2007; Kennett et al. 2010; Killion 2013; VanDerwarker 2006). Most scholars of the Early Formative agree that "horticulture" refers to the limited use of domesticated plants as part of a diet based largely on wild resources, while "agriculture" connotes reliance on domesticates as staples, albeit ones usually supplemented with wild resources. This distinction is important because recent evidence suggests that Mesoamericans practiced horticulture based on maize, beans, squash, and probably manioc and malanga (*Xanthosoma violaceum*) for thousands of years before they became reliant on them, a transition that did not occur in many regions until the Middle Formative (P. Arnold 2009; Bronson 1966; Clark et al. 2007; Kennett et al. 2010; Killion 2013; Pohl et al. 1996; Rosenswig 2007; Sheets et al. 2012). For some authors (e.g., Y. Cohen 1974), agriculture also carries a connotation of more intensified use of the land than does horticulture. For the purposes of my discussion here, I will follow these established definitions for "horticulture" and "agriculture."

Historically, archaeologists have argued that the transition from Archaic period seasonal mobility to settled Early Formative villages was predicated on agriculture (e.g., Coe and Flannery 1967; Flannery 1968b; Marcus and Flannery 1996). Kent Flannery (Coe and Flannery 1967; Flannery 1968a, 1973, 1986) proposed a highland origin of domesticated maize and the sedentism and social complexity he felt agriculture promoted. Kent Flannery (1968b:79–81, 1986) argued for a causal link between agriculture and sedentism when he claimed that increased reliance on domesticates in the highlands prompted more permanent settlement as traditionally mobile groups remained stationary for increasing intervals to tend their crops. According to Joyce Marcus and Kent Flannery (1996:79–80), areas with the best agricultural land attracted the largest and most nucleated populations. These authors suggested, for example, that the Etla arm of the Valley of Oaxaca attracted dense settlement at sites such as San José Mogote because it possessed a greater allotment of fertile "Class I" land than did other areas of highland Oaxaca (Marcus and Flannery 1996:79–80).

In a study addressing the origins of agricultural villages in Pacific coastal Guatemala, Michael Coe and Flannery (1967:5) argued that "an effective maize-beans-squash agriculture" was a "prerequisite to fully settled village life in Mesoamerica." Drawing on Joseph Caldwell's (1958) interaction sphere model, Coe and Flannery (1967:7) proposed an ecological approach that viewed human populations as reacting not to entire "biomes," but rather to specific microenvironments within them. According to these authors (1967:102–5), agriculture originated in the highlands, where it promoted sedentism, population growth, and social complexity. Coe and Flannery (1967:103) sought to explain social complexity on the Guatemalan coast (suggested by apparently permanent villages and earthen architecture indicative of labor management by about 1400 cal BC) in light of their belief that farming, sedentism, and social complexity originated in highland areas such as the Tehuacán Valley. Causal mechanisms proposed by Coe and Flannery (1967:102–5) for such social and economic diffusion included the exchange of goods, products, and ideas among societies in different ecological settings and the "efficiency" of adopting maize in coastal zones. Pacific coastal social complexity, for these authors (1967:103), was the result of fortuitous ecological circumstances in the highlands, followed by diffusion of maize agriculture to the coasts. According to this model, coastal communities played a passive or secondary role in developing their own social complexity. Although ambitious in their attempts at synthesis, early studies such as that by Coe and Flannery (1967) were hampered by a lack of extensive information about the Early Formative at the time of their writing. For example, the Barra phase, now recognized as the earliest ceramic assemblage in the Soconusco, was not yet identified at the time of Coe and Flannery's study (G. Lowe 1975, 2007). As Lesure (2009:15) noted, these early studies remain important for their reconstructions of Early Formative subsistence, paleoenvironmental analysis, and modeling of daily life in Early Formative communities.

Recent evidence suggests that the timing of full-blown agriculture and permanent sedentism was not consistent across Mesoamerica. While many archaeologists (e.g., Coe 1981; Coe and Diehl 1980a, 1980b; Coe and Flannery 1967; Flannery 1968b; Sanders and Price 1968) have claimed that agriculture was a prerequisite to sedentism in both inland and coastal settings, this conclusion is under revision. Several scholars (e.g., P. Arnold 2009; Blake et al. 1992; Clark 2004a; Kennett et al. 2010; Killion 2013; Rosenswig 2007) have argued that coastal subsistence from the Late Archaic through the Middle Formative was based on wild floodplain and estuarine resources supplemented by horticulture. According to these models, agriculture appeared *after* sedentism on the coasts, thus contradicting Coe and Flannery's (1967:5) claim that agriculture was "prerequisite" to sedentism. Underlying these contrasting interpretations are the basic assumptions that agriculture is either necessary for sedentism or an independent social and economic phenomenon often affiliated with sedentism but not strictly *necessary* for it. Of these two basic approaches, I favor the latter because it takes the association among discrete social and economic variables as a point of departure for further inquiry rather than assuming an *a priori*, causal relationship between them.

Survey and excavation in the Mazatán region over the last two decades have demonstrated that sedentism, ceramics, and monumental architecture were all well developed on the Pacific coast as early as, or *earlier than*, they were in many highland areas (Clark 1991; Lesure 2009; Love 2002:193; G. Lowe 2007; Voorhies 1989). These various lines of evidence cast doubt on models proposed by authors such as Richard MacNeish (1969, 1972, 1992; MacNeish and Eubanks 2000; MacNeish and Nelken-Terner 1983), William Sanders (1956, 1965, 1968; Sanders and Nichols 1988;

Sanders et al. 1979; Sanders and Price 1968), and Flannery (Coe and Flannery 1967; Flannery 1968b; Flannery and Marcus 2003; Marcus and Flannery 1996) for the highland origins of agriculture and social complexity. Research in highland Mexico suggests that sedentism in certain areas significantly predated reliance upon domesticates (e.g., Niederberger 1979; see also Pohl et al. 1996:335). This pattern corroborates the Soconusco settlement data, indicating that sedentism need not imply an agricultural subsistence economy.

According to Mary Pohl and colleagues (1996:367–68), the two main competing explanations for the transition to agriculture in Mesoamerica are the "economic buffering" hypothesis (e.g., Coe and Flannery 1967; Flannery 1972a, 1972b, 1973, 1986; Flannery and Marcus 2003; Marcus and Flannery 1996) and models proposing that agriculture was "a strategy to manipulate social relationships in the context of emerging political hierarchies" (e.g., Clark and Blake 1994; Hayden 1990). The former approach is based on ecological systems theory and implies that maize agriculture was an adaptive response that promoted homeostasis with local environments as sedentism produced population increase and the threat of ecological imbalance. The latter explanation is part of a broader set of agency models (e.g., Blake and Clark 1989, 1999; Clark and Blake 1994; Hayden 1990; Hill and Clark 2001) focusing on the social maneuvering of individual "accumulators" or "aggrandizers" employing domesticates for purposes of social mobility or the development and maintenance of prestige, especially through competitive feasting. Importantly, the material implications of these two apparently contradictory models may not be mutually exclusive, given varied evidence for the establishment of agriculture in different regions. In their discussion of predominantly palynological evidence for early agriculture in northern Belize at wetland sites such as Cob and Pulltrouser Swamps, Pohl and colleagues (1996) determined that the transition to maize agriculture occurred at various times in different regions. Furthermore, these authors (1996) suggested that this varied sequence of horticulture and early agriculture indicates that maize served different purposes in different areas. In places prone to food shortage, such as central Panama, maize may have been an important component of the diet since its introduction to the region. This evidence does not contradict the use of maize as a feasting food employed as part of a social strategy by emergent elites in Mazatán, where wild estuarine products could provide dietary staples (Blake et al. 1992; Blake and Clark 1999; Clark 2004a; Clark and Blake 1994; Pohl et al. 1996:367). Differences in social organization among regions may have been just as important to the adoption of agriculture as were environmental factors. In Belize, communities near centers of elite political power appear to have shifted to maize reliance before their rural counterparts. Despite evidence from Belize, Panama, the Río Balsas drainage, and highland Mexico for plant manipulation by the Early and Middle Archaic (ca. 5000 cal BC), Pohl and colleagues (1996:368) concluded that significant agricultural intensification did not occur in northern Belize until 1500–1000 cal BC.

In Mazatán, Clark and colleagues (2007:35) found settlement evidence suggesting that by 1900 cal BC, the Mokaya people were selecting village sites according to the distribution of interfaces between "well-drained soils and humid soils." This combination of soil types would have been ideal for a variety of crops, including legumes, tubers, and maize. On the basis of macrobotanical finds and a lack of tooth wear from metate grit, it may have been beans, rather than maize, that was the real staple crop for the Early Formative Mokaya (Clark et al. 2007:25, 34–35). Use of legumes as a dietary staple might explain not only the distribution of early villages in the Soconusco but also the increase of ceramic cooking vessels and the relative lack of isotopic indicators for maize reliance in human remains (Clark et al. 2007:35). In perhaps their most revolutionary refinement of traditional subsistence models, Clark and colleagues (2007:35) suggested that it might have been sedentism that eventually prompted maize dependency (by about 1000 cal BC), rather than the other way around. In other words, these authors indicated that a combination of factors, including the availability of estuarine resources and the potential productivity of legume horticulture, might have spurred the development of sedentism. The establishment of cultivated fields at the interfaces between different types of arable land, coupled with increasing social pressures emanating from status competition associated with ritual feasting, later drove the transition toward maize agriculture. If the earliest sedentary peoples of the Soconusco focused heavily on domesticates other than maize, it suggests that archaeologists may need to reassess their methods of identifying horticulture versus agriculture. Other considerations beyond isotopic markers of maize dependency could include microbotanical remains such as starch grains, pollen, and phytoliths (e.g., Morell-Hart et al. 2014) and skeletal indices of the health effects of different dietary regimes (see Hodges 1987; Larsen 1987; chapter 6). Furthermore, as Clark and colleagues (2007:37) concluded, it is difficult to determine the social significance of a given domesticate, since even a calorically insignificant crop could have "tipped the balance" toward increased sedentism and/or social complexity despite a subsistence economy generally based on other resources. I am sympathetic to Clark and colleagues' interrogation of traditional models for maize dependency in Early Formative Mesoamerica. As dietary evidence from La Consentida demonstrates, however, maize consumption on the western Oaxaca coast may have been more significant than it was in the Soconusco region during the Early Formative period (Blake et al. 1992; chapter 6).

Ancient peoples of the Soconusco undertook the labor of crop cultivation despite an apparent abundance of nearby estuarine resources (Kennett et al. 2010:3401–2).

For example, Michael Blake and colleagues (1992:89–90) identified isotopic indicators in human remains for limited Early Formative maize consumption in the Soconusco accompanied by faunal evidence for a diet largely composed of marine, brackish, freshwater, and terrestrial wild game. These hunted animals included gar, mojarra, turtles, iguanas, crocodiles, deer and various snakes. Domesticated dogs also represented a significant portion of the diet (Blake et al. 1992:90). Limited macrobotanical remains also suggest some maize use in the Early Formative Soconusco, but not as a dietary staple (Blake et al. 1992:91).

On the basis of isotopic data and the abundance of wild fauna in the Early Formative Soconusco diet, several authors (Blake et al. 1992; Clark et al. 2007:25, 32) have argued that it may have been for *social* rather than strictly *dietary* reasons that maize grew in popularity. Such social factors could have included the production of alcoholic beverages brewed from the sugary stalk of the maize plant. Feasting beverages may have been integral to competitive generosity among aggrandizers and to the establishment of social complexity despite the relative insignificance of maize in the diet (Blake et al. 1992; Clark and Blake 1994; McGovern 2009; Rosenswig 2007:22; Smalley and Blake 2003). The interpretation that Barra phase ceramics were used for serving food or drink at feasts rather than for cooking is supported by the absence of evidence for burning on these early vessels (Clark and Blake 1994; Clark et al. 2007). Clark and colleagues (2007:25) suggested that tecomates in particular likely held fermented maize or cacao beverages. This interpretation is supported by residue analysis demonstrating cacao consumption from Barra phase vessels (Powis et al. 2007, 2008). Based on increased evidence for burning on later ceramics, along with an apparent decrease in cooking with boiling stones, the use of Mazatán pottery for cooking apparently increased after the Ocós phase (1550–1400 cal BC), likely in tandem with the transition to maize use as a dietary staple (Clark et al. 2007:29). Clark and colleagues (2007:33–34) pointed out, however, that the steady increase in cob and kernel size in domesticated maize beginning in the Archaic indicates that the harvesting and selective planting of the kernels themselves was a consistent factor in the plant's cultivation (though see Webster 2011). By subsisting on wild resources supplemented by maize and other domesticates, the Chantuto and Mokaya (among other coastal peoples) demonstrated both the productive wealth of coastal resources and the social significance of feasting foods (Blake et al. 1992; Clark 2004a; Clark and Blake 1994; Coe and Flannery 1967; Cyphers et al. 2013; Kennett et al. 2010; Rosenswig 2007; VanDerwarker 2006; Voorhies 2004:342–43).

One limitation of many subsistence models for Early Formative Mesoamerica is researchers' tendency to focus on the importance of maize at the expense of other domesticates. As Gareth Lowe (1967, 1975, 1977) and Clark and colleagues

(2007:25, 35) have argued, Archaic and Early Formative Pacific coastal horticulture likely included the cultivation of beans and manioc. Maize preserves well in the archaeological record because its hard kernels carbonize readily and its pollen and phytoliths are distinctive. The opposite is true of manioc. Lowe (1967, 1975, 1977) proposed that the apparently random obsidian flakes common among Early Formative lithics might indicate the use of obsidian graters for manioc processing. Such processing is necessary for the use of bitter manioc in order to avoid cyanide poisoning (see Isendahl 2011:455). Although experimental archaeology (Davis 1975) suggested that the use of obsidian for manioc graters is consistent with Early Formative lithic microwear and debitage morphology, other research indicated that archaeological obsidian chips bear little resemblance to ethnographically recorded manioc graters (DeBoer 1975:431; Lewenstein and Walker 1984). In particular, DeBoer (1975:430–31) noted that ethnographically documented manioc-processing flakes tend to be much smaller (less than 1 cm in diameter) than most archaeological obsidian flakes. Another potential drawback of the obsidian chip grater hypothesis is the probability that microscopic fragments of brittle obsidian could render the processed root pulp dangerously inedible, as Dee F. Green and Gareth W. Lowe (1967:128) acknowledged (see also Lewenstein and Walker 1984). Therefore, it seems unlikely that the Early Formative informal obsidian flakes came from manioc graters. Furthermore, graters would likely only be necessary for bitter varieties of manioc, and thus their identification would be irrelevant in tracing the use of the sweet varieties identified in the archaeological record of Central America (Payson Sheets, personal communication, 2015; see also Sheets et al. 2012). Recent research has suggested, however, that some obsidian tools *were* used to process manioc in ancient Mesoamerica (Sheets et al. 2012). Shanti Morell-Hart and colleagues (2014:74), for example, identified starch grains indicating the use of prismatic obsidian blades in ancient manioc processing. It also may be possible to identify diagnostic wear on obsidian tools heavily used for the removal of manioc cortex (Sheets et al. 2012:271–72).

Despite the difficulty of demonstrating ancient consumption of manioc (which does not preserve well in hot coastal climates), discussions of the tuber as a possible Early Formative cultigen persist. Pohl and colleagues (1996:362) identified fossil pollen comparable to domesticated manioc in northern Belize by approximately 3400 cal BC. The analysis of macrobotanical remains, as well as DNA sequence variation of modern manioc, has demonstrated an early date for the crop's domestication (likely from a single wild ancestral subspecies of *Manihot*) and the presence of domesticated manioc in central Pacific Panama by about 5700 cal BC (Olsen and Schaal 1999; Piperno 2011:S454, S458). Natural volcanic casts from El Salvador demonstrate that manioc was farmed in Central America during the

Classic period (Sheets and Woodward 2002). At the site of Cerén, Sheets and Woodward (2002:189–90) found that manioc farming formed part of a "zoned biodiversity" in Classic period kitchen gardens, which also contained chiles, guayaba, maize, maguey, and cacao. Given Lowe's (1967, 1975, 1977, 2007) convictions regarding late Archaic and Early Formative contact between Central America and the Soconusco, evidenced by the abrupt arrival of sophisticated Barra phase ceramics with no local precursor, the diffusion of manioc from Central America should not be disregarded. In fact, archaeobotanical evidence is mounting that manioc was a component of Mesoamerican diets by 4600 cal BC (Isendahl 2011; Pohl et al. 1996; Pope et al. 2001). The answer to correcting preservation biases in the study of ancient domesticates may lie in the analysis of absorbed residues in cooking vessels and on grinding stones (Clark et al. 2007:300; Isendahl 2011). Following excavation, most of the ground stone and chipped stone artifacts recovered at La Consentida were not washed. Ongoing analyses of residues collected from some of these artifacts may help to identify microbotanical evidence for their use (see Clark et al. 2007; Morell-Hart et al. 2014).

Early studies of coastal subsistence in the Soconusco (e.g., Coe and Flannery 1967; Green and Lowe 1967), while not benefitting from recent methodological advances such as isotopic and residue analyses, nonetheless provided a rough outline of dietary practices in Early Formative villages. At Altamira, Dee Green and Lowe (1967:31) found evidence for the consumption of deer, rodents, and possibly fish. The scarcity of faunal remains in Barra and Ocós deposits at Altamira suggested to Lowe (1967:58; see also Clark 1994:228) that occupants focused more on plant cultivation (perhaps that of manioc) than on fishing or hunting. Green and Lowe (1967) suggested that their faunal evidence for small mammal consumption corresponded with similar finds at La Victoria (Coe 1961:12, 141). The small Barra phase occupation identified at Altamira Mound 19 does not necessarily promote subsistence reconstructions but does establish continuity of occupation at the site since the beginnings of the Early Formative (G. Lowe 1967:56). Although Lowe (1967:56–57) found it "impossible to conclude much . . . about the Barra phase economic base," he did note a lack of evidence for extensive fishing or shell fishing in comparison with Central and South American cultures from which he proposed that Barra phase ceramics diffused.

What most of the diverse reconstructions of Early Formative subsistence discussed here have in common, it seems, is their focus on agriculture and their attempt to establish its chronology relative to sedentism. In a critique of traditional reconstructions of Early Formative subsistence on the Gulf Coast, Phillip Arnold (2009:398; see also J. Arnold 1996; Killion 2013) has challenged several researchers for relying on "agricentrist" models that depend too heavily on agricultural explanations for social change. Arnold (2009:397) proposed that competition over floodplain resources such as fish promoted social complexity in coastal settings. This model is supported by faunal remains at San Lorenzo, where aquatic resources may have composed 60 percent of meat in the diet (Wing 1978; also see discussion in P. Arnold 2009:400). Elizabeth Wing (1978:31) estimated the relative importance of faunal resources in the San Lorenzo diet by calculating the number of identified specimens (NISP) and minimum number of individuals (MNI) of identifiable taxa, and then applying "correlations between a linear dimension in the skeleton and live body weight using a least squares regression analysis." Arnold (2009:398) suggested that such aquatic resource exploitation was also likely important in the Soconusco and that Early Formative peoples in both coastal regions were fisher-forager-horticulturalists before a Middle Formative adoption of agriculture. Similarly, Amber VanDerwarker (2006:195; see also Cyphers et al. 2013) argued that maize horticulture was only one aspect of a broad-based Early Formative Gulf coastal diet that included exploitation of fruit trees, hunting, and fishing. I agree with Arnold and others (e.g., Garcea 2006; Marshall 2006) that the relationship between early village communities and agriculture must be questioned rather than assumed as components of a "Neolithic Package." For example, in areas with ample wild resources, such as north African rivers and coastal regions of the Pacific Northwest, with fish communities abundant enough to support an intensified subsistence regime, foragers may form sedentary or semisedentary, socially complex communities (J. Arnold 1993, 1996; Garcea 2006).

To summarize the findings of many of the dietary studies discussed above, Early Formative communities on the Soconusco and Gulf coasts appear to have subsisted upon estuarine and floodplain products (such as freshwater and saltwater fish, waterfowl, shellfish, ungulates, beans, and palm fruits) supplemented by limited horticulture of maize and perhaps other domesticates such as beans, squash, and manioc (P. Arnold 2009; Blake et al. 1992; Clark et al. 2007; Jones and Voorhies 2004; Killion 2013; Lesure 2009; Voorhies 1976, 2004). Consistent themes among many coastal subsistence models include the importance of individual and group agency and the impact that diet, and especially the shift to agriculture and use of domesticates in feasting, may have had on social organization. These results differ from earlier studies in the Mesoamerican highlands (e.g., Flannery 1972b, 1973; Flannery and Marcus 2003; MacNeish 1992; Marcus and Flannery 1996; Sanders 1968; Sanders and Webster 1978) that applied an ecological systems approach and that inferred a direct relationship between agriculture, sedentism, and eventual social complexity.

Obviously, La Consentida's coastal environment implies different ecological relationships than were present in the highlands. Even other coastal studies, however, may not be wholly appropriate for comparison with La Consentida. While most of the coastal sites discussed in this chapter are located near estuaries, sediment cores from the lower Río Verde Valley indicate that estuaries in the region are geologically recent and probably formed by around 400 cal BC (Goman et al. 2005). The relationships between domestic mobility, natural resource exploitation, and social organization may thus have been unique at La Consentida in comparison to Early Formative sites elsewhere. Compared to mobility and diet, social organization may be the most complicated and elusive element of ancient life to understand. It is to the discussion of social organization (particularly hierarchical and hereditary inequality) in Early Formative Mesoamerica that I now turn my attention.

## EARLY FORMATIVE PERIOD SOCIAL COMPLEXITY

Archaeological and ethnohistoric evidence indicates that Mesoamerican social organization of the Middle Formative through Postclassic periods included hierarchical inequality inherited according to lineage or familial affiliations. Hierarchical differentiation between nobles and commoners and networks of interaction among nobility in different regions promoted a rich elite culture exemplified by the monumental public buildings and complex political systems of the Late Formative through early Colonial periods (e.g., R. Adams 1966; Drennan 2009; Feinman and Nicholas 1989; A. Joyce 2000, 2010; Kowalewski 1990; Lesure and Blake 2002; de Sahagún 1974; Sanders and Nichols 1988; Sousa 1998; Spores 1997; Taylor 1979; Terraciano 1994, 2000, 2001). Due to the pervasive social influence of commoner/nobility distinctions of later times (and the traditional social models used to explain that complexity [e.g., Morgan 1877]), the type of Early Formative Mesoamerican social complexity most frequently discussed is both *hierarchical* and *ascribed*. This is not to say that heterarchical differences in social roles were insignificant, as I discuss below (Fried 1967:11–14). Rather, it was the establishment of hierarchical differences that many researchers (e.g., Blake and Clark 1999; Clark 2004a; Clark and Cheetham 2002; Flannery and Marcus 2003; MacNeish 1992; Parsons 1974; Sanders and Nichols 1988; Sanders and Webster 1978) associated with the Early Formative period changes in subsistence and settlement discussed in this chapter.

Traditional explanations for the advent of social complexity in Mesoamerica have tended to focus on ecology and the economics of interregional interaction. For instance, human ecology models have viewed social complexity as arising in areas with particular environmental characteristics such as fertile agricultural land (Flannery and Marcus 2003; MacNeish 1992; Marcus and Flannery 1996; Sanders 1956; Sanders and Nichols 1988), aridity requiring the importation of goods and the regulation of irrigation agriculture (Sanders 1968; Sanders and Nichols 1988;

Sanders and Price 1968), or even agricultural shortfall (Symonds et al. 2002). Economic models for social complexity (e.g., Rathje 1971) have focused on interaction and exchange among communities in diverse ecological zones. William Sanders and Deborah Nichols (1988) argued that the rise of social complexity has often occurred in semiarid regions of the world, such as the highlands of central and southern Mexico. Basing their argument in part on Julian Steward's (1955) cultural ecology, Sanders and colleagues (1956, 1965, 1968; Sanders and Nichols 1988; Sanders et al. 1979; Sanders and Price 1968; Sanders and Webster 1978) suggested that the risk of agriculture in arid regions, coupled with the necessity to import goods otherwise unavailable in such areas, promoted the development of "central places." These central places in marginal environments become trade hubs, centers of population nucleation, areas for the accumulation of agricultural surplus, and (through such emphasis on social interaction and increasing population density) loci of developing social complexity. These authors argued that highland Oaxacan sites, such as San José Mogote and Monte Albán, fit the description of such centers. For Sanders and Nichols (1988), natural environmental conditions and human agricultural responses to them provide the main causal influences on cultural development. According to these authors, for example (1988:72), the cultural development of the Oaxacan Mixteca Alta experienced "retardation" in comparison to that of the Valley of Oaxaca because the former lacked strains of domesticated maize properly resistant to frost.

Recent research in the Soconusco, Gulf, and Honduran coastal regions considers the importance of ecological factors such as estuarine productivity, but primarily focuses on the activities of social actors in the establishment of complexity (P. Arnold 2009; Blake and Clark 1999; Clark 2004a; Clark and Blake 1994; Clark and Cheetham 2002; Clark and Pye 2006; Joyce and Henderson 2001, 2007; Lesure and Blake 2002; Love 2007). Rather than focusing on the community or regional level of analysis, such actor-based approaches consider smaller social units, such as individuals and inter- and intra-community alliances. Some of this recent coastal scholarship has demonstrated that key social transformations occurred in variable sequences in domestic, economic, and mortuary contexts in Early Formative sites such as Paso de la Amada. The "aggrandizer model" is one application of agency theory that has been particularly influential among researchers of the Mazatán region. Authors applying variants of this model (e.g., Blake and Clark 1989, 1999; Clark 2004a; Clark and Blake 1994; Hill and Clark 2001; Lesure and Blake 2002; Love 1999, 2002; see also Hayden 1990, 1995, 2009) have considered interregional interaction, exchange and redistribution, competitive generosity, and the management of public events as strategies employed by aggrandizers in promoting the development of social complexity. Community events organized by aggrandizers, according to these authors, included feasts, sporting events, gambling, and communal labor projects. This model essentially proposes that relationships of social indebtedness (in areas of ecological abundance, where resource surplus was possible) sparked the development of hierarchical complexity. The aggrandizer model has been applied mostly in the Soconusco region, though it has also received refinement from archaeologists working in other areas such as the Caribbean coast of Honduras (e.g., R. Joyce 2004a; R. Joyce and Henderson 2001, 2007).

"Transegalitarian" is a heuristic category used to describe societies transitioning between relative egalitarianism and ascribed status distinction (Blake and Clark 1989, 1999; Hayden 1995, 2009). Brian Hayden (1995) identified three principal types of transegalitarian societies: despotic, reciprocator, and entrepreneurial. He identified increasing evidence of social inequality among those three types, respectively. Archaeological implications of transegalitarian societies, according to Hayden (1995:41–42, 49–50, 60–61), include increasing evidence for competitive feasting, exchange of prestige goods, public architecture, procurement and storage of surpluses, ancestor veneration, burial offerings indicating a transition from achieved to ascribed status, population nucleation, and differences in house structures. Although transegalitarian societies do not require agriculture, Hayden (1995:61) argued that the most complex groups among them tend to be agriculturalists, as the social necessity to produce, store, and exchange surplus goods amidst increasing population density becomes extreme. Transegalitarian social organization represents a key theoretical bridge between egalitarian and ranked societies, which are too often viewed as diametrically opposed modes of social organization (Hayden 1995:18).

Michael Blake and John Clark (1999:67) analyzed the development of transegalitarian society as an element of their aggrandizer model for social complexity in Mazatán. These authors determined that trajectories of social change differ among regions but tend to follow the establishment and maintenance of internal and external alliances, the procurement of surplus goods, the sponsorship of craft specialists producing elite wealth items, and the elaboration of mortuary practices. A central aspect of Blake and Clark's (1999) discussion of transegalitarian society is the redistributive economy of surplus goods and prestige items that allows aggrandizers to accumulate indebtedness and acquire influence over their contemporaries. Blake and colleagues (1992) and Clark and Blake (1994:28) argued that Early Formative maize use, for example, consisted primarily of consuming prestige foods used in shows of "competitive generosity" such as feasting. Beyond the redistribution of material wealth, the social capital collected by aggrandizers who organized events such as feasts and ball games left them at the pinnacle of an increasingly institutionalized social hierarchy. By sponsoring feasts and encouraging gambling debts, according to the model, aggrandizers accrued indebtedness convertible into social capital such as control over labor for monumental construction projects. It was only in the context of rising ecological abundance (especially in places such as fertile estuaries, where r-selected resources are plentiful, reproduce quickly, and are difficult to overexploit and where people can accumulate surpluses) and human population growth during the Holocene that such social dynamics sparking complexity could develop.

Material evidence for hereditary inequality in Mazatán suggests that the origins of hierarchical social complexity in Mesoamerica may be sought in the initial Early Formative period and at the very end of the Archaic (Blake and Clark 1999; Clark 2004a; Clark and Blake 1994; Hill and Clark 2001; Lesure and Blake 2002; Love 2007). Diversity in the size and architectural elaboration of domestic structures at sites such as Paso de la Amada demonstrates growing differentiation among households (Flannery 2002; Lesure and Blake 2002). The construction of a public plaza and a large ballcourt at Paso de la Amada by around 1650 cal BC suggests organized communal labor and public events (Clark 2004a:62; Hill and Clark 2001). Although it may be difficult to demonstrate whether these hierarchies were hereditary rather than achieved, Warren Hill and John Clark (2001:333) sought to do so through the study of Structures 2 through 6 at Paso de la Amada's Mound 6. These structures increased in size and preeminence among houses at the site beginning around 1600 cal BC. Hill and Clark (2001) proposed that the ball game became a focal point of communal identity and loyalty, or *communitas* (see also Turner 1967). These authors (2001:331) argued that the Mound 6 construction sequence indicates one of Mesoamerica's first episodes of "ascriptive" social differentiation. Successive generations born into the Mound 6 kin group began to inherit their role as event organizers rather than earning it during their lifetimes as their ancestors had done.

According to Hill and Clark (2001), social actors engaged in playing, ritualizing, observing, and even gambling on ball games produced a self-perpetuating cycle of increasing social differentiation. Ethnohistoric evidence of ball games played in Mazatán at the time of European arrival helped the authors establish a link between the form of Paso de la Amada's ballcourt and the "hipgame" likely played there (Hill and Clark 2001:334, 336). Nineteenth-century reports of a North American Choctaw ball game suggest that at least in some New World contexts, hundreds or even a thousand participants might play in a ball game, while an even greater number observed (Catlin 1953; Hill and Clark 2001:339). Early Spanish accounts of Aztec ball games focused, according to Hill and Clark (2001:339), at least as much on the importance of spectator gambling as on the game itself. Ethnographic accounts of Amazon Basin societies further demonstrate that gambling is sometimes considered sacred, rather than secular or unethical (Gabriel 1996; Hill and

Clark 2001:339). What these examples suggest is the need for facilitators at lively and well-attended sporting events. Hill and Clark (2001) concluded that the role of sport in Mesoamerican social complexity (further evinced by the presence of possible ballplayer helmets on the colossal Olmec chiefly portraits of slightly later Mesoamerican history) supports the aggrandizer model as it pertains to the management of large-scale social events (see Clark 2007:30). The ethnographic analogies imply that pre-Hispanic sporting and gambling sometimes had significant ideological connotations that may have justified the growing prestige of individual facilitators. These interpretations rely, however, on the assumption that a single kinship group occupied Mound 6 throughout its use.

Demonstrating a slightly different perspective than Clark (2004a, 2004b), Lesure and Blake (2002) found that some lines of evidence for the establishment of hierarchical social inequality at Early Formative Paso de la Amada are conflicting. For example, the construction of probable elite residences atop earthen platforms, which contrast with the majority of nonelevated residences at the site, was contemporaneous with homogenous sitewide distributions of obsidian, greenstone ornaments, incense burners, rattles, figurines, celt and animal bone offerings, and possible bloodletters (Lesure and Blake 2002:2, 12). Mortuary evidence of hereditary inequality, such as greenstone objects and a mirror made of mica, is weakly associated or unassociated with the elevated houses, which likely required control over communal labor for their construction (Lesure and Blake 2002:12). These authors argued that such apparently contradictory material traces of complexity indicate that Early Formative social transformations affected some realms of social life (e.g., architectural elaboration) before others (e.g., wealth distribution). This conclusion suggests that evidence for initial social inequality should be sought in discontinuous and variable patterns potentially unique to individual communities or regions. Lesure and Blake (2002) questioned a uniform or monocausal development of social complexity but did find some support for Clark and Blake's (Blake and Clark 1999; Clark and Blake 1994) aggrandizer model. It thus appears that the emergent elites of the Soconusco had the ability to convince their neighbors to build structures and participate in public events, but perhaps could not yet levy this influence for the accumulation of notable personal wealth. I agree with Lesure and Blake that hierarchical complexity is historically and socially contingent rather than the predictable outcome of a simple set of environmental and economic variables. As I will discuss in chapter 9, I also feel that the cultural and ecological variability of Mesoamerica makes the use of any single causal explanation for social complexity in the region untenable.

The aggrandizer model for the origins of Mesoamerican social complexity is not without caveats, and critiques from scholars working in the Soconusco and elsewhere have promoted its refinement. Rosemary Joyce (e.g., 2004a; R. Joyce and Henderson 2001, 2007), Lesure (1997b, 1999b; Lesure and Blake 2002), and Frederick Bove (1989) suggested that the activities of social factions, rather than just those of ambitious individuals, may have been a source of complexity. R. Joyce (2004a) argued that many of the most impressive architectural achievements, and even the development of elite culture itself, likely began as the unanticipated consequences of groups seeing to their daily concerns. According to this perspective, such communal endeavors initially promoted group solidarity and heterarchical social distinctions, without foreknowledge of what would eventually become hierarchical social inequality. Although John Clark and Michael Blake (e.g., Blake and Clark 1999; Clark and Blake 1994) suggested a similar lack of clear predictions about the outcomes of aggrandizer activities, the key difference that R. Joyce (2004a) proposed was the importance of agency on the part of social collectives. Many architectural features that would later become impressive mounds, platforms, and pyramids, for example, may have begun as domestic foundations or as modest stages for public performance. These features were gradually elevated through resurfacing, thus creating a demarcation of social space and thereby opportunities for influential people to appropriate those spaces. I agree with R. Joyce in that assuming Early Formative emergent community leaders were fully aware of the ramifications of their actions produces a teleological fallacy. I would add, however, that assuming those social actors were totally *unaware* of the significance of their aspirations is too simplistic as well. In a society so focused on kinship and ancestor remembrance, emergent Mesoamerican social elites would have considered the legacy they left to their descendants (Clark and Cheetham 2002:292; Hepp and Joyce 2013; Love and Guernsey 2011:181; Marcus 1998; Marcus and Flannery 1996:78, 95).

Specific critiques of the aggrandizer model have included assertions that the social agency of aggrandizers is assumed to have been the exclusive privilege of atomistic, even "westernized," individuals with some foreknowledge of the social hierarchies their activities would produce (e.g., R. Joyce 2004a:16–17). Joyce (2004a:17) claimed that many aggrandizer arguments (e.g., Blake and Clark 1999; Clark and Blake 1994; Clark and Gosser 1995) assume that these hypothetical social actors were independent, ambitious, and usually male. According to such critiques, the aggrandizer model overlooks the activities of not only women and children, but also of collectivities (perhaps organized according to kinship or nonkinship corporate principles) of people working together to modify their social and geographic landscapes. Such critiques suggest that agency must be considered on different scales, with everything from individuals to collectives and entire communities engaging in activities that promoted social complexity. My response to such critiques is twofold. First, Clark and Blake (1994) do not claim that the aggrandizers of the Soconusco were fully aware of the eventual results of their activities. In fact, they state that "the development of permanent social inequality is an unanticipated consequence of individuals pursuing self-interests and personal aggrandizement" (Clark and Blake 1994:28). Second, I do agree that Clark and Blake's model is overly focused on the social maneuverability of individuals at the expense of the agency of social collectives, as the quote I have just provided demonstrates.

I suggest that one potential limitation of the aggrandizer model is that it may underemphasize the negotiated status of aggrandizers' identities and the social contexts, constraints, and norms within which agents operated (Giddens 1979; R. Joyce 2000a, 2004a; R. Joyce and Henderson 2007; Sewell 1992). In other words, agents cannot be considered atomistic individuals divorced from their surroundings. Also, aggrandizer arguments tend not to account for changes in ideology that must have accompanied the transition from achieved to ascribed status. Given that Mesoamerica developed from egalitarian societies of the Archaic, such reformulations of ideology and social interaction must have been significant (Clark 2004a; Clark and Cheetham 2002; R. Joyce and Henderson 2007). The relative diminution of social and geographic independence that the transition to sedentism must have brought (in comparison to seasonally mobile communities, whose members can more easily relocate when displeased with some aspect of their surroundings) would have made for novel modes of interaction with one's increasingly permanent neighbors and may itself have been a factor in establishing hierarchical complexity. I will develop this idea further in chapter 7, particularly regarding the discussion of anthropomorphic iconography at La Consentida.

According to R. Joyce and Henderson (2007:642), social groups, rather than just aggrandizing individuals, built status by sponsoring communal feasts and by using foods to highlight social distinctions. These authors argued that beginning in the Ocotillo phase, the consumption of cacao in coastal Honduras formed part of an environment in which successful feasts increased the social capital of factions sponsoring them. R. Joyce and Henderson (2007) argued that Early Formative cacao use at Puerto Escondido shifted from feasting events centered on the consumption of alcoholic cacao beverages to the consumption of nonalcoholic cacao as an element of elite "cuisine." Supporting evidence for this interpretation includes ceramic residue containing theobromine, a distinctive alkaloid marker of cacao. This perspective differs from the aforementioned aggrandizer model in that it focuses on community constructions of elite "cuisine" rather than on feasts sponsored by ambitious individuals. For a food to be transformed from an element of communal feasting to one of elite cuisine, however, implies that its availability at some point became restricted. This limiting of access to a good, similar in some ways to the demarcation of space discussed previously by R. Joyce (2004a), indicates a process of restriction that must have been central to increasing hierarchical social distinctions. I find Joyce and Henderson's discussion of shifting cultural significance attributed to food compelling and will discuss this further in chapter 6 with regard to what I believe were changing practices of maize consumption at La Consentida.

Clark (2004a) later reformulated his version of the aggrandizer model in light of the critiques summarized above. Explaining his (with Blake's) earlier hypothetical construction of aggrandizers as "frenetic and single-minded individuals" to be "descriptive excesses involved with initial stages of model-building," Clark (2004a:47) reenvisioned them as diverse social actors with negotiated and socially constructed identities. These agents sought their ambitious aims of self-promotion within the confines of an egalitarian or transegalitarian ethos retained from the Archaic period. Such aggrandizers were able to transform their social landscapes only in settings in which resources such as labor and consumable goods were abundant (Clark 2004a:47). The actions and strategies of aggrandizers, according to Clark (2004a:47), included "ritual feasting and drinking, sponsorship of craft specialists, long-distance exchange, gift-giving, competitive sports, and communal construction projects." Archaeological evidence for these strategies includes ornate Barra phase ceramics, which appear to have been used to serve feasting beverages such as maize beer or chocolate, rather than for daily domestic meals (Clark and Blake 1994; Clark et al. 2007). Of such proposed strategies, Clark (2004a:67) has viewed the organization of community labor for constructing monumental architecture and public spaces, which symbolized a growing community identity, as the most important factor in the birth of Mesoamerican social complexity. Clark's revisions have partially met the challenges of detractors by reimagining the social unit of Early Formative period agency but have done less to address explicit gender bias in the model. The best contribution of the aggrandizer concept as applied in Mesoamericanist research, in my opinion, is the diverse activities it proposes (such as organizing feasts, labor projects, sporting events, and gambling) as causal to social change.

Reformations of the aggrandizer model have not quieted all critiques. Although Love (2002) supported the model in the past, he appears to have grown more critical of it over time. Michael Love (2007:285–86) noted that Clark and Mary Pye (2006) and Lesure and Blake (2002) continued to support the aggrandizer model despite its "weak" "empirical base." Without significant variation among household middens in ceramic styles and vessel forms at Paso de la Amada, argued Love, one must question the interpretation that feasts permitted aggrandizers to accumulate indebtedness or to display wealth. He further stated that evidence for obsidian redistribution at the sites of Tajumulco, El Chayal, and San Martín Jilotepeque, which Clark and Tamara Salcedo Romero (1989) originally saw as evidence for a "petty chiefdom," has more recently proven inconclusive (Clark and Pye 2006). Finally, Love found mortuary evidence for Locona phase social differentiation to be unsatisfactory. While Clark (1991, 1994) interpreted differences in mortuary goods in juvenile burials as indicative of ascribed inequality, Love (2007:285) found this evidence insufficient to rule out "differential regard for those who die young." In general, while Clark's model predicts wealth inequality as a major indicator of growing social differentiation, the archaeological evidence for such inequality is mixed.

Love (2007) acknowledged the presence of a regional settlement hierarchy in Mazatán, the status of Paso de la Amada as a likely regional ceremonial center, the interpretation of Structure 6 as a possible elite house, and the presence of the earliest known ballcourt in Mesoamerica (Clark 2004b; Hill and Clark 2001; Lesure and Blake 2002). In terms of proposing his own explanations for Early Formative social complexity, Love (2002:199–200) has tended to emphasize interregional social interaction, of which he discussed two main types. Interaction sphere models, according to Love, tend to consider the Gulf Coast and other regions of Mesoamerica as relatively equal partners in exchange and early social complexity (e.g., Flannery and Marcus 1994; Grove 1989; Marcus 1989; [cited in Love 2002:199]). Core-periphery models, including the Olmec "mother culture" concept, tend to view the Gulf Coast as a primary source of early Mesoamerican social complexity, iconographic traditions, and monumental sculpture and architectural production, perhaps after an initial fluorescence in the Soconusco (Clark 1997; Coe 1968, 1989; Diehl and Coe 1996; Tolstoy 1989b; [cited in Love 2002:200]). Recent research regarding early Maya monumentality (e.g., Inomata et al. 2013) suggests that the Olmec heartland was not the only source of Formative period hallmarks of social complexity such as ceremonial architecture.

Although Love (2002:201–4) concluded that the evidence is insufficient to choose between these models, he pointed out important problems with each, including that San Lorenzo chronologies are too imprecise to demonstrate early dates for all monumental sculptures and that symbols on ceramics belonging to the "X Complex" appear unassociated with social rank. This discrepancy between indicators of interregional interaction (such as the ceramic X Complex) and other archaeological evidence of social hierarchy is reminiscent of Lesure and Blake's (2002) findings regarding contradictory traces of complexity in Mazatán. Ultimately, Love (2002:202, 204) implied that the only consistent causal mechanism behind Early and Middle Formative social complexity in different regions is that of population size and nucleation. Regardless of the as-yet poorly understood mechanisms that drove Early Formative interregional interaction, Love (2002:202) found more decorated, exchanged ceramics at larger sites that arguably served as "nodes for interregional interaction and economic exchange." He did not explain what caused sites to grow larger or better connected to interaction networks in the first place, though favorable ecological conditions seem a probable candidate. Such contradictory interpretations of the Mazatán and Gulf coastal data underscore the importance of both ecology and agency in explanations of social complexity. Despite conflicting conclusions such as those discussed here, several authors (Blake and Clark 1999; Clark 1991, 2007; Hill and Clark 2001; Love 2007:285) describe the Soconusco as "precocious" in its development of social complexity, regardless of later Gulf coastal phenomena that eclipsed it. I am generally supportive of Love's shrewd comparison of the material records of different regions, though I find that ethnographic evidence is inconsistent with some of his specific critiques of the aggrandizer model, as discussed below.

Divergent interpretations regarding the advent of social complexity in regions such as the Soconusco stem from conflicting data as well as from differing theoretical perspectives (P. Arnold 2009; Blake and Clark 1999; Clark 1991, 1994, 2004a; Love 2002, 2007; Marcus and Flannery 1996). The case for Early Formative social complexity is tantalizing at Paso de la Amada, with its 3,600-year-old ballcourt and a possibly associated residence (Clark 2004a; Hill and Clark 2001; Lesure and Blake 2002). In the Río Naranjo subregion of the Soconusco, however, significant settlement hierarchy and public architecture did not appear until the Middle Formative, and even then, most likely as the result of contact with Mazatán (Love 2002:19–26, 199–202). I believe that it is only through careful study and comparison of various lines of evidence, such as those discussed by Lesure and Blake (2002), that archaeologists can hope to reconstruct Early Formative social organization. This comparison of multiple lines of evidence is important because initial social complexity is likely to have left diverse material signatures in different sites and regions, as I will discuss more fully in chapter 9 (R. Joyce 2004b; R. Joyce and Henderson 2001; Lesure and Blake 2002).

Love's (2007) critique of the evidence for social complexity in the Early Formative Soconusco is similar to arguments by Blanton and colleagues (1999:39), who suggested that late Tierras Largas and San José (1450–1150 cal BC)-phase highland Oaxacan burial practices, though indicative of burgeoning wealth discrepancies, might imply moieties rather than elite/commoner differences. In this vein, Arthur Joyce (2010:113–15) argued that mortuary, settlement pattern, and architectural data for Early Formative highland Oaxacan social complexity are weak in comparison to those in the Soconusco and Gulf coastal regions. Variation in quantities and types of probable prestige goods (such as worked shell and imported stone) in domestic refuse at sites such as Santo Domingo Tomaltepec in the Valley of Oaxaca has been interpreted as evidence for early social complexity (Blanton et al. 1999:34–42; Flannery 2002; Marcus and Flannery 1996:76–110; Whalen 2009:76–77; see also

Cervantes Pérez et al. 2017). Richard Blanton and colleagues (1999:35) discussed burial evidence for San José phase social differentiation, which included patterning in flexed versus extended body positioning, quantities of grave goods, and stone slabs over apparently "elite" burials. Mixed evidence for complexity in highland Oaxaca, such as a lack of obvious "chiefly" houses despite mortuary evidence for the beginnings of achieved wealth accumulation (Blanton et al. 1999:37), resembles the aforementioned mixed signal of formalized hierarchy in the Soconusco.

Many researchers (Carballo 2009; Flannery 2002; Hill and Clark 2001; Lesure 1997a, 1999b; Pool 2007:124; Whalen 1983:24) have argued that public architecture in Early Formative population centers suggests group ritual and political activities developing from earlier domestic practices and that these public activities may have been organized by emerging elites. Although public spectacle itself does not denote inequality, the organization of labor for constructing public architecture such as platforms and ballcourts (in Mazatán, for example) to house those events may do so to a certain extent (Clark 2004a:67). Also, though examples may be found elsewhere in the world for monumental and public architecture made by ostensibly "egalitarian" groups (e.g., Göbekli Tepe on the Anatolian Plateau), such examples must be used with the caveat that there is still much we do not understand regarding the social dynamics of the communities who produced them (Banning 2011; Peters and Schmidt 2004). Several authors (e.g., Drennan 2003b; Marcus and Flannery 1996) have argued that San José Mogote in the Valley of Oaxaca attracted a large population during the Early Formative in part due to its fertile agricultural land. This site may have later developed public ritual architecture indicative of initial social complexity. It is worth noting that while some authors (e.g., Drennan 2003b:47; Flannery 2009b; Marcus and Flannery 1996:109–10) interpret a set of buildings at Tierras Largas and San José phase San José Mogote as public ritual structures, others (e.g., Winter 2002:69) suggest instead that they were high-status houses. Although Flannery and Marcus (1994:129–32; Marcus and Flannery 1996:87) interpret these "Men's Houses" as public structures repeatedly razed and reconstructed over time in the same spots, other authors disagree. Based on stratigraphic reinterpretation, Clark (2004a:50) argued that the structures were a set of neighboring buildings used concurrently. Clark (2004a:52) concluded that the possibly simultaneous use of these relatively small buildings suggests political competition among contemporaneous social factions rather than diachronic maintenance of community ritual. If Clark is correct, and the Early Formative highland Oaxacan structures are a collection of redundant community buildings rather than a long-used "Men's House," a major component of the argument for hierarchical social complexity in these Early Formative communities is lost.

Differing interpretations of evidence for initial social complexity in Mesoamerica may sometimes relate to basic theoretical differences such as degrees of adherence to General Systems Theory. Critiques of the application of systems theory in archaeology (e.g., Berlinski 1976; Salmon 1978) have suggested that archaeologists have imprecisely used many of the specific terms and ideas developed within systems ecology, but which are inappropriate for application to anthropological questions. As Merrilee Salmon (1978:182) argued, even the best uses of systems theory by archaeologists (her example being the work of Kent Flannery) could have borrowed equally productive theoretical influence from other realms of study. Responses to Salmon's critique (e.g., Lowe and Barth 1980; see also Salmon 1980) were swift, and in some cases seemed to imply that a rejection of the principles of ecological systems thinking was tantamount to a rejection of any form of systematic or scientific thinking in archaeology. In a sense, the debate seems to have descended into accusations of scientistic versus antiscience approaches to archaeology. Salmon (1980:576) acquiesced that General Systems Theory (GST) may represent a useful heuristic device for thinking about the social dynamics studied by archaeologists but maintained that most attempts by archaeologists to utilize the terms and approaches developed within GST have been "confused, jargon-ridden, and full of grandiose claims." Perhaps most damningly, Salmon (1980:578) pointed out that the use of GST principles to formulate replicable models for ancient social dynamics (as in the case of computer simulations) must necessarily include the assumption that certain variables of ecological relationships and social dynamics are causally linked to one another (i.e., in order to promote "homeostasis"). The potentially "high degree of arbitrariness" in such determinations of causality, argued Salmon (1980:578) has led proponents of GST in archaeology to produce misleading explanations of social change that may do more to obfuscate causality than to clarify it. In short, Salmon argued (and I agree) that human societies cannot be reduced to the mathematical models or computer simulations favored by GST-inspired archaeology. Attempting to explain social dynamics in this manner may lead to models that appear "logical" or that seem to find causal links for social change, but that bear little relationship to reality.

One of my main critiques of the ecological systems approach that Flannery and colleagues have applied to the highland Oaxacan evidence relates to the authors' overly straightforward associations between the purported quality of agricultural land and the advent of social complexity. I agree with recent findings of Mesoamerican subsistence research (see discussion above) that contradict proposals that social complexity can be predicted by something as simple as the distribution of fertile "Class I" land (Marcus and Flannery 1996:79–80), though I do acknowledge the importance of soil variation to agriculturalist communities. In a similar vein, archaeologists applying ecological systems theory in Oaxaca have tended to rely on overly strict dichotomies between masculine and feminine activities and between the public and domestic spheres of community activity (e.g., Flannery and Marcus 1994:129–32; Marcus 1998, 1999; Marcus and Flannery 1996:87). As I have argued elsewhere (Hepp and Joyce 2013), I feel that "domestic," as applied to Formative Mesoamerica, is in some ways a misnomer. When viewed through a modern, Western lens, "domestic" seems to imply "private," or "hidden within the household." My research indicates that Formative period domestic life was defined by public and social interaction, a circumstance in part driven by the necessity of cooperating with ever-more-permanent neighbors in increasingly sedentary communities. No longer as free to "vote with their feet," as their Archaic forebears, Formative period peoples gathered their domestic lives into communal scales of collaboration and contestation in a process I have referred to as "communal domesticity." Furthermore, historical analyses of early colonial Mixtec gender concepts (Sousa 1997, 1998:108; Spores 1997:186; Stern 1995:242, 248; Taylor 1979:108; Terraciano 1994:393, 2000:16), linguistic study (Terraciano 1994:176–77), and Zapotec ethnography (Stephen 1991:76–77, 2002:41–59) suggest that gender roles in ancient Oaxaca existed within a framework of complementarity rather than one of strict segregation. These data contradict interpretations that figurines were always used in homes by women (Marcus 1998) or that groups of men gathered in public spaces to make all decisions for the community (Flannery and Marcus 1994:129–32; Marcus and Flannery 1996:87).

Several archaeologists (e.g., Blanton et al. 1999; Clark 2004a; A. Joyce 2010; Winter 2002) have generally been critical of attributions of hierarchical social inequality to Early Formative highland Oaxacan communities. Arthur Joyce (2010:74) argued that no highland Oaxacan settlements of the Early Formative, with the exceptions of San José Mogote and Yucuita, had populations above "a few dozen people." Even in these larger towns, a similarity in size and form of houses (averaging between 18 and 24 m$^2$ in size) and bell-shaped domestic storage pits suggests that neighborhoods were organized according to kinship rather than social status. Neighborhoods at San José Mogote were remarkably similar to those in the region's smaller, more "typical" villages (A. Joyce 2010:75). According to Joyce (2010:77), Tierras Largas and San José phase burials indicate an emphasis on domestic community organization, variability according to kinship, and "minimal wealth and status distinctions." This relative lack of hierarchical distinction suggests an emphasis on kinship over ascribed status. Although they acknowledged the evidence for some social differentiation in the region by the San José phase, Blanton and colleagues (1999:39–42) suggested that Early Formative highland Oaxacan society was organized not according to inherited hierarchical status, but instead pursuant to moieties dedicated to the earth and the sky. The evidence for the moiety model comes from imagery interpreted as referring to "lightning" and "earthquake" on ceramic vessels (Blanton et al. 1999:40; Marcus 1999). Even if arguments for social complexity in Early Formative highland Oaxaca are accepted, the earliest dates for that complexity (ca. 1350 cal BC) postdate those in Mazatán by several centuries (Clark 2004a:62; Marcus and Flannery 1996).

The debates summarized above indicate that archaeologists have considered many material correlates for identifying and understanding social complexity in Mesoamerica. One line of evidence that I have so far only briefly discussed is that from mortuary data. Differential burial practices, by which certain individuals were interred with elaborate offerings or with different body treatment, provide evidence of the diverse ways Mesoamericans treated their dead (Gillespie 2001; R. Joyce 1999; Spencer and Redmond 2004; Whalen 1983:30–33; Winter 2002:68). For decades, archaeologists have inferred degrees of social difference through the analysis of burial practices. Some (e.g., Binford 1971; Gillespie 2001; Saxe 1971) have argued that differential burial of children may indicate hereditary status distinction. Clark (1991, 1994), for example, interpreted offerings interred with juveniles at the Early Formative Chilo "site cluster" in the Mazatán region, such as a mica forehead mirror with a juvenile at Locona phase Vivero, as indicative of ascribed inequality. This interpretation is partially based on anthropomorphic figurines with forehead mirrors recovered at the Ocós phase site of Cosme (Clark 1994:126). Other scholars have voiced concerns about the use of mortuary data to infer social organization, in part because it is difficult to determine what variations in the quantity and type of grave goods indicate status rather than other social affiliations or idiosyncratic variation (Carr 1995; Love 2007). As mentioned previously, Love (2007:285) warned, for example, that mirrors interred with juveniles in the Soconusco might simply reflect the social importance of youth in ancient communities. This interpretation may contradict Amerindian ethnographic evidence for infancy and personhood, however, particularly in societies with traditionally high infant mortality. Among the Wari', for example, it is not the act of being born, but rather the establishment of social connections, that makes a person (Conklin and Morgan 1996:681; see also R. Joyce 2000a; Parsons 1936).

Carr (1995:188) called on archaeologists to recognize that the determining factors of differential mortuary treatment include not only social organization but also "philosophical-religious" beliefs. Following Jos Bazelmans (2002), Rosemary Joyce (2005:143) has argued that funerary offerings, accoutrements, and attire represent a complex interplay of social meanings, and are "not simply ... reflections of a coherent 'identity'" or status. Joyce (2005:143) referred to these complex meanings of burial goods as "the enactment of embodiment in mortuary contexts." Strengthened by these evaluations, mortuary analysis remains an important step in understanding social relations in ancient communities (Morris 1991; Rakita et al. 2005). Mortuary analysis is important for studying social organization not just because it promotes comparison of burial offerings, but also because ancient human bodies themselves varied in ways that sometimes relate to status or to other dimensions of identity. Markers of skeletal health indicative of dietary and lifestyle differences include porotic hyperostosis, suggesting malnutrition; linear enamel hypoplasia, indicating juvenile growth interruption (Cook 1981; Goodman 1991; Skinner and Goodman 1992); cribra orbitalia, suggestive of anemia; and Harris lines in skeletal growth plates that indicate arrested growth (Lallo et al. 1977; Martin et al. 1985; Mays 1995; Roberts and Manchester 1995; Stuart-Macadam 1989). Although variable health among individuals is not exclusively a result of unequal status, skeletal health indicators that correlate with evidence from offerings or body treatment may indicate that people led dissimilar lives (Danforth 1999:17–18; Márquez Morfín et al. 2002; Paynter 1989; M. Pearson 1999:81, 210; Santley and Rose 1979; Storey et al. 2002). For that reason, mortuary analysis is appropriate as one avenue (accompanied by other types of data such as those from iconographic study and the comparison of domestic contexts) for inferring modes of social organization.

What most of the models for social complexity discussed above have in common is their emphasis on ascribed hierarchical inequality. I believe this to be problematic because many anthropologists (e.g., J. Arnold 1996; McGuire 1983; Pauketat and Alt 2003; Pauketat and Emerson 2007) recognize that inherited status distinctions are only one type of social complexity, which may also take the form of heterarchical specialization without formal hierarchy. Furthermore, heterarchical and hierarchical social distinctions may be negotiated, contested, and variable throughout one's lifetime according to such factors as age, gender, roles enacted in ritual cycles, specialized knowledge, and myriad social affiliations (Brumfiel 2003, 2006; R. Joyce 1999, 2000a, 2002; Stockett 2005). Heterarchical and hierarchical distinctions may be ascribed or achieved during one's lifetime. Although all societies have some heterarchical and hierarchical differences and may practice division of labor according to variables such as sex and age, certain forms of economic specialization are less common (Crumley 1995, 2004; McIntosh 1999; Pauketat and Emerson 2007; Vega-Centeno Sara-Lafosse 2007:169). I argue that acknowledging the significance of heterarchical complexity is not the same as stating the facile truism that "all human groups are socially complex." Instead, groups with marked economic or ritual specialization despite a relative lack of ascribed hierarchical inequality are *different* (perhaps as a matter of degree rather than of kind) in their modes of social organization than other egalitarian groups (see J. Arnold 1996; Fried 1967). Material indicators of complex heterarchy might include anthropomorphic iconography (see R. Joyce 2000b; Lesure 1997b, 1999a, 2011c) or grave goods suggesting that community members fulfilled diverse social roles or possessed different kinds of crafting abilities (perhaps suggesting emergent specialization), ecological knowledge, or ritual knowledge or skill (see Carr 1995; Gillespie 2001).

Most anthropologists agree that groups marked by inherited status display a different form of complexity from those in which social distinction is attainable only through achievement (e.g., Blake and Clark 1999; Hendon 1991; A. Joyce 2010; Kowalewski 1990; McGuire 1983; Paynter 1989; Spores 1997). It is widely accepted (Banning 1998:229; Clark and Cheetham 2002; Feinman 1995; Hayden 1990:32–33; Sanders and Price 1968) that hierarchically ranked society is a recent development in comparison to the relatively egalitarian societies that comprised most of human history. Egalitarian societies have traditionally been defined as those in which there are "as many positions of prestige in any given age-sex grade as there are persons capable of filling them" (Fried 1967:52). Nevertheless, the recognition that status can vary according to gender, age, and achievement in all societies has led many to argue that nowhere are people truly equal (Blake and Clark 1999; Blanton 1998; Clark and Blake 1994; Feinman 1995:261; Flanagan 1989; Paynter and McGuire 1991). This conclusion is not new, however. As Morton Fried (1967:11–14) argued, true "egalitarianism" is a fallacy because humans will always be somewhat self-interested. The key difference between "egalitarian" and ranked groups may lie in the relaxation of leveling mechanisms that worked against the formalizing of hierarchy during our egalitarian past (Feinman 1995:262; Hayden 1995).

As discussed above, authors such as Blake and Clark (1989, 1999; Clark and Blake 1994) and Brian Hayden (1995) have proposed "transegalitarian" as a category for the transition between egalitarian and hierarchical social organization. In my opinion, the concept of transegalitarian society is important for understanding the origins of Mesoamerican social complexity because it combines extrasocietal factors such as the availability of rapidly reproducing r-selected resources with the agency of real people operating according to societal freedoms and constraints, while recognizing that transitional communities leave variable and often ambiguous material evidence (Blake and Clark 1999; Clark 2004a). As I will argue in subsequent chapters and conclude in chapter 9, "transegalitarian" is a useful descriptive concept for much of the archaeological evidence at La Consentida. One should remember, however, that many results of aggrandizing practices were likely unintentional outcomes of more immediate and personal goals and that aggrandizing agents need not always have been ambitious (or male) individuals, but instead could be collectivities of diverse scale and composition. For that reason, I am generally sympathetic to the refinement of the aggrandizer model to accommodate agency on multiple social scales (Clark 2004a; R. Joyce 2004a).

Debates discussed above regarding such lines of evidence as architectural differentiation (e.g., Clark 2004a:50–52; Flannery and Marcus 1994:129–32; Marcus and Flannery 1996:87; Winter 2002:69), scales of social agency (such as that of individual aggrandizers versus that of social collectives) (e.g., Blake and Clark 1999; Clark and Blake 1994; R. Joyce 2004a:16–17), and mortuary analysis (e.g., Carr 1995; Clark 1994:126; Love 2007:285) indicate that no one line of evidence can supply a satisfactory indication of past social organization. Put simply, there may be many *necessary* types of evidence for social complexity, but there appears to be no single *sufficient* indicator. The possible permutations of extracommunal and intracommunal influences on Early Formative period social change leave a convoluted and regionally variable material record. As Lesure and Blake (2002) discussed, the search for material indicators of Early Formative complexity often yields mixed results, and archaeologists should expect some forms of evidence in the absence of others. One should thus not anticipate that certain kinds of evidence for social complexity would appear in unison across domestic, economic, and ritual milieus of Early Formative society. Furthermore, though clear evidence of domestic, mortuary, and economic differentiation at a site might suggest the presence of fully developed hereditary social inequality, the very beginning stages of status differentiation will likely be much more difficult to recognize (Clark and Blake 1994:29). In such cases of incipient complexity, even faint hints regarding the different roles of individuals may be significant as precursors of fully institutionalized inequalities of the later Formative and Classic periods (R. Joyce 2004a). It may be that some Early Formative period individuals were preeminent community administrators, but that such roles were achieved through personal accomplishment or endowed through participation in collective action with corporate partnerships rather than being ascribed by birthright (see Blanton et al. 1996; Carballo 2013; Fargher et al. 2010). Perhaps also these nascent elites were initially unable to use their influence for *personal* gain but could instead enlist it on behalf of the community. Such considerations might be particularly applicable to Early Formative period sites, where long-held leveling mechanisms of the Archaic likely instilled resistance to the novel status reorganizations of the Formative or constrained the degree to which influential people could institutionalize hereditary status distinctions.

## SUMMARY

As reviewed in this chapter, ongoing debates in Early Formative period archaeology concern the establishment of sedentism, the subsistence economy supporting early village life, the origins of institutionalized social complexity, and the relationships among these transformations. While Early Formative sedentism and social complexity in the highlands may have been based on agriculture (see Marcus and Flannery 1996:79–80), recent scholarship suggests that some coastal populations founded villages before they were reliant on domesticates such as maize and that sedentism was adopted gradually in some areas (P. Arnold 2009; Kennett et al. 2010; Killion 2013; Lesure 2009; Rosenswig 2007). One potential limitation of subsistence evidence from coastal contexts is that it typically comes from sites near estuaries (Blake et al. 1992; Kennett et al. 2010; Lesure 2009). This pattern begs the question: what was the economic basis for communities in coastal regions without estuaries? Such a lack of evidence represents a limitation to understanding Early Formative period diversity. The interplay between ecology and social organization in Early Formative coastal communities is still not well understood in many regions, but likely had significant influences on practices of mobility and community development (P. Arnold 2009; Clark and Cheetham 2002; Coe 1981; Joyce and Goman 2012; Stark and Voorhies 1978; Wing 1978). Patterns of material culture variation among ecologically distinct zones, such as differences in ceramic vessels between estuarine and slightly inland sites in the Soconusco, emphasize this point (Lesure 2009:2).

In light of continued debates in Early Formative archaeology, it is clear that there was no single set of conditions or sequence of events behind transitions in settlement, subsistence, and social organization in Mesoamerica. Initial sedentism began at different times in various regions and seems (at least in some coastal zones) to have been based on a mixed subsistence economy of horticulture and foraging, rather than on a truly agriculturalist diet. Early social complexity seems to have been historically contingent and regionally variable. Its development was based on agency at different scalar levels and on ecological circumstances promoting the accumulation of resource surplus that could be employed for communal events. Debates regarding these transformations suggest that individual causal mechanisms fail to explain the transformations across diverse regions and also signify the value of evidence from sites in previously unstudied or understudied regions, which may provide new insight into this era of significant social change. Investigations at La Consentida represent an opportunity to address the relationships between sedentism, subsistence, and social organization in a region that has seen very little previous study of the Early Formative period. Coastal Oaxaca may have been a significant but previously unrecognized participant in the roots of Mesoamerican culture. As I will seek to show in the remaining chapters of this book, research at La Consentida will help to illuminate that role.

# 3

## Methods and Mapping for the La Consentida Archaeological Project

This chapter summarizes the LCAP's phases of research, field and laboratory methods, mapping results, and definitions of some terminology used throughout the book. Research at La Consentida in 2008 included ground-penetrating radar (GPR), informal site reconnaissance, and an analysis of ceramics collected during the excavation of a test pit at the site in 1988 (Barber 2009; A. Joyce 1991a:116–17; A. Joyce et al. 1998:105–6; Winter 1989). In 2009, the research team returned to La Consentida for limited total station mapping and a pilot excavation project during which we removed about 15.4 m³ of sediments in two operation areas at the western edge of Platform 1, on its northern and southern slopes. A 2010 laboratory project helped to begin the classification of La Consentida's ceramic and chipped stone artifact assemblages. The results of these preliminary studies proved useful for demonstrating the early dates of site occupation and for securing grant funds to support a longer project in 2012. The 2012 field season consisted of nine months of research divided into three phases. The first phase (January–February) focused on mapping. The second phase (February–June) focused on excavations. The final phase (June–October) included preliminary laboratory study and processing of artifacts and samples collected during excavations. A 2013 laboratory season focused on the analysis of ceramics, figurines, and ground stone, and a 2014 laboratory project included the study of faunal remains. In 2017, laboratory studies focused on the lithic assemblage, the production of digital artifact models using photogrammetry, and sonication to sample microbotanical remains from probable food-processing artifacts. These latest studies are still in progress. Samples of carbon, obsidian, ceramics, human bone, animal bone, human teeth, and microbotanical remains have been exported from Mexico with INAH permission in 2010, 2012, 2014, and 2017 to undergo specialized laboratory analyses. Studies completed so far have included AMS radiocarbon dating of charcoal, carbon-rich sediment, and human bone processed with XAD purification (a special procedure using resin to remove older or younger carbon that has invaded the bone's porous matrix), X-ray fluorescence (XRF) of obsidian, and the analysis of stable isotopes in the dentin and enamel of human teeth and in human long bones and faunal bone. All La Consentida materials that have not been exported are stored at the INAH facility in Cuilapan, Oaxaca.

## REMOTE SENSING

Analysis of GPR in 2008 was undertaken in two long transects along a modern road bisecting the western edge of La Consentida's Platform 1, as well as in three rectangular areas atop and at the northern and southern margins of the western end of Platform 1 (Barber 2009; figure 3.1).[1] Due in part to interference from plant roots, the rectangular areas of GPR produced little useful information. Analysis of the data produced by the long transects, however, indicated as many as six to eight subsurface anomalies along the modern road. As discussed below, these anomalies formed the basis for the 2009 excavations at La Consentida.

## MAPPING AND SURVEY

Much information about the location and setting of La Consentida can be learned from publicly available remote sensing imagery. Figure 3.2, for example, shows the site's location as visible on a Microsoft Bing Maps™ image. The map demonstrates the densely vegetated location of the site within the Chacahua National Park, as well as the site's proximity to modern coastal estuaries.

In May of 2009, fieldwork at La Consentida began with the clearing of underbrush to expose key areas for excavation, which were chosen based on the GPR results from the previous field season. A site datum point was set, and a north/south baseline established to facilitate mapping. Excavation areas were placed adjacent to the disused road bisecting Platform 1. Although GPR results indicated anomalies directly under the road, excavations were performed slightly west of the road to avoid potential compaction interference produced by vehicle traffic that crossed over the platform as recently as the 1990s. An arbitrary Cartesian grid was set for the site, with the main datum designated as 5000N 5000E, in order to record the

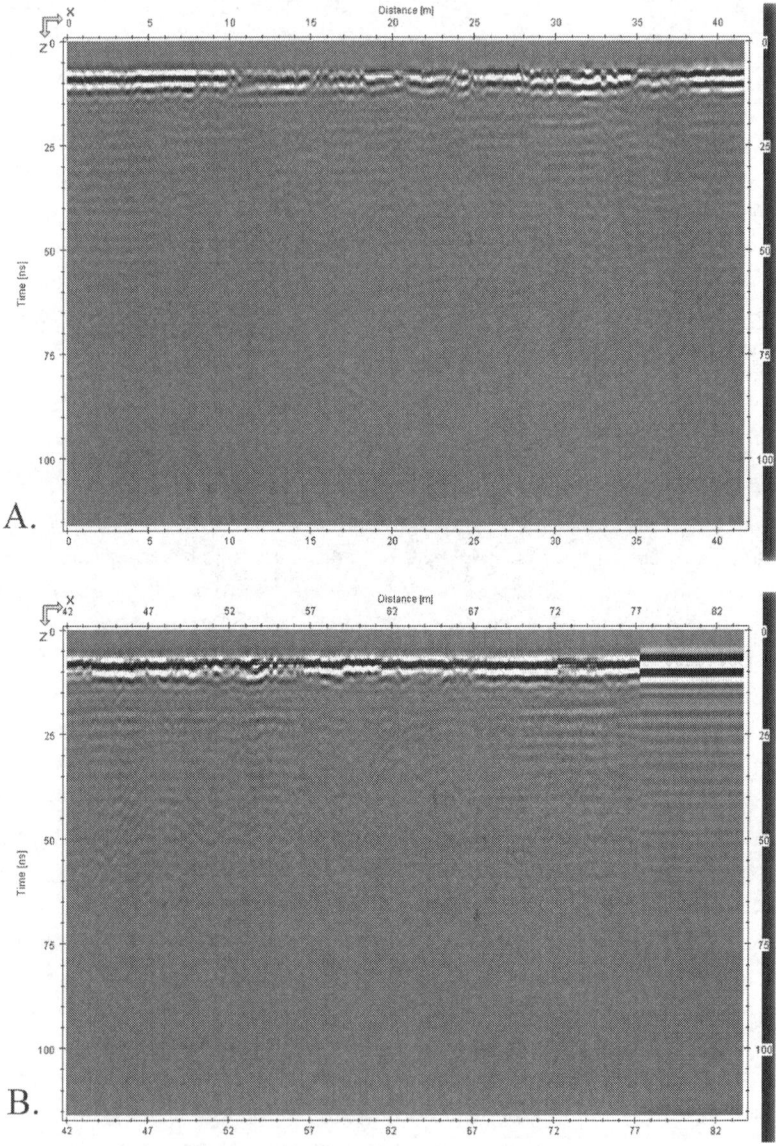

**FIGURE 3.1.** GPR results from La Consentida: (A) results from the southern margin of Platform 1—Anomalies detected at 10 m, 15 m, 18 m, and 35 m (Left: South; Right: North); (B) results from the northern margin of Platform 1. Anomalies detected at 54 m, 61 m, 66 m, and 72 m (Left: South; Right: North)

**FIGURE 3.2.** Bing Maps™ image showing La Consentida in relation to modern estuaries in the lower Río Verde Valley

location of the datum, mapping points, and the southwest corner of excavation operations. A casual walking survey in 2008 and 2009 (performed beyond the boundaries of an initial GPS mapping project undertaken in 2000) indicated that the site was larger, and its topography more complex, than initially thought (see Joyce et al. 2009b:524). Several earthen substructures were identified atop Platform 1. Survey also located the collapsed remains of a historic brick structure near the northwestern edge of Platform 1. Approximately 50 m west of Platform 1 is a dry streambed that may relate to the deposition of sand and silt in the natural strata beneath earthen architecture at La Consentida (A. Joyce 1991a:166, 408; Mueller 1991:826; Winter 1989). Terrain, time constraints, and vegetation limited the extent of mapping during the 2009 season. The task was completed as an initial component of the 2012 study.

The first step of the 2012 field season was to use machetes to clear as much vegetation from Platform 1 and a surrounding buffer area as possible. Because the site is located in a national forest, no large trees were damaged during this process. Following site clearing, the research team spent three weeks mapping with a total station. We recorded a total of 1,507 three-dimensional data points. This mapping process revealed the location of seven earthen mounds (Substructures 1–7) atop Platform 1 (figures 3.3 and 3.4). We discovered that another modern road crosses the eastern edge of the site. As of 2012, this road was providing local farmers with access to large colonies of bees kept in the forest. The better understanding of site topography afforded by these mapping results, as well as surface finds identified during mapping and survey, helped to refine plans for excavation during the following phase of research. As demonstrated by figures 3.3–3.5, several types of maps can be generated from these data. Figures 3.3–3.5 are topographic maps using 20 cm intervals. Figures 3.3 and 3.4 are identical plan view maps, with the exception that the former indicates the locations of earthen features at the surface, while the latter indicates the locations of excavation operations at the site. Figure 3.5 provides an oblique view of the site surface with the vertical dimension exaggerated for greater visibility of Platform 1 and the various substructures.

## EXCAVATIONS

In general, excavations at La Consentida followed procedures established by previous projects in the region (e.g., Barber 2005; A. Joyce 1991a; Levine 2007). This was largely done so that project results would be comparable to findings from later sites in the lower Río Verde Valley. The 2009 LCAP excavations were undertaken in two operation areas, which were respectively labeled LC09 A and LC09 B. The 2012 excavations occurred in eight new operation areas (LC12 A–LC12 H). These excavations also reopened Op. LC09 B, in order to access previously identified human burials, some of which could not be fully excavated in 2009 due to time constraints (Hepp 2011a). Each operation was divided into 1 × 1 m excavation units in order to maintain horizontal control. In some cases, exposing features and stratigraphic associations required the excavation of units outside the original grids. To preserve vertical control, excavations proceeded in levels of 5–20 cm in both arbitrary and natural lots wherever stratigraphic changes were identified. The relatively low quantity of artifacts in some construction fill contexts made the thicker 20 cm lots practical in some cases. In areas with high densities of artifacts, features, or burials, excavation lots were reduced to 5–10 cm in either arbitrary or natural stratigraphic levels. In most excavation areas, surface sediments were heavily disturbed by bioturbation. Tree roots and *camote de agua* tubers were among the most destructive culprits, and their size is a good reminder of what can cause artifact disruption and reverse stratigraphy, particularly near the surface (figure 3.6). All excavated sediment was passed through 1 cm mesh screens. In areas of highest sensitivity (e.g., structure floors, burial fill, and an offering), we used .4 cm mesh to prevent losing small items from good contexts.

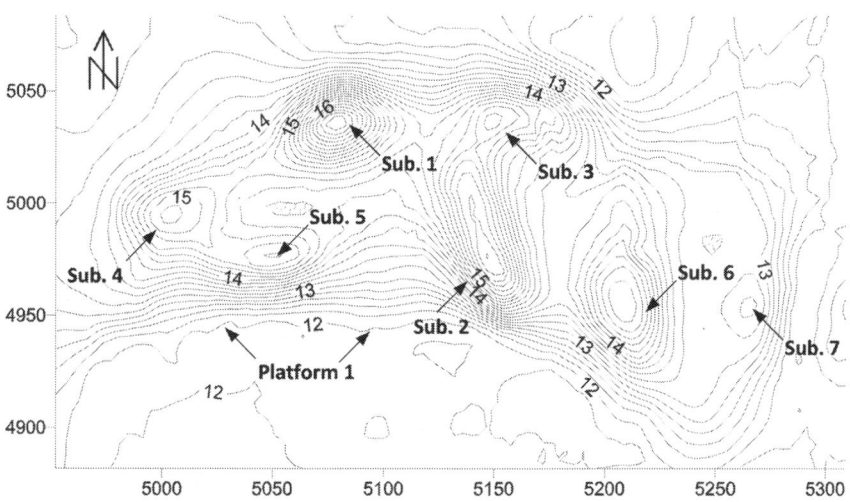

**FIGURE 3.3.** Topographic map of La Consentida showing Platform 1 and Substructures 1–7

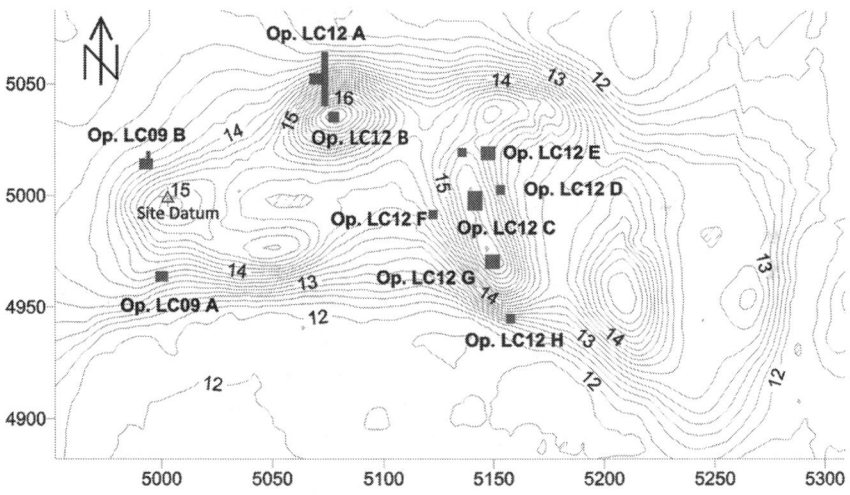

**FIGURE 3.4.** Topographic map with locations of 2009 and 2012 excavations (dimensions of operation areas are approximate)

The few exceptions to this screening procedure included some redeposited fill contexts that were deemed appropriate for excavation without screening because they were so disturbed. All units were backfilled and manually compacted after excavation and illustration.

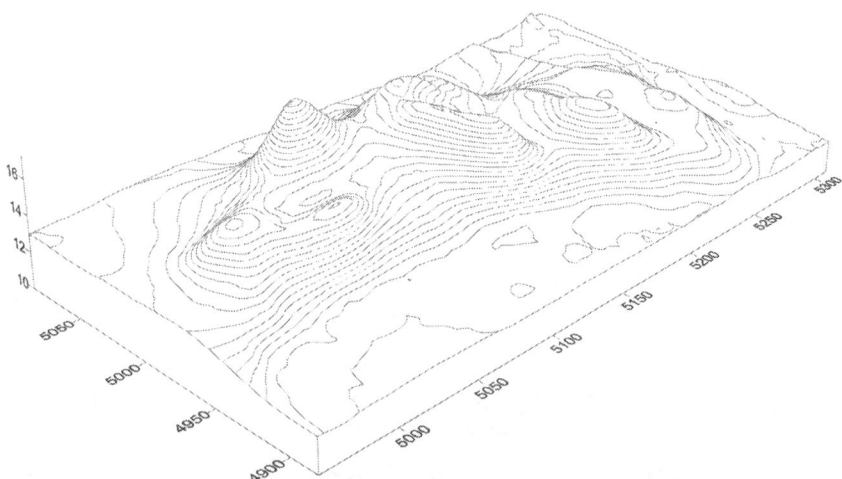

**FIGURE 3.5.** Topographic map of La Consentida with oblique view elevations exaggerated approximately ten times

Excavation teams collected sediment samples for basic flotation and soil chemistry analysis from certain contexts such as occupational surfaces, fill from human burials, hearths, and middens. Most samples intended for flotation have been processed, and heavy fractions from six samples have been studied for faunal remains (see chapter 6). These samples typically consisted of 2 to 4 l of sediment, depending on the size of the feature and the amount of material available. During the field investigations, project archaeologists completed 1:10 scale excavation profile and feature plan drawings. The research team also took well over 2,000 photographs of excavated units and of important discoveries such as stratigraphic changes, features, and in situ artifacts. Investigators described each stratum by color, consistency, plant activity, artifact density, and stratigraphic relationships. Project archaeologists organized and assigned field specimen numbers to all artifacts according to provenience and material type. Following previously established conventions in the region (e.g., Barber 2005; A. Joyce 1991a; Levine 2007), provenience coordinates for each excavated context included operation, unit, and lot designations according to the Cartesian grids governing LCAP field investigations. Either during excavation or during sediment screening, the research team collected all identified ceramic artifacts, lithics, shell, and bone. All sherds were counted and weighed by paste type. Decorated or diagnostic sherds were curated for typological analysis, while many undecorated body sherds were discarded due to INAH regulations governing very limited storage space.

**FIGURE 3.6.** Examples of camote de agua tubers that disturb surface deposits at La Consentida

Excavations in 2009 were intended to explore the aforementioned GPR anomalies. Excavations in 2012 built on results of the 2009 pilot study (Hepp 2011b, 2011a). Goals of the 2012 excavations included bisecting part of the northern edge of Platform 1 and Substructure 1 (especially in Op. LC12 A) with a long trench to reveal the stratigraphy produced by platform and mound construction episodes (figure 3.7). To test hypotheses regarding social complexity, excavations returned to Op. LC09 B, an area with several human burials initially identified in 2009 (Hepp 2011b). Because numerous burials were identified here and in Op. LC12 A, mortuary analysis became another major focus of the 2012 investigations. As described in chapter 7, investigation eventually determined that these two areas represent possible cemeteries. Mortuary studies included the excavation, photography, and detailed illustration of each burial. Human skeletal remains were then collected for analysis by bioarchaeologist José Aguilar, as described in more detail below (see also Aguilar and Hepp 2015; Hepp et al. 2017). The research team recorded and collected artifacts interred as burial offerings (such as ceramic vessels, figurines, and stone tools) for analysis and comparison with similar artifacts from other contexts.

Finds from the very earliest site occupations were uncovered in multiple operations in 2012, and included stratified midden deposits and a few human burials, along

**FIGURE 3.7.** Operation LC12 A trench overview, facing north

with their mortuary offerings. Additional platform construction layers, slightly different ceramics, and several human burials were deposited at La Consentida later in the site's occupation and were uncovered in the excavations at Op. LC09 B, as well as in several areas excavated in 2012. Chronological ceramic variation is subtle and does not yet justify the establishment of separate formal subphases, though future analyses may help to further quantify these variations (see appendix 2). Early Formative period occupations at La Consentida produced large quantities of lithic debitage. Prismatic blades, which were introduced to much of Mesoamerica in the late Early Formative and Middle Formative periods, are rare at the site. These occur mostly in shallow deposits dating to shortly before site abandonment and/or to later reuse of the site. Although this pattern may suggest to some that La Consentida was occupied into the Middle Formative, Early Formative period prismatic blades identified in other regions (see Hirth et al. 2013), as well as the single ceramic phase identified at La Consentida (see appendix 2), are consistent with a shorter occupation that took place entirely within the Early Formative period. As discussed in chapter 4, the site later experienced a small Early Classic period reoccupation atop Substructure 1.

To test hypotheses regarding sedentism and social organization, some 2012 excavation areas (including Ops. LC12 B, LC12 C, and LC12 G) were intended to expose probable domestic contexts atop substructural mounds. These excavations targeted the substructures in part because such mounds often contain domestic contexts at later sites in the region (Barber 2005:140–41, 235; A. Joyce 1991a:292). The density of surface artifacts atop these mounds, identified during the mapping phase, was also useful for locating probable domestic areas. Once such contexts were identified, the research team employed shallow, horizontal excavations in order to expose structure floors and occupational surfaces. After exposing the remains of structures, excavation teams used a few smaller, penetrating excavations to identify superimposed occupational surfaces, explore wall fall and the remains of foundations, and investigate probable postholes. Operations such as LC12 C, LC12 E, LC12 F, and LC12 H searched for middens at the base of Platform 1 and its substructures. The ceramic and lithic artifacts, as well as faunal remains from middens, provided evidence of the site's vessel assemblage, iconographic artifacts suggesting interregional relationships and aspects of social organization, and dietary practices.

## ARTIFACT PROCESSING AND LABORATORY ANALYSIS

Preliminary artifact processing and some laboratory analysis of the artifacts and samples collected during 2009 and 2012 excavations took place at field houses / laboratories rented in the coastal Oaxacan town of San José del Progreso. Local workers and project archaeologists were instrumental in these phases of research. Formal laboratory analysis projects in 2010, 2013, and 2014 took place at the INAH storage and research facility in Cuilapan, Oaxaca. Bioarchaeologist José Aguilar analyzed the human remains collected during the 2009 and 2012 field seasons. Aguilar's assessment of each set of remains focused on basic information such as age, sex, pathology, and evidence of habitual activities (Aguilar and Hepp 2015; Hepp et al. 2017; chapter 7). Aguilar and I collected M2 and M3 molars for analysis of stable isotopes as evidence of ancient diet. Stable isotope expert Paul Sandberg and I then prepared the samples at the University of Colorado for further analysis at specialized laboratories. Additional human teeth and long bones, as well as faunal bones, were exported in 2014 and have recently been part of a regional stable isotope study (Joyce et al. 2017; chapter 6). David Williams (2012) analyzed lithics recovered during the 2009 excavations. Of those hundreds of lithic artifacts, forty obsidian samples were submitted for XRF analysis to reconstruct the ancient exchange networks in which La Consentida was involved (see chapter 8). Biologist Silvia Pérez Hernández and I organized the study of faunal remains from the 2012 excavations during a 2014 laboratory season at Cuilapan (see Pérez Hernández and Hepp 2015). Our study focused in particular on the comparison of animal remains from the Ops. LC09 B, LC12 D, LC12 E, and

LC12 H middens. We also analyzed the faunal remains from the LC12 A-F15 ritual cache (see chapters 4 and 7). Pérez Hernández undertook the majority of species identification, while I chose appropriate contexts, provided reference materials and species research, and sorted the heavy fractions of six flotation samples from the Op. LC12 D, LC12 E, and LC12 H middens. We followed standard faunal analysis procedures for estimating the number of identified specimens (NISP) and the minimum number of individuals (MNI) of each taxon identified (see, for example, Banning 2000:187–211). Methods of species identification and data reporting benefitted from the example of a previous study on animal remains in the region by Deepika Fernández (2004; see also A. Joyce 1991a).

I analyzed the ceramics (appendix 2), figurines and musical instruments (chapters 7 and 8), and ground stone (chapters 5 and 6) recovered from the site. Methods of ceramic analysis followed previous examples in the lower Río Verde region (A. Joyce 1991a:121–73) and in the valley of Oaxaca (Caso et al. 1967; Martínez López et al. 2000). Ceramic analysis began with sorting sherds by paste. The two primary paste categories identified were medium brown ware and coarse brown ware. Although excavations recovered a small collection of fine brown and gray wares, these more recent ceramics occurred near the modern surface and postdate the site's primary occupation.

Temper in the paste of the ceramics consisted of sand, grit, gravel, and possibly grog (crushed, recycled pottery) and shell. Some sherds also contained voids and carbonized materials indicating the use of organic temper. Nearly all La Consentida ceramics have a micaceous paste, which is naturally occurring in clay sources within the Río Verde fluvial system, rather than representing intentional additives (Mueller 1991:834–36). Comparative count and weight analysis of ceramic paste types indicates a predominance of medium brown wares among all ceramics from the site (Hepp 2011b; appendix 2). Medium brown ware ceramics are rare in the lower Río Verde Valley assemblage except during the late Middle Formative Charco phase (A. Joyce 1991a:126–29; Urquhart 2010). The occurrence of this paste type at La Consentida is consistent with an occupation in the Middle Formative period or earlier, as indicated by the lack of medium brown wares in later ceramics in the region (A. Joyce 1991a). Preservation of surfaces of ceramics from the earliest occupations at the site was generally superior to that of later ceramics. Due to the apparently low firing temperatures, however, few large sherds from the earliest vessels survive. It is possible that the better surface preservation of the earliest sherds as compared to later examples is due to their heavy slips and burnished surfaces. Flooding events caused by the seasonal variability of the Río Verde may also have influenced differential artifact preservation, as the condition of ceramics varies strongly according to stratigraphy (see chapter 4 and appendix 1). Geochemical processes may also be responsible for some of the ceramic deterioration. La Consentida sediments are mildly acidic with a pH of 6.2, whereas most sediments in the lower Río Verde valley are neutral with a pH of 7.0–7.2. This slight difference in acidity is unlikely to have a significant impact on ceramic preservation, however (Raymond Mueller, personal communication, 2009). I feel that the most likely cause of ceramic erosion at La Consentida is low firing temperature used to produce the region's first ceramics, which generally have a soft, sandy paste, leaving them susceptible to erosion from redeposition.

Following the analysis of sherds according to paste, various vessel forms were identified within each paste category. This process was difficult in many cases due to the fragmentary condition of the collection. Although many rim sherds were present, for example, rarely did they include an entire vessel wall. Only a handful of complete vessels, mostly interred as mortuary offerings, have been recovered at the site. Identified vessel forms among each paste type were numerous. Some vessel forms, especially various types of globular jars and conical bowls, make up the majority of the collection. Other vessels—such as tecomates, bottles, and semispherical bowls—were present and relatively consistent in form across paste categories but were not as common as jars and conical bowls. Grater bowls suggest food preparation or some other crafting activity such as pigment processing, while worked sherd discs may have been used as grinders or perhaps as lids (see chapter 6 and appendix 2).In order to investigate patterns of artifact production and symbolic representation at La Consentida, I illustrated, photographed, and analyzed the best-preserved figurines, musical instruments, decorated ceramic vessel fragments, and other types of figural artifacts. Examples of these artifacts can be found in chapter 7, chapter 8, and appendix 2, among other locations. Analysis of chipped stone artifacts indicated the use of gray and black obsidian, chert, quartz, and chalcedony (Hepp 2011c; D. Williams 2012). Lithics are dominated by gray obsidian debitage, some of which has been utilized. Although few formal tools were identified, exceptions include bifacially flaked chert knives (one example of which was found near the cranium of burial B2-I3), scrapers, and especially stone drills (see chapter 6). Other lithics include a few probable exhausted chert cores and a variety of ground stone tools. Obsidian XRF results indicate that La Consentida was part of a complicated exchange network including six obsidian sources located hundreds of kilometers away in Central Mexico (see chapter 8).

Stable carbon ($^{13}C/^{12}C$) and nitrogen ($^{15}N/^{14}N$) isotope composition of teeth has been recognized an important indicator of diet in many archaeological contexts (Blake et al. 1992; Boyd et al. 2008; Katzenberg 2000; Price et al. 2008; Schwarcz and Schoeninger 2011; Sealy 2006; Tykot and Staller 2002; Webster et al. 2005). The results of dental isotopic analysis of human and faunal remains from La

Consentida are discussed in chapter 6. A key factor when considering nitrogen isotope levels as indicators of marine resource use is the comparison of $^{15}N/^{14}N$ levels between human remains and those of local terrestrial animals. Such animals, if they did not consume marine products, can provide a "baseline" for nitrogen isotope levels, which can help to differentiate between marine and maize dietary indications in human remains (Sealy 2006:578). Human teeth and long bone samples from numerous individuals, as well as fifteen marine and fifteen terrestrial animal bones, have thus been exported and analyzed for their isotopic indicators (Hepp et al. 2017; Joyce et al. 2017; chapter 6).

Chipped stone, ground stone tools, boiling stones, and ceramic vessel forms can also provide evidence of ancient food preparation and diet. Manos and metates are often associated with maize processing, while mortars and pestles have been more frequently associated, due to their portability, with wild resource exploitation and horticulture practiced by mobile groups (Clark et al. 2007; Rosenswig 2006:339). It is noteworthy that some small manos were likely used for hide processing and other crafting tasks (J. Adams 1988), but the presence of metates at La Consentida suggests that at least some ground stone was for food processing. The study of probable food-processing artifacts provides evidence for reconstructing both ancient diet and domestic mobility patterns. Stone tools and ceramics may also be useful for identifying the presence of domesticates additional to maize, particularly if cooking jar styles or stone tool use-wear suggest the relative unimportance of maize or the processing of domesticates not demonstrated by a dental isotopic analysis (Clark et al. 2007). I performed a basic typological analysis of ground stone tools according to established methods of categorizations employed by other researchers working in Oaxaca (e.g., Winter and Mateos 2010). I photographed, weighed, and measured ground stone tools recovered at the site. I illustrated many of the complete or otherwise diagnostic examples (see chapters 5 and 6). I inspected each artifact for use-wear from activities such as grinding, hammering, polishing, and chopping. In many cases, the tools appeared heavily used for a variety of tasks. As discussed above, and though it hampered some use-wear analysis, ground stone and chipped stone artifacts were generally left unwashed to allow for ongoing residue analyses of pollen, phytoliths, and/or starch grains (see Morell-Hart et al. 2014). To date, the primary goal of studying La Consentida's ground stone has been to identify basic tool types (such as manos, metates, grinders, hammer stones, pestles, etc.) and to determine if there are any obvious chronological or spatial patterns present in the kinds of tools found in different contexts. As described in chapter 5, these artifacts represent secondary evidence of the community's practices of domestic mobility. In chapter 6, I discuss these artifacts as an indication of changing dietary practices.

## A COMMENT ON RADIOCARBON DATING

Throughout this text, I make reference to dated contexts and to individual AMS radiocarbon dates that allow me to make specific chronological statements about La Consentida's Early Formative period occupation. The carbon samples I have chosen for processing have been selected out of the larger number originally collected in the field for their stratigraphic significance and for the integrity of the deposits in which they occurred. For instance, a piece of carbon "floating" in fill is not as useful as one associated with a hearth or directly from human remains, because it may have been redeposited from an earlier occupation layer. Table 1.1 briefly reports the contexts from which I have collected dated samples, which include hearths sandwiched between layers of platform fill, burned food (likely from an annual plant such as maize) adhering to a jar fragment collected from a midden, occupational surfaces, and a direct date from human bone. I have endeavored to elaborate on the exact contexts from which each of the reported dates was collected. Many of these descriptions occur in appendix 1, where I describe the excavated deposits at La Consentida, or in chapter 4, which contains discussion of the site's occupational history.

A special discussion is worthwhile regarding the AMS radiocarbon date of human bone collagen from the B12-I14 burial (see table 1.1, figure 1.2, chapter 7). Radiocarbon dating of bone has traditionally been considered problematic because bone, as a porous material, absorbs exogenous, or "background" carbon from its surrounding matrix. Purification via XAD is a process designed to remove that erroneous carbon in order to date the collagen from the bone itself (Waters et al. 2015). Despite poor collagen preservation in bone from the lower Río Verde Valley generally, collagen extracted from a femur fragment from the best-preserved burial thus far excavated at La Consentida (B12-I14) produced enough material for AMS dating following XAD purification. The collagen from B12-I14 was actually dated twice. The resulting dates were 3310 ± 25 (PRI-5423A [H6]; human bone) or 1660–1510 cal BC and 3375 ± 30 (PRI-5423B [H6]; human bone) or 1750–1610 cal BC (Cummings 2017). When these two samples are combined, the resulting date is 3335 ± 20 (PRI-5423A/B [H6]; human bone) or 1690–1530 cal BC.[2] This dated human bone thus represents direct evidence of human occupation (as well as additional evidence for pottery production, the consistent orientation of human remains, stone beads used in burial offerings, and the production of cemeteries) at La Consentida during the Early Formative period (see Aguilar and Hepp 2015; Hepp et al. 2017).

## TERMINOLOGY

Throughout this book, I use technical terms to refer to certain types of features, artifacts, and contextual relationships. I try to use these terms in a way consistent with standard archaeological practice (e.g., Banning 2000; Kipfer 2007; Roskams 2001) and as established by previous research in the lower Río Verde Valley (e.g., A. Joyce 1991a). Although I define many of these terms the first time I use them, I will take this opportunity to clarify some of the most common examples. I use the term "occupational surface" to refer to an interface between strata that seems to have been a stable surface on which daily activities were carried out in antiquity. Evidence for identifying such surfaces includes refitting ceramic sherds in a horizontal position at a stratigraphic interface and other indications of occupation such as thin bands of ash and alignments of architectural stones. I use this term independently from "floors," which I consider to have been inside ancient structures, and "soils" or "paleosols." In some cases, "paleosols" (ancient soils identifiable by their dark color and prismatic structure) are evidence that a stratum was a stable surface for an amount of time sufficient to allow soil formation. "Domestic" contexts are identified as living areas atop substructures that are associated with small buildings (probably houses) and lenses of artifacts such as ceramics, animal bone, and ground stone. Sharp chipped stone debitage is frequently not found in domestic zones, because ancient people were no fools and recognized the danger of stepping on it. Of course, postdepositional processes can muddle such patterns, if they were ever clear to begin with. I differentiate "domestic contexts" in the general sense from "domestic middens," which are related to the former but were the locus of trash accumulation away from actual living and daily activity areas. In general, I make reference to possible or probable domestic zones, rather than definitive domestic zones. I do this because it is possible (though unlikely, for reasons I explain in detail for each circumstance) that some buildings atop La Consentida's mounds were public in nature. "Daub" refers to fired earthen material used to seal the walls of structures made of "wattle" (woven sticks) and likely with thatched roofs and is useful for identifying probable domestic contexts (see A. Joyce 1991a). It is generally assumed that most daub was fired either unintentionally or as part of a practice of retiring buildings, rather than as a step in construction. "Fill" contexts are composed of redeposited sediment used to construct earthen platforms and mounds. "Middens" are refuse heaps deemed to be in primary context, unless described as otherwise, as in the case of a few likely redeposited middens. They tend to contain ceramics, lithics, animal bone, and often shell and ash. In some situations, high artifact density and poor artifact preservation have led to the interpretation that redeposited midden has been used as fill. The term "in situ" is used to refer to features or artifacts that appear not to have been moved or redeposited after their formation by either cultural or postdepositional processes. In some places, such as in chapter 8, I use the term "diagnostic" to refer to certain artifacts such as ceramic vessels or vessel fragments. Depending on the context, I use this term to denote materials recognizable and/or specific to a given place, time period, or category. In this usage, "diagnostic" may be taken as an antonym of "generic."

I employ an abbreviation system to refer to operation areas, excavated units, and features. I have endeavored to keep this system consistent with previous studies in the region (e.g., Barber 2005; A. Joyce 1991a). Operation areas are described first by year (i.e., LC09 or LC12, depending on whether they were excavated in 2009 or 2012), and then by letter according to the order in which they were established. In some cases in which I have already stated the year, I avoid repetition by omitting it. The research team labeled excavation units according to a Cartesian grid in which numbers increase from west to east and letters increase from south to north. Refer to figures in appendix 1 for visual examples. Excavated strata were labeled either natural (N) or features (F) and were assigned numbers. These numbers increase from the surface downward, though they are described chronologically, beginning with the deepest/earliest deposits (see appendix 1). I frequently refer to elevations "asl" (above sea level) for excavated features. Where contextual data permit, I report these elevations to the nearest centimeter (for example, 15.15 masl refers to 15.15 meters above sea level). Where contextual information is approximate (such as with the elevation of strata that vary somewhat in their size and location), I refer to approximate elevations "asl" with 0.1 m accuracy. Approximate elevations asl can be found in all profile drawings. Burials are considered cultural features and are independent of individual sets of skeletal remains. This is an important distinction in Mesoamerican archaeology, as burials often contain multiple individuals (e.g., A. Joyce 1991a:app. 5). Following convention in the lower Río Verde region, burials at La Consentida are numbered (e.g., B1, B2, B3, etc.) separately from individual sets of remains (e.g., I1, I2, I3, etc.). A standard reference to a set of human remains would thus indicate both the burial and the individual (i.e., B1-I1).

# 4

## La Consentida's Occupational History

This chapter presents a summary of information available to date for how La Consentida was settled, occupied, and modified through time. These results are based on the detailed description of excavated deposits, which I present in appendix 1. For that reason, figure references in this chapter will frequently direct the reader to the appendices. Throughout this discussion, I use radiocarbon dates and artifact comparison to indicate vertical and horizontal relationships among features. I emphasize vertical stratigraphic relationships and chronological change, though I also discuss horizontal/synchronic patterns where the data allow. When it is impossible to identify stratigraphic crossties among excavated contexts, I divide the discussion according to operation areas. In total, the research team undertook the 2009 and 2012 excavations in one hundred and eight 1 × 1 m units and a single .5 × 1 m unit, organized into ten operations. These excavations analyzed approximately 146.0 m$^3$ of sediment. The locations of operations were selected with the goal of answering key project research questions.

### PREOCCUPATIONAL STRATIGRAPHY AT LA CONSENTIDA

Preoccupational strata were uncovered in Ops. LC09 A, LC09 B, LC12 A, LC12 D, LC12 F, and LC12 H, most of which were located near the edges of Platform 1. In many areas, alternating strata of natural alluvial and fluvial silt and sand (e.g., LC09 A-N1 through N3, LC09 B N-1 through N-5, and LC12 A-N1 through A-N2) indicate the migrations of a stream across the site prior to the first stages of platform construction (e.g., figures A1.1–A1.3, A1.5, A1.8, A1.12, A1.15, A1.16). These natural strata tend to occur at about two meters below modern ground surface (10–11.5 masl), depending on the depth of fill at a given location. In one area (Op. LC12 H), excavations were halted at 9.97 masl after they encountered both fluvial sands (H-N1) and the modern water table. La Consentida's natural deposits include sands of various grain sizes, indicating that the depositional force varied over time, likely as the channel of the river or stream migrated. Tiny grains of volcanic glass diagnostic of sediments transported by the Río Verde support that river's identification as the ultimate source of these materials, though it was likely a smaller stream that finally brought the sediments to La Consentida (Mueller 1991; Mueller et al. 2014). It may have been due to the unpredictable flooding and/or migration of the river that the first layers of architectural fill were constructed, thus raising the site above floodwaters. Fluvial sands at La Consentida often contain carbonized wood fragments, which have not been collected for radiocarbon dating due to the possibility that they were redeposited from earlier burning events upstream. Frequently occurring in these natural strata, and even in some of the deepest cultural strata above them, are rhizoliths, which are calcium deposits formed around the roots of ancient plants. Ceramics and bone fragments from both the natural and the deepest cultural strata are sometimes found within similar calcium carbonate concretions. In some cases, bone from cultural deposits appears to be mineralizing.

Test auguring performed by Raymond Mueller and a field assistant helped to assess deposits below the area of some excavations (Mueller et al. 2014). An auger core below the deepest excavation in Unit LC09 B.1H, for example, identified gravel under the fluvial sand at about 10.10 masl, indicating that deeper strata resulted from higher energy channel deposits. The auguring also identified groundwater at about 9.80 masl. At Op. LC12 A, Mueller performed sediment coring beneath an area of human burials at the very edge of Platform 1 and Substructure 1, at the boundary between Units −1R and −2R. He found fluvial sand at about 10.90 masl and groundwater at about 9.95 masl (Mueller et al. 2014). The composition of the deep, natural strata in Op. LC12 A (e.g., A-N2), suggests actions of the river similar to those that produced the Op. LC09 A strata (A-N1 and A-N3) and those elsewhere at the site.

Some of the sands directly below cultural occupations at La Consentida contain artifacts likely brought down from shallower deposits by postdepositional processes such as bioturbation. A small number of artifacts were perhaps interred in natural strata through cultural processes such as excavation of sediment for use as fill material, but these cases are difficult to demonstrate and are probably rare. Artifacts from the deepest contexts include ceramics, burned bone, and black or gray obsidian flakes.

These materials are similar to those from the first identified occupation layers deposited immediately prior to, and during, initial Platform 1 construction.

## INITIAL SITE OCCUPATION

The earliest occupation surfaces provide a glimpse of the material culture of the people who first occupied the site prior to the construction of Platform 1. In some cases, as in Ops. LC09 A and LC09 B, no clear occupational surfaces were identified between natural strata (e.g., LC09A-N1 and LC09B-N1) and early platform fill layers (e.g., LC09 A-F5 and LC09 B-F16; see figures A1.1–A1.3 and A1.5). It is likely that areas such as the extreme western edge of Platform 1, where these operations were carried out, were not intensively occupied prior to platform construction. It is also possible that parts of the site were occupied before the construction of Platform 1, but that the river washed cultural deposits away. Although some of La Consentida's ground stone tools, such as small one-handed manos, are "Archaic period" in style (Clark et al. 2007; MacNeish et al. 1967:101–21; Winter and Mateos 2010; Winter and Sánchez Santiago 2014a:10–11), it is noteworthy that no preceramic or Archaic component has been identified at the site. Based on the artifacts recovered in the deepest cultural layers, it appears that the site's first occupants were already producing ceramics.

The clearest example of a prearchitectural occupational surface was uncovered in the Op. LC12 A trench, which was excavated at the northern margin of Platform 1 and Substructure 1 (figures 3.4 and 3.7). This thin cultural zone, identified at the interface between A-N1 and the A-F19 platform fill layer, included in situ ceramics (including a tecomate rim), obsidian, burned daub, animal bone, and charcoal (figure A1.8). This zone suggests a brief occupation atop natural sediments prior to initial construction of Platform 1. The few artifacts were lying flat, suggesting that they were deposited atop the A-N1 surface rather than within fill. Ceramics identified at this depth include highly burnished medium brown wares with black, orange, and red slipped surfaces. They are generally representative of the earliest of the site's Tlacuache phase ceramics, and indeed of the earliest known ceramics in the entire lower Río Verde region (see appendix 2). The ceramics were not appreciably different in their paste or surface treatment from those associated with the first phases of Platform 1 construction. The few animal bones identified on this occupational surface included a large deer vertebra.

## EARLIEST EARTHEN ARCHITECTURE: PLATFORM 1 CONSTRUCTION

The first levels of mounded earthen architecture (e.g., LC09 A-F5, LC09 B-F18, LC12 A-F19, D-F11, F-F6, and H-F4-s3) tended to consist of a compact, yellowish silty clay or silty clay loam fill. Due to the relative consistency of these earliest fill strata across the site, and of the redeposited artifacts they contain, it appears that Platform 1 already covered much of its greatest horizontal area by its first phase of construction or that initial earthen architecture consisted of several small mounds made of similar sediments. Some initial fill strata (e.g., LC09 A-F5) contain inclusions of granodiorite and gneiss, which are components of the region's natural bedrock, and are commonly found near the surface in coastal piedmont zones (Raymond Mueller, personal communication 2009). It is likely that the earliest fill layers at various areas of the site differ slightly in clay versus silt content because they include redeposited natural sediments that vary due to the fluvial energies of the ancient river or the position of collected sediments relative to that river. Despite these minor fluctuations, the uniformity of earthen construction across the site, along with early carbon dates for various initial fill deposits, suggests a foundational settlement of considerable dimensions and which probably entailed significant labor investment (see tables 5.1–5.4). Artifacts associated with early fill included slipped and burnished black, red, and orange medium or coarse brown ware pottery, essentially identical to artifacts from prearchitectural occupation surfaces in the deepest cultural layers of Op. LC12 A (see appendix 2). Also recovered from early fill were anthropomorphic figurines, animal bone, black and gray obsidian flakes, and marine shell.

The tops of several early platform fill layers (e.g., LC09 A-F5) were used as occupational surfaces before being covered by further construction. Hearths (LC09 A-F4-s1 and LC09 B-F15) that intrude into initial fill layers demonstrate these early platform occupations (figures 4.1–4.4). The LC09 A-F4-s1 hearth was constructed with a circular arrangement of stones and contained burned earth and shell. The A-F4-s1 hearth was 34 cm deep and had a diameter of 125 cm. A smaller possible auxiliary hearth (A-F4-s2), located on the eastern edge of the large hearth, had a diameter of 50 cm. It is possible that A-F4-s2 was used for food preparation activities supplemental to those in A-F4-s1; similar auxiliary hearths are used today in the lower Río Verde Valley. Because this hearth intrudes into the A-F5 fill stratum (at an elevation of about 11.88–12.22 masl), its construction must have postdated initial platform construction and have been contemporaneous with occupation atop A-F5. Although few artifacts were found in direct association with A-F4-s1, several medium brown ware sherds were found in and around the hearth. One large sherd appeared either to have been a piece of construction material for making the hearth or to have become affixed in place by the burning of the A-F5 sediments composing it. Although shell is not especially common in Platform 1 fill layers, A-F4-s1 was filled with a compact shell deposit that was about 25 cm thick and was mixed with an ashy silt matrix (figures A1.1 and A1.2). The A-F4-s1 hearth dates to 1900–1690 cal BC (table 1.1).

**FIGURE 4.1.** LC09 A-F4-s1–s3, Units A.3B and A.4B

Another hearth, similar to LC09 A-F4 but slightly postdating it according to the radiocarbon dates (see tables A1.1 and A1.2), was identified at the northern edge of Platform 1 in the main portion of Op. LC09 B. In this area, B-F18 was likely the initial platform fill layer. B-F16 must have served as an occupational surface, as evidenced by the large B-F15 hearth (figures 4.3 and 4.4) intrusive into it. B-F15 had something of an irregular form in its northwestern quadrant due to rocks and pieces of burned earth that may have fallen away from the rest of the hearth. This feature had a diameter of 120 cm and consisted of a circular arrangement of stones and burned sediment that may have hardened during firing episodes. The hearth was similar in composition, stratigraphic relationships, and perhaps function to LC09 A-F4-s1, though it lacked shell and instead contained dark, carbon-rich sediment. As discussed above, the B-F15 hearth dates to 1755–1525 cal BC.

Early platform fill deposits uncovered in Op. LC12 A, at the northern edge of Platform 1 and Substructure 1, are similar to those in the LC09 A and LC09 B areas. At this northern operation, LC12 A-F19 was the initial stratum of platform fill. Occupation atop this layer is indicated by in situ refit sherds, daub, and burned animal bone at the interface between A-F19 and A-F18-s2 and produced an AMS date of 3443 ± 35 (AA101267; plant charcoal; $\delta^{13}C = -27.2‰$), or 1885–1665 cal BC. In addition to providing a date for architectural construction, occupation atop

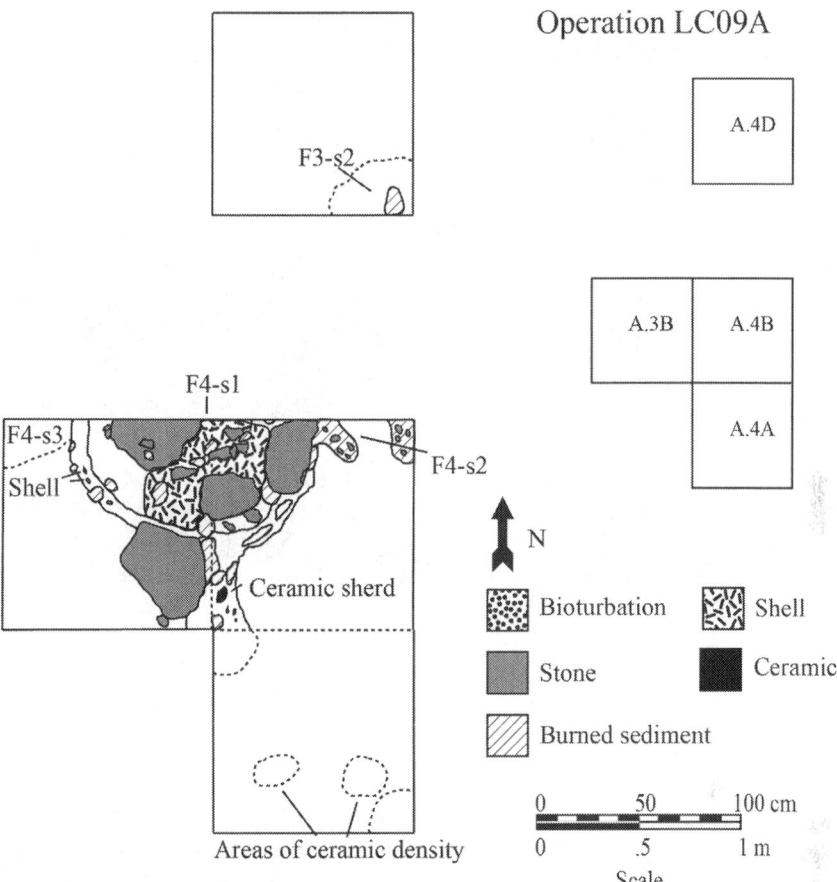

**FIGURE 4.2.** LC09 A-F4-s1 hearth plan map

A-F19 is indicative of a broad trend, wherein Platform 1 experienced sometimes-brief periods of occupation and stability between construction episodes. In terms of chronological crossties within the site, it is likely that this level was roughly contemporaneous with the LC09 A-F4 hearth.

At Op. LC12 D (figure A1.12), the initial platform fill (D-F11) was intermingled with substrata consisting of D-N1 fluvial sands, suggesting that the river continued to flood the area or perhaps that a hurricane occurred during an early construction phase. A narrow band of probable occupational debris (D-F10-s1) was then deposited atop D-F11. This brief occupation layer is consistent with other areas of the site, such as in Ops. LC09 A, LC09 B, and LC12 A, where fill layers topped with

**FIGURE 4.3.** B-F15 hearth in Units 1B and 2B of Op. LC09 B

occupational deposits or early hearths (e.g., LC09 A-F4-s1, LC09 B-F15 [figure A1.5], and LC12 A-F19) indicate occupation of the initial platform before subsequent construction. In the area investigated by Op. LC12 F (figure A1.15), initial fill deposits (F-F6 and F-F5) consisted of dense layers of clay with low artifact density. These early fills produced some interesting artifacts, including a complete ground stone tool and an early figurine head bearing an ear spool and what may be part of a headdress or banded hairdo (see figures 5.4.D and 7.8.B). Because they were incorporated in the initial fill, these artifacts suggest an early, perhaps preplatform occupation nearby.

In the area of Op. LC12 E (figure A1.13), occupation in the eastern portion of Platform 1 produced a sequence of midden deposits (E-F16 through E-F9) containing ash, dense shell lenses, well-preserved animal bone, and sherds from vessels such as decorated serving bowls and bottles (see chapter 6, appendix 2). The calcium from the shell and the ash (probably from hearth-cleaning events) likely reacted with natural rainwater percolation to produce an extremely hard concretion layer (E-F11-s1) within the midden. Although this layer appeared to have an intentionally shaped, domed form with a consistent, step-like lip, it probably resulted from natural postdepositional processes. The stratum may indicate the presence of a stable surface within these sediments, or perhaps represents the depth to which rainwater has been absorbed into the sediments from subsequent overlying surfaces.

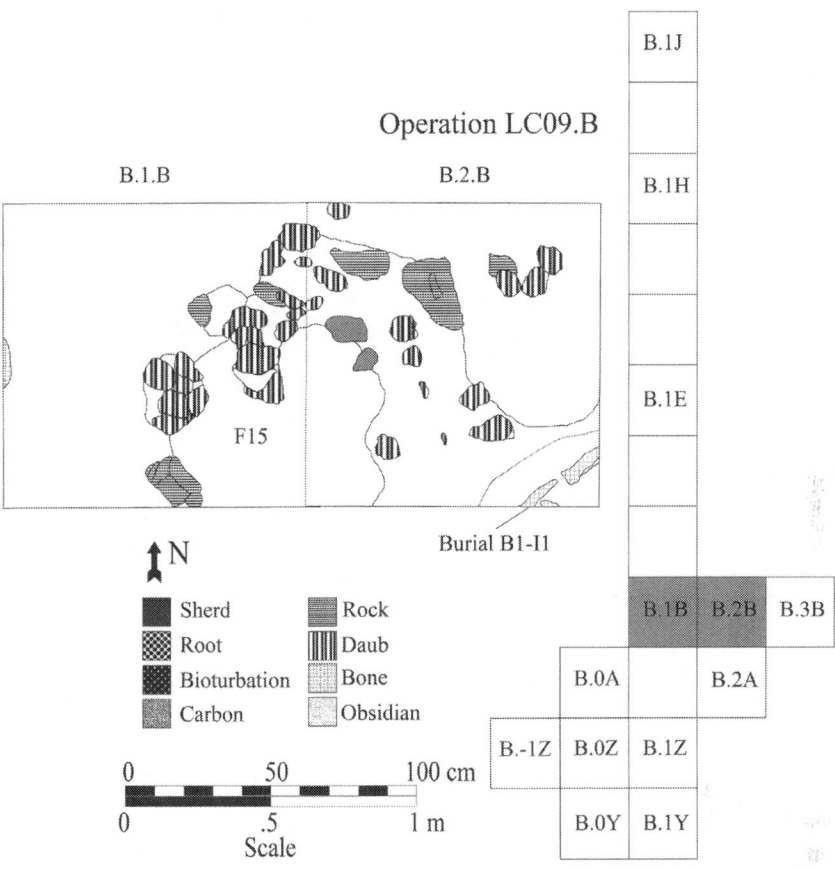

**FIGURE 4.4.** Plan map of B-F15 hearth

Intrusive into the E-F11-s1 concretion layer is a possible hearth or burning feature (E-F10; figure A1.13). The contents of this feature returned an AMS date of 3435 ± 44 (AA101269; carbon-rich sediment; $\delta^{13}C = -25.5‰$), or 1885–1635 cal BC. This date suggests that the Op. LC12 E area was in use at roughly the same time as the LC09 A-F4-s1 hearth and the LC12 A-F19 occupation surface. Also recovered from E-F10 was part of an anthropomorphic figurine (figure 7.7.A) reminiscent of Cruz A phase examples from the Mixteca Alta (Jeffrey Blomster, personal communication, 2015). Above E-F10, nearly a meter more of ashy, silty clay (E-F9) containing dense lenses of *Mytilidae* mangrove mussel shell (E-F9-s2) accumulated. In some spots, these deposits are mostly shell and contain little sediment. Ceramics found in the midden included decorated bowls and bottles (see chapter 8, appendix 2, and especially table A2.28). The interpretation that the decorated vessels in this deposit suggest feasting is consistent with results of research elsewhere in Mesoamerica and beyond, notably regarding the Barra phase ceramics of the Soconusco region (e.g., Clark and Cheetham 2002:294; Hayden 1990; G. Lowe 1975; Rosenswig 2007). The presence of refitting fragments of vessels such as a decorated bottle, which were spread horizontally over four 1 × 1 m excavation units and vertically over 20 cm of sediment or more, suggests that much of this midden was deposited quickly, perhaps resulting from just a few events. Such rapid deposition of food-related garbage suggests a large public event or events. The general rarity of shell at La Consentida, and the occurrence of finely finished, decorated wares in two of the contexts containing the most shell (in Ops. LC12 D and LC12 E) suggests that shellfish was perhaps often a public feasting food rather than a dietary mainstay for the community. Also present in the midden (E-F9 through E-F16) were faunal remains, especially of fish (see Pérez Hernández and Hepp 2015; chapter 6).

Whereas the LC12 E midden appears to have been a product of public feasting and the use of decorated serving vessels, a small but dense midden deposit uncovered in Op. LC12 H appears to be the result of different practices. Op. LC12 H was located at the base of Substructure 2, where Platform 1 meets the surrounding natural floodplain (figure 3.4). In this area, the research team discovered the remains of a midden consisting of an ashy clay matrix and a primary deposit of broken ceramic vessels. The lowest layers of this midden (H-F4-s3) occurred at a depth of about 10.3–9.9 masl. These deposits likely date to some of the earliest occupations at La Consentida (roughly contemporaneous with the Op. LC12 E midden), and may relate to domestic contexts atop Platform 1. In addition to ceramics, other finds from the deepest sediments at Op. LC12 H included a canid mandible fragment from a probable coyote (figure 6.3.D), other faunal bone, a small amount of marine shell, and a few obsidian flakes.

The densest layers of the Op. LC12 H midden (F4-s1 through F4-s3) contained large cooking jars of a style similar to the Tierras Largas phase vessels from the Valley of Oaxaca (Flannery and Marcus 1994; Marcus Winter and Cira López Martínez, personal communication, 2013; Ramírez Urrea 1993). The jars vary in style from in-leaning neck examples to out-curving neck jars with large diameters (as much as 53 cm), suggestive of food preparation for communal feasting events (see chapter 8 and appendix 2). The emphasis on jars over other vessel forms such as bottles and bowls (with the notable exception of one large, hemispherical bowl) suggests that the midden resulted from a special-use event. As was the case with the Op. LC12 E midden, large Op. LC12 H vessel fragments from depths of approximately 60 cm apart came from the same vessels and had sharp, uneroded edges. This pattern indicates the rapid deposition of large cooking jars from a single, or very few, events (see

Barber 2005:179). The event or events that produced the Op. LC12 H midden may have been related to the adjacent domestic area atop Platform 1, suggesting that even communal feasts could be hosted by one or a few households. As discussed below, the floor of Structure 2 was identified in Op. LC12 G (figure 5.3), at the southern end of Substructure 2. Though Ops. LC12 H and LC12 G were close to one another, Substructure 2 and the domestic structure atop it (Structure 2) occur late in the straigraphic sequence, and likely postdate the initial Op. LC12 H deposits (figure 3.4). Very few decorated or serving wares of any kind were recovered in Op. LC12 H, suggesting that the feast itself took place elsewhere, perhaps in a more central location near the middle of Platform 1, such as the area near Op. LC12 E. Carbonized food adhering to a jar fragment from LC12 H-F4-s2 has been dated to 3419 ± 36 (AA104836; carbonized food; $\delta 13C$, −15.5), or 1880–1625 cal BC (see table 1.1).

Taken together, the seven most reliable radiocarbon samples from secure contexts at La Consentida (one excavated in 1988, two in 2009, and four in 2012) have produced a calibrated AMS radiocarbon date range of 1950–1525 cal BC for the site's earliest architecture (Hepp 2015; A. Joyce 1991a; Winter 1989; table 1.1). Initial occupations almost certainly predated these deposits. Because the Op. LC09 A, LC09 B, and LC12 E hearths or probable hearths were intrusive into platform fill, their dates provide conservative chronological estimates for the earliest earthen architecture at La Consentida.

## SUBSEQUENT PLATFORM AND MOUND CONSTRUCTION

Following the initial construction of Platform 1 (or perhaps of several smaller mounds later subsumed into Platform 1), subsequent building episodes added to the platform and produced several substructures atop it. At Op. LC09 A, Platform 1 construction continued after the disuse of the A-F4 hearth with the deposition of a thick (up to about 80 cm) fill stratum (A-F3-s1 through A-F3-s3). In the area of Op. LC12 A, platform construction atop A-F19 (the first layer of Platform 1 fill with a probable occupation surface on top) resulted in the deposition of A-F18-s2, a fill layer that is nearly 1.8 m thick in some spots (figure A1.8). In the southern half of Op. LC12 A, several artifacts found near the interface between A-F18-s2 and A-F17-s2—including ceramics, bone, burned daub, shell, and a figurine head (figure 7.6.A)—may indicate an occupation surface between these strata. The burnished and slipped medium brown ware ceramics from this deposit and the style of the associated figurine fragment support an early date for this context, even if some of the artifacts were redeposited as part of the A-F18-s2 fill event.

Platform 1 construction episodes were punctuated by the interment of human remains and ritual deposits. The LC12 A-F15 ritual cache (figures 4.5 and A1.7) was

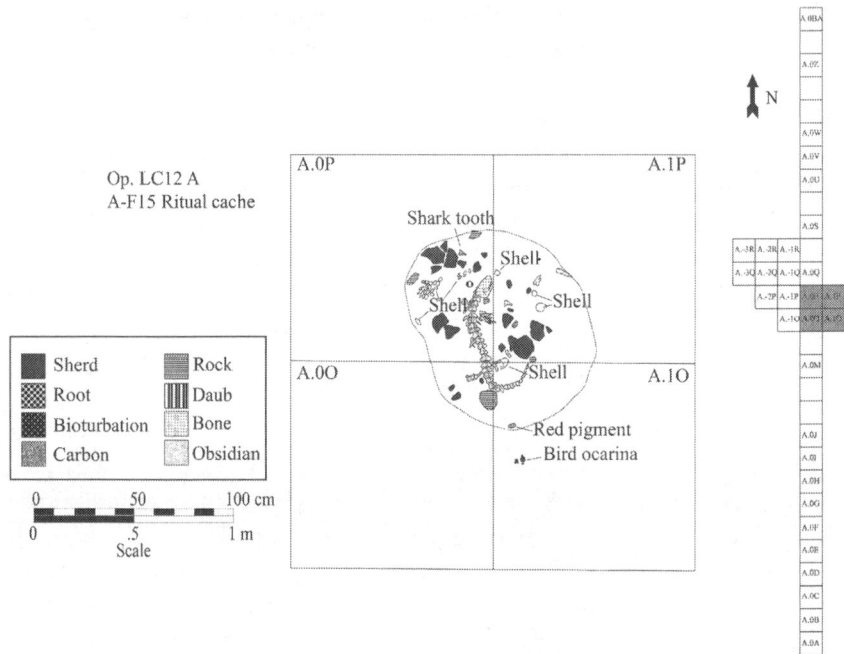

**FIGURE 4.5.** A-F15 Ritual cache near human burials in Op. LC12 A

deposited within a pit feature (A-F14) intrusive into the A-F17-s2 fill. The cache contained faunal remains, ceramics, and probably a musical instrument (figure 7.16.A). The proximity of this ritual offering to an area containing at least nine sets of human remains may be significant, as may be its location at the base of Substructure 1. The cache, perhaps deposited as a bundle, might have been a dedicatory offering to the burials and/or to the Substructure 1 architecture (Hendon 2000; Zedeño 2008; see chapter 7). The early style ceramics recovered within A-F15 (see appendix 2), as well as the stratigraphic position of this feature, are consistent with an Early Formative period date.

The placement of a playable bird ocarina (figure 7.16.A) at the edge of the offering merits comment. In part because of the artifact's good state of preservation, I believe it was carefully deposited as part of the cache or bundle rather than as part of the surrounding fill. If so, the A-F14 pit may have been larger than its outline appeared during excavation. If the ocarina came from the A-F17-s2 fill instead, it may have a slightly earlier date of manufacture. In either case, this bird ocarina apparently stands as one of the earliest examples of a musical instrument thus far recovered in Oaxaca or, for that matter, in Mesoamerica (Hepp et al. 2014, 2017).

Some of La Consentida's earliest human burials appear to be those deposited in the area of Op. LC12 A. B12-I14, B11-I13, and B9-I11 were interred in early fill layers such as A-F17-s2, which was subsequently capped by later fill episodes and several additional burials (B5-I6 through B10-I12). One of these burials, B12-I14, has been directly dated to 1690–1530 cal BC (table 1.1, figure 1.2, chapter 3) by extracting collagen from a femur fragment and obtaining an AMS radiocarbon date from that collagen following XAD purification (Cummings 2017; see also Waters et al. 2015). Other burials in the LC12 A mortuary area can be dated by stratigraphic comparison with B12-I14. Ceramics and other offerings associated with these burials also evince their early date (see appendix 2).

In the southern portion of the Op. LC12 A trench, the relatively thick (approximately 50–70 cm) A-F17-s2 fill deposit was overlain by strata A-F12 and A-F13, which likely represent another occupation surface atop the platform. In situ artifacts demonstrating occupation of A-F12 and A-F13 included ground stone, burned daub, obsidian, marine shell, bone, and some figurine fragments. Based on the size of daub pieces and the occupation surface identified, burned daub associated with A-F13 may actually be in situ architectural remains or wall fall. Ceramic sherds were recovered adhering directly to the daub, indicating the use of recycled ceramics in architecture. Overlying LC12 A-F12 and A-F13 was stratum A-F11-s1. Given the thickness of this deposit, it is apparent this stratum represents a concerted fill episode that raised the height of Platform 1 as much as 115 cm in some areas. Also noteworthy is the shape of A-F11-s1, which increases in thickness to the south, indicating that the Substructure 1 mound was already under construction at this time. Because all the artifacts found in these sediments (and in several overlying strata) are Formative rather than Classic period in style, it is most probable that Substructure 1 was a Formative period construction and that its brief Classic period reoccupation (discussed below) represents a minor addition to a substantial and preexisting Early Formative period mound.

Later fill deposits (e.g., LC12 A-F10-s1, A-F4-s1, and A-F2) were often modest in their alteration of Platform 1 and overlying substructures. Some strata (e.g., LC12 A-F10-s1) taper noticeably, suggesting that they were used to increase or lessen the angle of substructures relative to Platform 1. Other deposits (e.g., LC12 A-F4-s1, A-F2) raised the overall height of architecture without marking any distinction between the substrcutures and the underlying platform. At Op. LC12 B (located atop Substructure 1), thick fill deposits (B-F6-s1 and B-F5) resulted in the final form of Substructure 1, the largest of the mounds atop Platform 1 (see figures 3.4 and A1.9; table A1.4). The style of the ceramics from these strata (many of which appear to be similar to the earliest ceramics at the site) support the interpretation that Substructure 1 was constructed during the Formative period. Some of the more recent construction layers were perhaps deposited later in the site's Early Formative period occupation, but utilized redeposited sediment containing artifacts of early style. Given the stratigraphic location of these deposits and the types of artifacts they contain, it is likely that strata B-F6 through B-F4 are related to fill deposits in the nearby Op. LC12 A area. B-F6-s1 probably corresponds to A-F11-s1, B-F5 likely corresponds to A-F10-s1, and B-F4 probably corresponds to A-F4-s1.

At Op. LC09 B, located in the northwest corner of Platform 1, the uppermost fill layers demonstrate periods of both construction and site abandonment (table A1.2, figures A1.4 and A1.5). Following the disuse of the B-F15 hearth, earthen architecture was modified by the deposition of the B-F14 fill and resurfacing layer. Next came a thick (up to about 90 cm) stratum of platform fill composed of redeposited midden (B-F12). Both within and atop B-F12, burned sediment and ceramics occurred in horizontal lenses, with refitted sherds immediately beneath burned sediment layers. The formation of a prismatic paleosol within these sediments indicates the stability of an ancient surface. Stratum B-F10 also contains redeposited midden materials and evidence for the formation of an ancient soil on a surface that remained stable for some time. It is likely that B-F10 and B-F12 represent a single ancient surface. This soil was one of the most strongly developed paleosols identified in the entire lower Río Verde region, indicating that it was a stable surface for a relatively extended period and that this area of the site was likely unoccupied during its formation. Intrusive pits (LC09 B-F3 through B-F9) later interrupted strata B-F10, B-F12, and B-F14. These pits appear associated with human burials (B1-I1 through B4-I5). In some cases, human remains were later covered by sediments including dense deposits of ceramic sherds.

In the area of Op. LC12 D, stratum D-F9 represents an early fill episode atop the D-F10-s2 occupational surface. The lack of any soil or occupational surface formation indicates that this layer was covered quickly. The subtle and gradual transition between D-F9 and D-F7 suggests that these two layers may have been deposited in quick succession, providing further evidence of a concerted community construction effort to modify Platform 1. Stratum D-F8 likely represents a preserved fragment of a compact occupational surface containing some stones. D-F4 was a thin platform resurfacing layer containing much more marine shell than surrounding deposits. D-F5 was a small pit intrusive into D-F4 that contained an extremely dense dump of crushed shell. These shells appear to mostly be from Mytilid mussels (known locally as *tichinda*), suggesting resource collection in a nearby mangrove environment. The presence of the D-F5 shell dump indicates an occupation atop the D-F4 surface.

The uppermost fill deposits in the area of Op. LC12 F (F-F4, F-F3, and F-F2) varied in thickness, demonstrating different scales of construction efforts to produce this central area of Platform 1. These deposits also contained high densities of eroded ceramics, suggesting the possible use of redeposited midden as fill. Stratum F-F1 likely represents the formation of a modern soil within the F-F2 fill. A transformational figurine head (figure 7.10.B) was recovered at the surface in Op. LC12 F. "Transformational" figurines are those that bear evidence of a blending of human identity with that of animals and/or deities (see Hepp and Joyce 2013; chapter 7). In total, approximately two meters of fill deposited above natural strata in this area suggest that even low-lying portions of Platform 1 represent considerable labor investment.

### LATER SITE OCCUPATION DURING THE FORMATIVE PERIOD

As discussed in appendix 2, ceramic evidence for subtle change over time in the Tlacuache phase assemblage suggests that occupation at La Consentida may ultimately be divided into two subphases. The results of excavation at Substructures 1 and 2 imply that Substructures 1–7 may also have served as domestic zones supporting ephemeral wattle and daub structures with earthen floors during later occupation at the site (figure 3.3). Based on the style of ground stone and ceramic artifacts found in these areas, the upper strata of the substructures (e.g., LC12 B-F4, C-F2, and G-F2) saw later phases of occupation than did earlier Platform 1 occupation layers (e.g., LC09 A-F5 and LC09 B-F16). Several superimposed occupational surfaces and Structure 1 in Op. LC12 C, along with Structure 2 in Op. LC12 G, indicate that the Substructure 2 mound saw numerous phases of occupation (see chapter 5, figure 5.3). The types of ceramics found in Ops. LC12 C and LC12 G suggest that Structure 2 postdated Structure 1 (see appendix 2). The final occupations of these contexts likely took place toward the end of the Early Formative occupation of the site.

Op. LC12 C was dug atop the northern portion of Substructure 2 (figures 3.3 and 3.4). In this area, layers of in situ ceramics atop fill strata (C-F9, C-F8, and C-F7), demonstrate that the area was subject to occupation between construction events (e.g., figure A1.10). Stratum C-F7, in particular, bears evidence of being an occupational surface, as the C-F4 structure floor overlies this deposit. The structure was small (measuring only about 2 × 1.5 meters in its preserved portions) and had large stone tool fragments, including those from broken metates, incorporated in its foundation or walls (figure 5.1). The lenses of refitting, in situ ceramic fragments, manos, and broken metates found in association with this structure suggest that its use may have been domestic. Despite the large, recycled ground stone fragments associated with this floor, the lack of much other preserved construction material suggests that the structure itself was ephemeral. Fragments of burned daub suggest that Structure 1 had wattle and daub walls and a thatch roof. The preserved portions of the structure appear small for a house and may instead represent a domestic outbuilding. After abandonment, occupational debris (C-F2) probably gradually covered this structure. The upper surface of C-F2 may also have been occupied, as evidenced by a relatively high concentration of ceramic artifacts near the modern surface.

In the area of Op. LC12 D, at the eastern edge of Substructure 2, the shallower deposits of platform fill (e.g., D-F3-s1, D-F2, and D-F1) represent more modest alterations of Platform 1 than did earlier deposits (e.g., D-F9). These shallower layers may represent resurfacing episodes. The Op. LC12 E midden (E-F16 through E-F9) was capped with a thin but consistent layer of fill and/or architectural debris (E-F4). Excavations in this western extension of Op. LC12 E recovered large pieces of daub with wattle impressions (E-F6), suggesting that a structure wall burned and collapsed at that spot. This deposit was also associated with a small ceramic dump that formed part of E-F4 (table A1.7). The style of associated ceramics, as well as their stratigraphic location nearer the modern surface of Platform 1, suggests that they date to shortly before site abandonment (see appendix 2). These architectural remains suggest that structures were placed toward the center of Platform 1 as well as atop the platform's substructures.

Excavations at Op. LC12 G (at the southern end of Substructure 2) uncovered the remains of Structure 2 (G-F2 through G-F15), which contained in situ ceramics dating to shortly before Formative period site abandonment (table A1.9). Although a few small excavation windows penetrated below the floor (G-F2) of Structure 2 to test for postholes, no deep vertical excavation was carried out here. It is thus possible that previous domestic occupation surfaces exist below the identified structure floor. The deepest deposit identified in this operation was stratum G-F16. This fill layer represents part of the construction of Substructure 2 atop the broad, early strata that compose Platform 1. Overlying G-F16 is the G-F2 structure floor (figures 5.2 and 5.3), which is associated with several pit features (G-F3 through G-F15). These features likely represent postholes in a ring around the structure and pits dug into the floor following the structure's construction. Based on its shallow depth and the lack of subsequent occupations atop it, Structure 2 represents one of the final occupation areas prior to Formative period site abandonment. A carbon sample from G-F2 returned a Middle Formative AMS date of 2433 ± 35, (AA101268; carbon-rich sediment; $\delta 13C$, $-21.6$) or 755–405 cal BC. Due to the shallow depth of this sample, possible contamination from later burning episodes and modern surface sediments, and an incongruously late date relative to associated ceramics, this date is considered suspect (table 1.1). A plateau in this part of the calibration curve may also affect the sample (Reimer et al. 2013). Following the abandonment of the

Op. LC12 G structure, a modern soil formed within G-F16, thus producing stratum G-F1(figure 5.3, table A1.9).

## FORMATIVE PERIOD SITE ABANDONMENT

The uppermost strata in most excavation areas (e.g., LC09 A-F1, LC09 B-F1, and LC12 A-F1) contained a mixture of Formative period occupational debris and modern materials. Modern soils have formed in these sediments, resulting in significant bioturbation. Based on the presence of medium brown ware ceramics and Formative period figurines at the surface of most of the site, and the lack of primary deposits from Late Formative or later occupation in all excavated areas except for Substructure 1, La Consentida was abandoned by the Middle Formative period. Prismatic obsidian blades have been considered a technological innovation made common during the Middle Formative period (Jackson and Love 1991), though in some regions they were present by the Early Formative. Kenneth Hirth and colleagues (2013:2789) found, for instance, that obsidian blades appeared at San Lorenzo by the Chicharras phase (1500–1400 cal BC), likely as specialized products produced by experts near the material source. The presence at La Consentida of a few prismatic blade fragments despite a complete lack of green (Pachuca) obsidian (which was heavily traded in Late Formative and Classic period times) is therefore consistent with site abandonment by the early Middle Formative, if not before (Cobean 2002:41 [cited in D. Williams 2012:112]; see chapter 6).

## CLASSIC PERIOD OCCUPATION AND LATER SURFACE ARTIFACTS

Excavations atop Substructure 1 uncovered the remains of a burned Early Classic building. This structure indicates a brief reoccupation of the site after its Formative period abandonment. The stratum of heaviest Classic period artifact density (LC12 B-F3) overlay Formative period fill layers (B-F6, B-F5, and B-F4). B-F3 contained large pieces of daub and lay directly below B-F2, a deposit of architectural refuse containing daub with post impressions. The uppermost layer of sediment in the Op. LC12 B area (B-F1) contained a mixture of Early Classic occupational debris and modern surface duff. Although the density of ceramic artifacts associated with the Early Classic structure was high, other artifacts indicative of occupation, such as faunal remains and lithics, were lacking. The layers of construction fill below the dense Classic period debris contained far fewer artifacts. All ceramic sherds identified at a depth greater than about 30–50 cm below the surface were Formative in style. This pattern suggests that much of the Substructure 1 fill was constructed before Formative period site abandonment. Another possibility is that the Early

Classic period reoccupation resulted in significant Substructure 1 construction using only redeposited earlier materials. The complete lack of Classic period artifacts at a depth of more than about 30–50 cm below the surface, along with available radiocarbon dates, make this interpretation unlikely, however (see table 1.1). It appears instead that the majority of earthen architecture at La Consentida was constructed during Formative period occupation.

The presence of a small number of Classic and Postclassic sherds, including at least one eroded polychrome and two Postclassic *malacates* (spindle whorls), at or near the surface in various parts of the site, indicates that La Consentida was known to later peoples in the lower Río Verde region. These people may have visited the site out of interest in and respect for the history of the region, in the process of collecting resources in the rich coastal zone the site occupies, or both. These more recent artifacts do not indicate more than a brief reoccupation of one part of the site during the Early Classic period.

# 5

## Settling Down

*The Shift to Sedentism*

As discussed in chapter 2, there are many lines of evidence that can support inferences about an ancient population's practices of domestic mobility. These may include the presence of monumental architecture or community buildings indicating organized communal labor (Drennan 2003b:47; Hill and Clark 2001; Marcus and Flannery 1996:109–10). "Durable" domestic architecture consistent in its placement through multiple construction phases (P. Arnold 1999:160) and formal storage features (Kent 1992; Smyth 1989:90, 92; Winter 2009:27–29) can also indicate settlement strategies. Other indices of mobility may include nonportable ground stone tools (Clark et al. 2007; McDonald 1991:85; Torrence 1983), the presence of cemeteries, and other evidence that a community held a specific spot on the landscape to be ritually significant and affiliated with memories, kinship ties, and a deep sense of place (Boyd 2006; Joyce and Goman 2012; Mitchell 2008). As I also outlined in chapter 2, identifying sedentism is a challenging task because all the material correlates discussed thus far have exceptions (e.g., Banning 2011; R. Bradley 1993, 1998, 2000, 2005; Pauketat and Alt 2003; Sherratt 1990; Tilley 1994) and because sedentism and mobile foraging are terms for two extremes within a continuum of domestic mobility (e.g., R. Kelly 1992; Kent 1992; Marshall 2006).

In this chapter, I discuss multiple types of evidence for the settlement practices of the La Consentida community. These sources of data include population estimates, earthen architectural sequences and associated labor estimates, attributes of probable domestic structures, and patterns identified among ceramic and ground stone artifacts. In many cases, I refer to other sections of this book because certain types of data, such as those from the analysis of ground stone tools, pertain to multiple interpretations about ancient life at La Consentida (e.g., domestic mobility and subsistence). After examining relevant evidence, I conclude that the population that initially founded La Consentida was likely somewhat mobile, and probably established the site as one of several seasonally occupied resource-gathering and domestic locales within the lower Río Verde region. By the time of Formative period site abandonment, the community appears to have become dedicated to permanent occupation at the site, though it likely maintained special resource-gathering locales in different ecological microregions nearby, as did Early Formative communities elsewhere in coastal Mesoamerica (see Blake et al. 1992:90; Lesure 2009:259; Voorhies 1989:116, 2004; Voorhies and Kennett 2011). In order to discuss shifting settlement practices over time, I begin with the evidence from La Consentida's mounded earthen architecture, which is among the earliest yet identified in Mesoamerica.

### EARTHEN ARCHITECTURE AT LA CONSENTIDA

Stratigraphic evidence (see chapter 4 and appendix 1) suggests that La Consentida was occupied only briefly before the community began to construct Platform 1. Dated hearths (LC09 A-F4-s1 and LC09 B-F15) and a possible hearth (LC12 E-F10) intrusive into the first fill layers (e.g., LC09 A-F5 and LC09 B-F16) demonstrate occupations atop those layers before the deposition of subsequent strata. The earliest of these hearths (1900–1690 and 1885–1635 cal BC) establish Platform 1, or perhaps the series of small mounds later subsumed into Platform 1, as among Mesoamerica's earliest examples of mounded earthen architecture (see tables 1.1 and 1.2). Specifically, La Consentida's earthen architecture appears to predate late Barra or early Locona phase (1700–1550 cal BC) mounds in the Soconusco and Bajío phase mounds in the Gulf Coast region (Clark 1994:141–42, 376, 462; Cyphers and Zurita-Noguera 2012; Hill and Clark 2001). The Soconusco's Chantuto B phase (3500–2000 cal BC) shell mounds, associated with late Archaic period mobile hunter-gatherers, predate La Consentida. While these Archaic shell mounds (e.g., at the site of Tlacuachero) may largely represent seasonal resource-gathering refuse, some deposits, such as a possible clay floor, suggest that they may also have been architectural (Voorhies and Kennett 2011:29). Although paleoenvironmental data (such as evidence of burning events for probable landscape clearance) indicate that a late Archaic period human population occupied the lower Río Verde region, no Archaic sites have been identified in the area (Goman et al. 2013; Joyce and Goman 2012).

In most parts of Platform 1, initial fill deposits are modest in comparison to later strata that imply increased labor expenditure, organization of communal labor, and perhaps a larger population (e.g., figure A1.8). This may indicate the mobilization of less construction labor than in subsequent times, when fill deposits reached nearly 180 cm thick. As discussed in chapter 4, the earliest fill layers, while not necessarily very thick, are relatively consistent in their sedimentary composition and associated radiometric dates across large areas of the site, such as between Operations LC09 A and LC12 E (see figure 3.4). This pattern suggests that the initial earthen architecture was broad and low. While it is difficult to quantify the labor responsible for a probable platform excavated only in ten discrete operation areas and never in complete cross section, rough estimates of the population and labor requirements are informative for the discussion of the community's settlement strategies.

## ESTIMATING LA CONSENTIDA'S POPULATION SIZE AND LABOR INVESTMENTS

At the surface, La Consentida is small, covering only about 4.5 ha. Excavations at the base of Platform 1 indicate that cultural deposits extend out from the platform farther than is apparent at the surface, likely because the level of the natural floodplain has risen since occupation and has covered the peripheries of the site (Joyce and Mueller 1992; Mueller 1991; Mueller et al. 2013, 2014; see chapter 4 and appendix 1). Adhering to the 4.5 ha size determination for La Consentida is thus conservative, but it suffices for a preliminary population estimate. Applying Paul Tolstoy's (1989b:95; [cited in Clark 1994:210]) Coapexco estimation technique (142–64 people per hectare) to La Consentida would produce the very high estimate of 639–738 people. This seems inappropriate for several reasons. First, the fourteen sets of human remains identified thus far do not seem to indicate a very large population. Also, the seven substructure mounds, if they were generally the exclusive location of houses at the site, seem scarcely capable of supporting such a large community. Applying the Valley of Oaxaca's population estimate guidelines (e.g., Feinman 1991; Feinman and Nicholas 1987, 1990, 1992; Nicholas 1989; [cited in Clark 1994:210]) would place La Consentida's population at 45–112, with a midpoint of nearly 80 people (see also J. Parsons 1971; Sanders 1965).

A population of 80 would fit comfortably in two to three small houses atop each substructural mound and could also have produced La Consentida's earthen architecture without requiring labor from outside the site. Any group occupying La Consentida would have included children, the infirm, and the aged. The presence of at least children and older adults is demonstrated by several of the burials recovered at the site (Hepp et al. 2017). Although children and the elderly can fulfill productive labor roles, La Consentida's population likely had no more than about two healthy adults per household unit of five people (assuming that the households also included two children and perhaps a grandparent) (see Flannery 2002 for a discussion of probable initial village family units). A workforce of no more than about thirty-two adults (40 percent of a population of 80 people) was thus probably undertaking the heavy labor at any given time. Although this may seem like a small group to produce Platform 1 and the substructural mounds, the young and the old likely supported these workers in incidental tasks.

Estimates of labor investments for the production of mounded earthen architecture in Mesoamerica have proven a useful measure for inferring lengths of site occupation, construction practices, and aspects of social organization including the mobilization of labor (e.g., Abrams 1994; Joyce et al. 2013; Rosenswig and Masson 2002). Joyce and colleagues (2013:149–53) discussed the energetics estimates and ethnographic analogies appropriate for inferring human labor practices in hot coastal tropical zones such as the western Oaxaca coast. These estimates assume five-hour work days (Erasmus 1965:283), a quantity of 2.6 $m^3$ of sediment excavated per person per day (Erasmus 1965:285), and the majority of labor for earthen architecture taking place during the dry season (Joyce et al. 2013:151).

Although some substructures at La Consentida are taller than others, and Platform 1 itself is not uniform in its dimensions, a good approximation of the earthen architecture at the site is that it measures 100 × 300 × 5 m, or 150,000 $m^3$ in volume (figures 3.3–3.5). Based on excavations in areas such as Ops. LC09 A, LC09 B, LC12 A, LC12 D, LC12 E, and LC12 F, it appears that the first iteration of the platform covered much of the maximal horizontal extent of the final platform, but was only about a meter tall (e.g., figures A1.2, A1.5, A1.8, A1.12, A1.14, and A1.15). An approximate volumetric assessment of the initial platform (incorporating the first one or two strata of fill) would thus be 100 × 300 × 1 m, or 30,000 $m^3$ of sediment. Although subtle variation is apparent in ceramics over time at the site, all levels of fill contain Tlacuache phase materials (see chapter 4 and appendix 2). Early Classic occupation apparently made minimal impact upon preexisting Formative period architecture. Probable borrow pits are located near Platform 1, some surrounding the platform in an arc to the northeast and in low-lying areas just outside the mapped site boundary (figure 3.3). I estimate that most of the fill sediments came from within 250 m of the site.

Following the lead of other energetics studies (e.g., Abrams 1994; Erasmus 1965; A. Joyce et al. 2013; Rosenswig and Masson 2002), I divide the labor to produce La Consentida's earthen architecture into three steps: excavation of fill material, transportation of fill, and construction (tables 5.1 and 5.2). Unlike in A. Joyce and colleagues' study (2013:151), I do not discuss the transportation of water or production of adobe blocks, as fill thus far excavated at La Consentida appears to result from the use of basket loads of sediment rather than from structured fill construction. As previous studies have done (e.g., Joyce et al. 2013), I adopt Stephen Aaberg and Jay Bonsignore's (1975:46) method for calculating person days of transport labor:

$m^3$ / person days = $Q * 1 / (L / V + L / V') * H$

Regarding the equation above, Q = container capacity and thus human transportation capacity (which Aaberg and Bonsignore [1975:47] and Joyce and colleagues [2013:152] estimate at 22 kg or .2 $m^3$), L = transportation distance, V = transport speed (estimated at 3 kilometers per hour), V' = return trips (estimated at 5 km per hour), and H = length of workday (Aaberg and Bonsignore 1975; A. Joyce et al. 2013). Given these estimated aspects of the workload, the completed formula is

$m^3$ / person day = .2 $m^3$ * 1 / (.25 km / 3 km hr + .25 km / 5 km hr) * 5
$m^3$ / person day = .2 * 1 / (.25 / 3 + .25 / 5) * 5
$m^3$ / person day = .2 * 1 / (.083 + .05) * 5
$m^3$ / person day = .2 * 1 / .133 * 5
$m^3$ / person day = 7.5

As A. Joyce and colleagues (2013:152) have also done before, I adopt Elliot Abrams's (1994:50) proposed rate of 4.8 $m^3$ of sediment per person, per day for the final stage of construction. Table 5.1 demonstrates the estimated labor investment, in person days, for the total amount of the La Consentida earthen architecture. Table 5.2 provides an estimate of the labor necessary for the first version of Platform 1, which was constructed shortly after initial site occupation.

As discussed above, several lines of evidence suggest that 80 is an appropriate average community size for La Consentida. Assuming that only 30–40 people were available for heavy labor and that the young, old, and otherwise unfit could help those laborers by procuring food, water, and other supplies, it is appropriate to propose a variety of possibilities for how long the earthen architecture took to construct. I have done so for the total quantity of architecture at the site (table 5.3), and for a proposed initial version of Platform 1 (table 5.4). When the site's population estimate is considered in conjunction with the carbon dates, which suggest an occupation of up to 425 years during the Early Formative (when the dates are reported with $2\sigma$ probability), one may further refine labor estimates for the site. When the dates are reported with $1\sigma$ probability, they suggest a 275-year period of occupation (1885–1610 cal BC [with a very small possibility of extending to 1565 cal BC]; table 1.1). During that occupation, a community with only 25–50 healthy workers could have produced La Consentida's mounded architecture without assistance from outside labor. It is also worth remarking that construction likely did not occur every dry season, as is suggested by the evidence for occupation (e.g., the LC09 A-F4-s1 and LC09 B-F15 hearths) atop some fill layers. Based on

TABLE 5.1. Estimated labor investment in all of La Consentida's mounded earthen architecture. Numbers are rounded.

| Labor step | Step 1: Fill excavation | Step 2: Fill transport | Step 3: Construction | Total |
|---|---|---|---|---|
| Person day calculation | 150,000 / 2.6 | 150,000 / 7.5 | 150,000 / 4.8 | N/A |
| Total person days | 57,692 | 20,000 | 31,250 | 108,942 |

TABLE 5.2. Estimated labor investment for initial version of Platform 1. Numbers are rounded.

| Labor step | Step 1: Fill excavation | Step 2: Fill transport | Step 3: Construction | Total |
|---|---|---|---|---|
| Person day calculation | 30,000 / 2.6 | 30,000 / 7.5 | 30,000 / 4.8 | N/A |
| Total person days | 11,538 | 4,000 | 6,250 | 21,788 |

the carbon date ranges and given that major construction would not have taken place annually, a group of 25–50 laborers likely constructed all of La Consentida's architecture over a span of about 250–300 years (table 5.3). The first version of Platform 1 may have been produced in only a few seasons (table 5.4), and the limited labor necessary for that construction is consistent with the community initially being semimobile. If pressured by impending floodwaters, the community could even have produced an incipient version of Platform 1 (or, instead, a series of low mounds) in a single season. Further excavation may help to refine the ceramic chronology and thus provide more temporal control for construction phases.

## DOMESTIC STRUCTURES

As discussed in chapter 2, the size and relative durability of architecture are two of several indices of a community's degrees of domestic mobility. The only clear examples of domestic buildings so far uncovered at La Consentida (Structure 1 in Op. LC12 C and Structure 2 in Op. LC12 G) date to relatively late in site occupation (see figure 3.3). While these probable domestic buildings were relatively ephemeral, the builders of Structure 1 incorporated large stones including recycled metates as construction material (figure 5.1). Narrow excavation windows below Structure 2 identified postholes associated with that building, but no obvious floors from previous structures (figures 5.2 and 5.3). It is possible that remains of earlier buildings exist in deeper, unexcavated deposits. Multiple, superimposed lenses of in situ artifacts such as ceramics and food-processing ground stone tools such as manos and metates identified in Op. LC12 C suggest that Substructure 2 was used as a domestic area over the course of

**TABLE 5.3.** Estimates of years necessary to construct all mounded earthen architecture at La Consentida, varying by season length and workforce size. Results are rounded to nearest year.

| Number of laborers | 30-day working season | 60-day working season | 90-day working season | 120-day working season |
|---|---|---|---|---|
| 25 | 145 | 73 | 48 | 36 |
| 50 | 73 | 36 | 24 | 18 |
| 75 | 48 | 24 | 16 | 12 |

**TABLE 5.4.** Estimates of years necessary to construct an early version of Platform 1, varying by season length and workforce size. Results are rounded to nearest year.

| Number of laborers | 30-day working season | 60-day working season | 90-day working season | 120-day working season |
|---|---|---|---|---|
| 25 | 29 | 15 | 10 | 7 |
| 50 | 15 | 7 | 5 | 4 |
| 75 | 10 | 5 | 3 | 2 |

multiple construction phases (see table A1.5 and figures A1.10 and A1.11). As demonstrated by Structure 2 (a square feature measuring 3.05 × 3.05 m in its footprint), some domestic buildings were ephemeral and could have been erased by subsequent construction or postdepositional processes, or simply remain overlooked in excavation.

While no other obvious Formative period domestic buildings aside from Structure 1 and Structure 2 have been fully uncovered at La Consentida, a few noteworthy finds suggest that other parts of the site were also domestic occupational areas. The LC09 A-F4 and LC09 B-F15 (figure 4.4) hearths, for example, indicate domestic practices such as cooking early in the occupation of Substructure 4 (see tables 1.1, A1.1, and A1.2; and figures 3.3, 4.1, and 4.3). Another possible hearth (E-F10) suggests similar practices in the north/central part of the site (see tables 1.1 and A1.7 and figure A1.13). Faunal remains (see chapter 6) and decorated ceramic serving wares (see appendix 2) suggest that the midden deposits (E-F9) overlying E-F10 result from communal feasting, however, rather than domestic cooking (also see discussion in chapter 7). In the area of Op. LC12 A, a fill deposit (A-F12) and a possible occupational surface (A-F13), were associated with an alignment of large, burned daub chunks. These finds may represent the edge of a building atop an early version of Substructure 1 (see table A1.3 and figure A1.8). Given the available evidence, it is unclear whether this probable building was domestic or public in nature.

Structures 1 and 2 on Substructure 2 and the LC09 A-F4-s1 and LC09 B-F15 hearths associated with Substructure 4 suggest that La Consentida's seven substructural

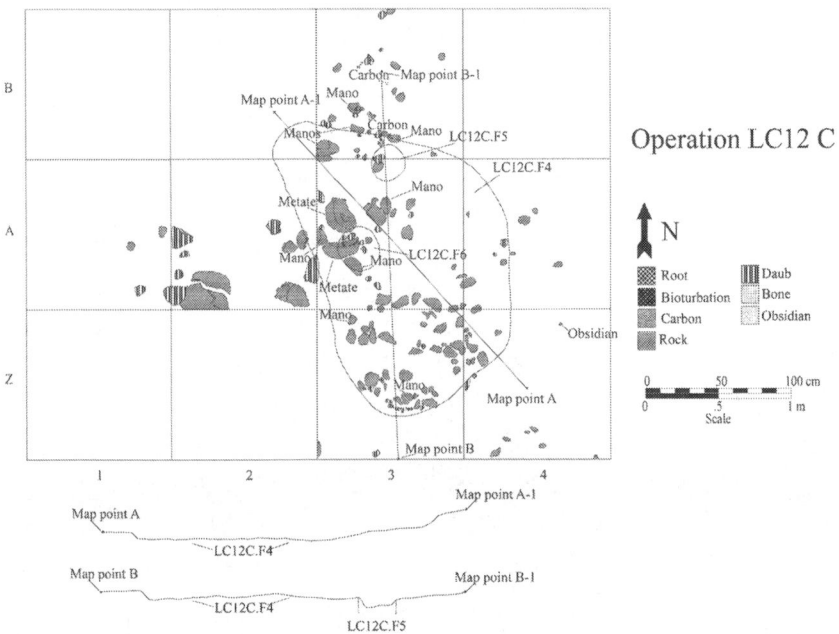

**FIGURE 5.1.** Plan and profile maps of Structure 1 (ceramic sherds removed for clarity)

mounds were largely domestic areas. In contrast, only a few midden deposits (e.g., in Op. LC12 E and in Op. LC12 H) suggest community gatherings beyond the household level. Both of these deposits occur at the edge of or between substructural mounds, suggesting that the households associated with Structures 1 and 2 (located atop Substructure 2) were involved with communal events. Specifically, individual households may have helped prepare for, or even have hosted, communal feasts. It is worth noting that clear evidence for permanent structures (including probable house floor stains, postholes, and large fragments of ground stone used as foundation or wall construction material) is not present until late in site occupation. Having discussed evidence from domestic structures, I will now turn to patterns identified among specific artifact types.

## CERAMIC AND GROUND STONE EVIDENCE

Like the architectural strata and domestic structures discussed above, La Consentida's subsistence-related artifact assemblage is consistent with a community in the process of shifting from seasonal mobility to committed sedentism. This interpretation

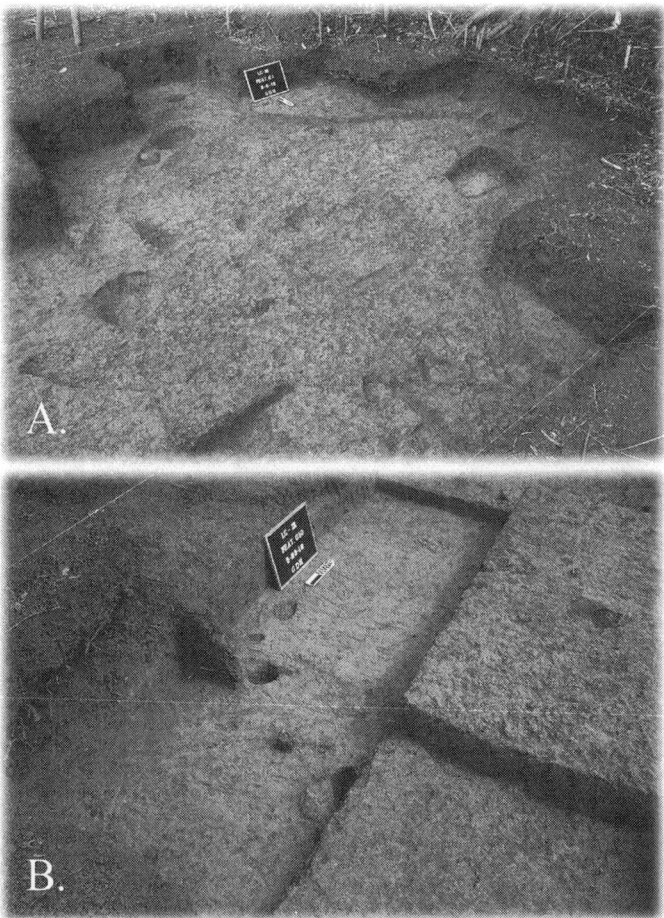

**FIGURE 5.2.** Structure 2: (A) floor and floor features of structure; (B) probable postholes associated with structure

requires careful assessment of the material evidence. In chapter 2, I discussed the use of ground stone and ceramic artifacts as indicators of mobility. Research from elsewhere in Mesoamerica (e.g., P. Arnold 1999, 2003; Rosenswig 2011) has suggested that the transition to sedentism in Mesoamerica was gradual and that pottery (for example) cannot be considered *a priori* evidence of a sedentary population. Ground stone tools changed from those emphasizing portability and multipurpose use to larger tools likely intended for more specific tasks as Mesoamericans established sedentary villages (e.g., Clark et al. 2007; MacNeish et al. 1967:101–21; Winter and

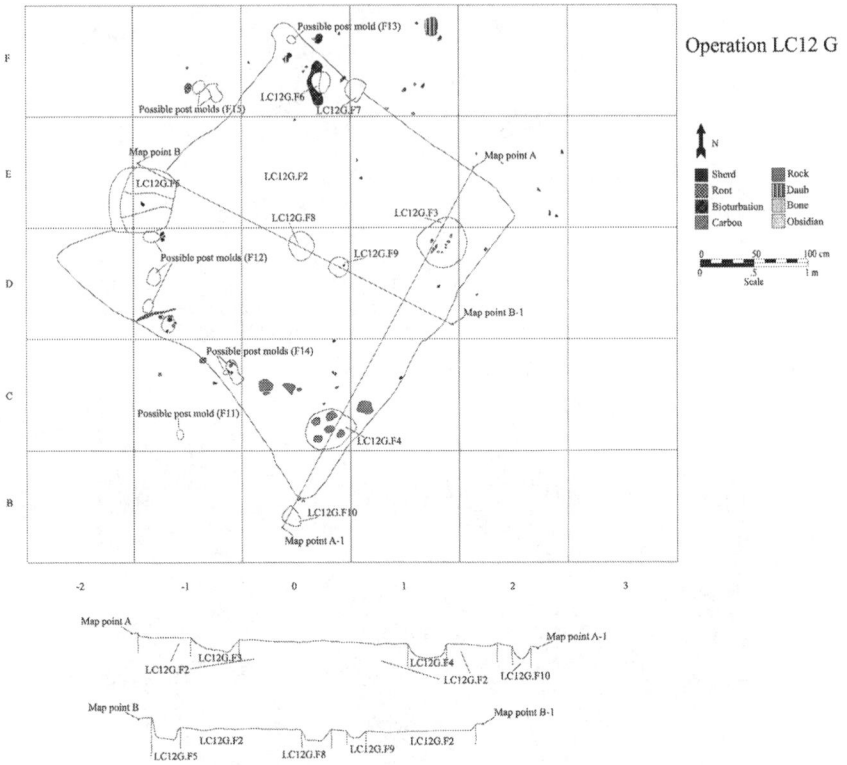

FIGURE 5.3. Plan and profile maps of Structure 2

Mateos 2010; Winter and Sánchez Santiago 2014a:10–11; see also J. Adams 1988). The earliest, prearchitectural occupation layers at La Consentida (see table A1.3 and figure A1.8) indicate that the community that established the site was already using ceramics. As discussed in chapter 8 and appendix 2, this vessel assemblage is not similar in its formal composition to the tecomate-heavy assemblages of the Barra and Tulipan phases (P. Arnold 1999, 2009; Clark and Blake 1994), which Arnold suggested might be evidence of a mobile occupational strategy. Regardless, caching practices or the use of a few portable and multipurpose vessels may explain how a semimobile group could employ ceramics, which further emphasizes that the presence or absence of ceramics is insufficient evidence for domestic mobility.

The ground stone tools identified in the deepest layers at La Consentida tend to be of a size and weight that would be relatively portable. As discussed below, the relative frequency of portable ground stone *decreased* over time, and nonportable

**FIGURE 5.4.** Early ground stone from La Consentida: (A) grinder or polisher and possible anvil from sheet midden atop early fill layer (LC12 D-F11); (B) grinder or mano and hammer stone or anvil with crystalline texture from probable occupation surface atop early fill layer (LC12 D-F10); (C) grinder or mano and possible hammer stone with crystalline texture from early midden (LC12 E-F14); (D) complete polisher or pestle from early fill context (LC12 F-F6)

ground stone *increased* over time (along with raw counts of ground stone), at La Consentida (see figure 5.7). In addition, use-wear marks on many of these earliest stone tools indicate their utility for multiple food-processing and/or crafting tasks. The tool shown in figure 5.4.A, for example, comes from a very early context (LC12 D-F11), where excavations uncovered thin layers of midden interspersed with fill. This tool bears sheen from grinding or polishing, in addition to probable impact pitting and fracturing from use as a hammer stone or anvil. When possible manos are recovered in La Consentida's earlier deposits, they tend to bear relatively little use-wear and are of a lower quality (i.e., grainier) granite than later manos at the site. The grinder or possible mano shown in figure 5.4.B is of a grainy material and has pitting from its use as a hammer stone or anvil. Figure 5.4.C shows a very similar mano or grinder and hammer stone from an early midden context (LC12 E-F14). In general, most of these earliest tools exhibit polish rather than grinding wear, which indicates they were probably not used with metates. The artifact shown in figure 5.4.D, for example, is a complete polisher or pestle with a smooth surface and a size and shape that appears to emphasize portability over grinding efficiency. Patterns of ground stone use-wear indicative of multifunctional use also suggest an emphasis on portability over efficiency and serve as supporting evidence of an initially semi-mobile population. Note that the rough-looking surface on some of the ground stone artifacts is the result of postdepositional calcium concretions or sediment that has not been washed off in order to permit residue analyses.

One kind of multifunction tool common in deeper deposits at La Consentida, but which can also be found (possibly redeposited) in shallower contexts, is that of the "polisher/hammer stone" (e.g., figure 5.5.A). These artifacts bear facets from their use as polishers as well as impact or flaking damage from hammering. Figure 5.5.B shows a similar multiuse tool that seems to bear attributes of a miniature mano, polisher, or even pestle, in addition to its impact scarring from use as a hammer stone. In general, a consistent pattern among such artifacts is that regardless of their other uses, most show evidence of being used for hammering. Such portable and multifunctional tools and one-handed manos are similar in form and likely in use to Archaic period manos, which were probably employed for "processing hard seeds of teosinte or primitive maize" (Winter and Sánchez Santiago 2014a:10–11, fig. 14.a [translation my own]; see also Clark et al. 2007; Winter and Mateos 2010). Small "manos" may also have been used for hide processing (J. Adams 1988). Small, "Archaic style" ground stone artifacts (e.g., the mano shown in figure 5.5.C) occur throughout excavated and surface contexts at La Consentida, and suggest an interest in portability and multifunctionality, which are emphases of mobile groups (P. Arnold 1999; Clark et al. 2007; McDonald 1991:85; Torrence 1983).

If Archaic-style tools remained in use after the community was sedentary and more reliant on agriculture (see chapter 6), their presence begs explanation. Such artifacts may indicate practices of caching, tool reuse, or the continued production of portable and versatile tools at the site. It is possible that tools best suited for Archaic period subsistence and crafting activities remained in use in order to conserve materials. A greenish stone axe or adze (likely made of a fine-grained basalt) demonstrates such conservation (figure 5.5.D). As discussed by Clark and Cheetham (2002:305), such artifacts can be used for a variety of tasks, including forest clearance and hoeing of dirt for planting crops. This artifact has been broken and refinished for continued use so many times that it has become very small and probably inefficient in comparison to its initial form. It is essentially an "exhausted" adze.

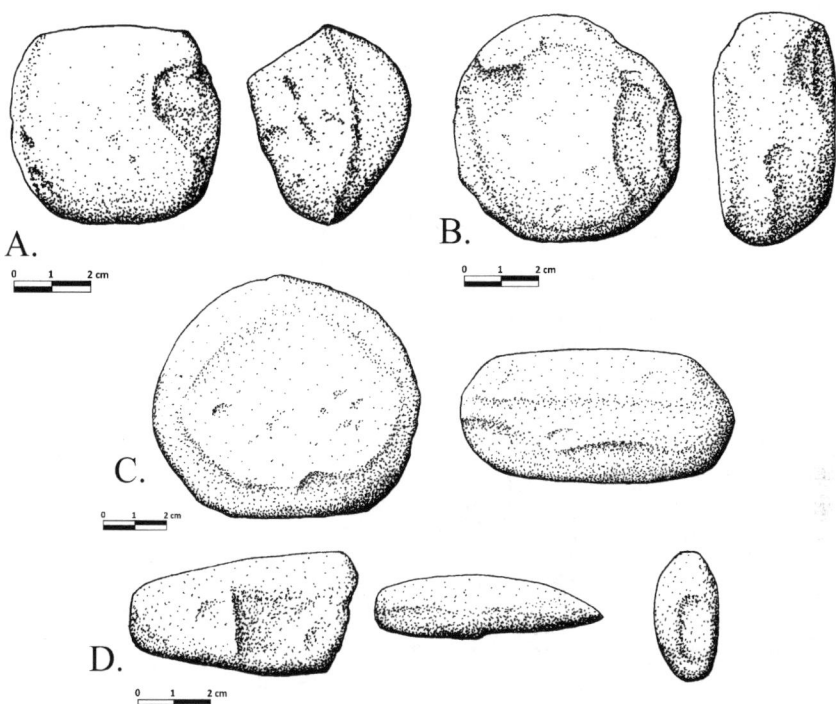

**FIGURE 5.5.** Multifunction ground stone tools from La Consentida: (A) polisher / hammer stone from fill (LC12 E-F2) above midden; (B) mano, polisher, or pestle and hammer stone from fill (LC12 G-F16) surrounding domestic context; (C) small, one-handed mano from surface at Op. LC12 B; (D) exhausted probable basalt axe or adze from fill with occupation layer (LC12 C-F7)

Although many of La Consentida's ground stone tools show evidence of use for multiple tasks (e.g., grinding, polishing, chopping, and pounding), a few examples appear to have had a more specific primary use. In shallower, later deposits, for example, manos are more common (figure 5.7.A). In one case, a mano even appeared as an offering with a burial (figure 6.9.C). Also of note for illuminating practices of domestic mobility are the metates, which (like the manos) remained small and of the "one-handed" variety throughout site occupation (see Winter and Mateos 2010; Winter and Sánchez Santiago 2014a:10–11). In later deposits, and particularly in and around Structure 1, large metate fragments were more common (figure 5.7.A). In fact, metate fragments seem to be exclusively located in more recent strata (see figures 5.1 and 5.6). Like the manos, these metates suggest an increasing emphasis on sedentism over time at La Consentida. Significantly, metates and manos do not connote sedentary

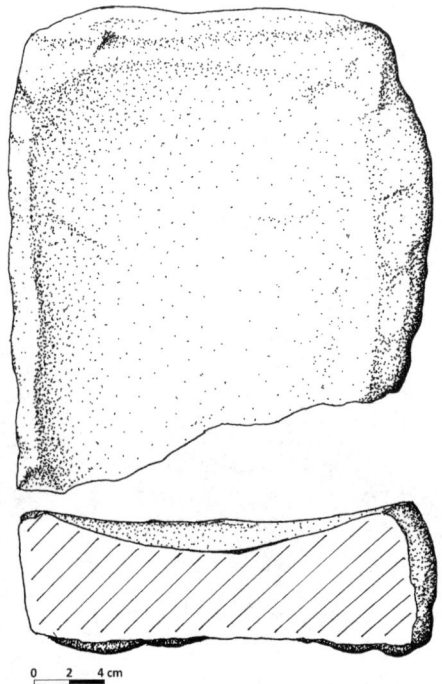

**FIGURE 5.6.** Partial metate from Structure 1 domestic context (LC12 C-F4)

agriculturalists, per se. As discussed by Pohl and colleagues (1996:365; see also Clark and Cheetham 2002:305), late Archaic deposits sometimes contain such artifacts and thus suggest some degree of maize processing in preceramic times. Although large metates and manos could hypothetically be cached at seasonal resource extraction sites used by a mobile group (see Mitchell 2008), their discovery in conjunction with a wide variety of ceramic vessels (see appendix 2), probable domestic buildings, and increasing production of mounded earthen architecture all suggest that a semimobile early occupation of La Consentida gave way to more constant settlement before Formative period site abandonment. Although La Consentida's later contexts demonstrate a shift toward nonportable grinding technology, likely for the processing of maize flour, it is worth mentioning that the metates never seem to have been very large. The best-preserved example, shown in figure 5.6, measures a maximum of only 245 mm in width and 75 mm in thickness. In cases where metate fragments are complete enough to determine original grinding trough dimensions, the metates seem well suited for use with early-style, one-handed manos rather than with the larger two-handed manos of later Mesoamerican history (see Clark et al. 2007; Winter and Mateos 2010; Winter and Sánchez Santiago 2014a:10–11). The fact that these metates were later recycled as construction material conveys a desire to reuse materials as much as possible, even after breakage. These recycling practices probably explain, at least in part, the evidence for multifunctional tool use at the site and should promote caution regarding arguments about changes over time in tool types, as tools were likely used for as long as possible.

Interpretations presented here about changes over time in residential mobility and the gradual adoption of material culture associated with agriculture are in part based on chronological changes in ground stone tools. Figure 5.7.A demonstrates chronological patterns in the relative frequencies of manos, metates, and polisher/hammer stones in groupings of excavated lots. For the purposes of this figure, excavation lots (which increase in number with greater depth) are taken as a general proxy of chronology. The majority (85.4 percent) of ground stone tools and tool fragments were collected within the upper ten lots of excavation throughout the site. Manos and metates were most frequent in about the first seven lots of excavation. Although part of this pattern is driven by the overall recovery context of ground stone, that circumstance alone cannot account for the shallow (i.e., "more recent") provenience of the manos and metates. For example, 74.0 percent of all ground stone was recovered from the first seven lots, while 87.5 percent of metates and 81.3 percent of manos were recovered in those shallowest seven lots. Manos peak in the second lot, while overall ground stone tools peak in lots four, five, and seven. Metate fragments are entirely absent from anything deeper than lot eight, despite the twenty-three pieces of ground stone recovered in those deeper (i.e., earlier) contexts. Polisher/hammer stones seem to be a particular kind of multiuse tool identified in some of the earlier contexts at La Consentida. These tools exemplify well the "Archaic-style" tradition of multiuse tools that emphasize portability over task-specific efficiency (Clark et al. 2007; McDonald 1991:85; Torrence 1983). These multiuse tools do not seem to follow the general pattern of increasing ground stone over time at La Consentida. In fact, only 57.1 percent of them occurred within the first seven lots, and 21.1 percent of them occurred in lots 19, 20, and 21. Polisher/hammer stones were located in earlier contexts than most manos and all metates. For the purposes of producing figure 5.7.A, I ignored any stone fragments that might be natural but included any "possible" manos, metate fragments, and polisher/hammer stones. This decision provides more artifacts for analysis but also means that some grinders or grinding platforms (for example) may dilute the patterns represented.

In general, ground stone tools at La Consentida follow the same chronological pattern as pottery remains. Figure 5.7.B displays relative percentages of ground stone tool counts (total = 123) as compared to grams of ceramic sherds (total = 499,199.9 grams), according to excavated lot. Ground stone and ceramic artifacts

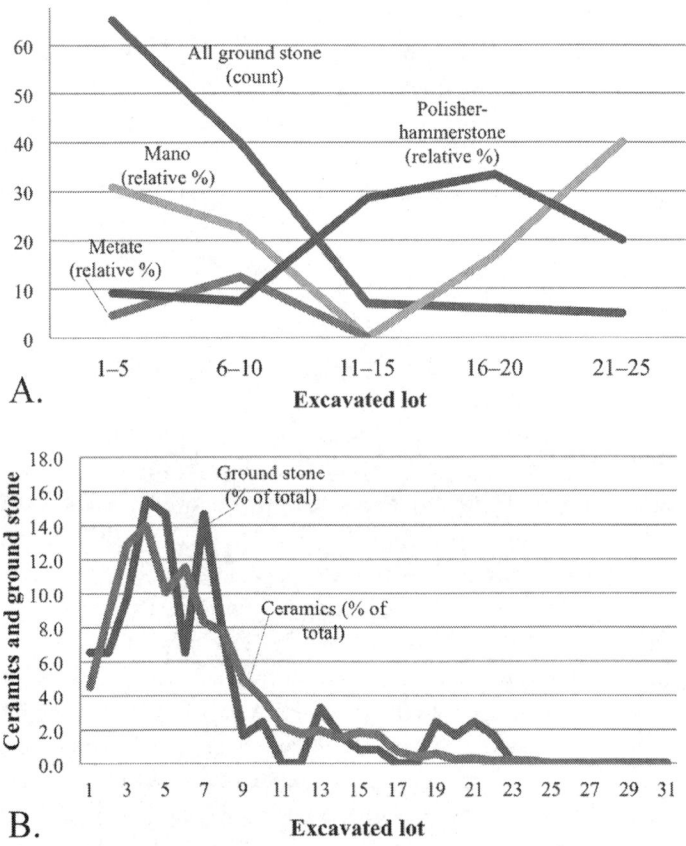

**FIGURE 5.7.** Material evidence for increasing sedentism: (A) relative percentages of selected ground stone tool categories by excavated lot; (B) percentages of total ground stone artifacts (count) and ceramics (g) by excavated lot

follow the same basic pattern, wherein they experience their greatest quantities between about lots three and seven. The ceramics experience a more gradual decrease in the earlier contexts, likely because their exponentially greater quantity provides for more normally distributed data. It is against the backdrop of these general trends that I contextualize my arguments about chronological patterns in the provenience of polisher/hammer stones, manos, and metates (figure 5.7.A).

As both figure 5.7 and table 5.5 demonstrate, the relative percentages of specific tool types change independently of the volume of sediment excavated. Polisher/hammer stones are common throughout the sequence but make up less of the total

**TABLE 5.5.** Changes over time in ground stone tools. Columns for manos, metates, and polisher/hammer stones include relative percentages by excavated level.

| Level | All ground stone | Manos | Metates | Polisher/hammer stones |
|---|---|---|---|---|
| 1–5 | 65 | 20 (30.8%) | 3 (4.6%) | 6 (9.2%) |
| 6–10 | 40 | 9 (22.5%) | 5 (12.5%) | 3 (7.5%) |
| 11–15 | 7 | 0 (0%) | 0 (0%) | 2 (28.6%) |
| 16–20 | 6 | 1 (16.7%) | 0 (0%) | 2 (33.3%) |
| 21–25 | 5 | 2 (40%) | 0 (0%) | 1 (20%) |
| Total | 123 | 32 | 8 | 14 |

ground stone collection in shallower deposits than do other tool types. Where polisher/hammer stones do occur in more recent layers, some may be redeposited. After a brief spike of two artifacts in the deepest levels (possibly accounted for by other crafting activities such as hide preparation), mano-like tools become most common in the shallowest deposits (see J. Adams 1988). This trend is stronger still among metates, which would be the most difficult tools for a semimobile population to transport and which are only found in shallower deposits. In general, larger (i.e., more difficult to transport) and more task-specific food-processing tools were disproportionately recovered from shallow excavation contexts. Smaller and more multipurpose tools were more likely to occur in earlier deposits. Overall, ground stone seems to have become much more common over time at the site. I find these data consistent with an increase in domestic sedentism and the use of task-specific tools associated with agriculture, and particularly the processing of maize flour.

## DISCUSSION: MOUNDS, MONUMENTS, AND THE BUILT LANDSCAPE

As discussed in chapter 2, monumental constructions at Göbekli Tepe (Peters and Schmidt 2004; though see also Banning 2011), in general across the megalithic landscape of ancient Europe (e.g., R. Bradley 2005; Sherratt 1990; Tilley 2007), and at North American sites such as Poverty Point (Gibson 2000) demonstrate the ability of semimobile groups to produce massive constructions that might have been viewed as evidence for sedentism according to traditional archaeological models (e.g., Childe 1950). Recent research on the European Neolithic has begun to tease apart these tacit associations between monumental architecture, sedentism, and social organization. Authors such as Alasdair Whittle (2003), Vicki Cummings and Whittle (2004), and Cooney (2007:560) have discussed the significance of monuments in Britain and Ireland as markers of "regional indigenous identity" during times of social upheaval or foreign incursion. In such circumstances, monuments may form elements of a group's cosmology or signify cultural affiliation in contradistinction to those of other peoples (Cooney 2007:560). Other researchers have explored the role of memory as it relates to monuments. With his concept of the "afterlife of monuments," Richard Bradley (1993) considered continuity and reinterpretations of the historical significance and symbolic meaning of human labor. Bradley (2000) argued that the maintenance of monuments and sacred areas of landscapes (such as long-standing loci for bog offerings and sacred cave shrines) indicates the importance of history, memory, and lineages associated with such places. Whether or not megalithic sites were also zones of domestic occupation is sometimes debated. The Anatolian site of Göbekli Tepe, for instance, has been the focus of debate regarding whether nonsedentary peoples were capable of such impressive architectural constructions (see Banning 2011; Peters and Schmidt 2004). What such discussions of ancient monuments have in common is an emphasis on the construction of symbolic landscapes. Although perhaps not sedentary farmers like their successors, ancient Old World communities occupied an environment constructed in part by anthropogenic modifications and the conceptual associations that they held (e.g., Tilley 2007).

Archaeologists interested in cultural memory have argued that landscape modification is a prime example of how humans build and maintain cosmologies and identity (Van Dyke and Alcock 2003). The intentional transformation of a landscape through such processes as earthen platform and burial mound construction has also been described in Marxian terms as the translation of the landscape from an "object" of human labor to its "subject" (Dunham 1999; Meillassoux 1972). As such, landscape modification is more than merely the epiphenomenal result of resource extraction and environmental degradation. Landscape modification and monumental construction serve to curate culture history and cosmology, preserve the memories of ancestors, and promote a sense of place (de Certeau 1984; Dunham 1999:120, 128). Such constructions as earthen mounds may be markers of territorial holdings signifying for a people (whether mobile, sedentary, or something in-between) shared aspects of ideology and physical markers of territory and/or identity, and perhaps meeting places for large social events. The labor and memories invested in anthropogenic earthen architecture and other monuments, therefore, were meaningful elements of ancient conceptions of landscapes and must be considered in discussions of the adoption of sedentism.

## SUMMARY

The case for changing practices of domestic mobility at La Consentida is based on stratigraphic, artifact, and contextual data, as well as population and labor estimates. Occupational surfaces below Platform 1, which contain in situ vessel fragments, demonstrate that people already producing ceramics founded the site (see chapter 4 and appendix 2). While some researchers (e.g., Clark and Cheetham 2002:311) consider ceramics and mounded architecture indicators of sedentism, others (e.g., P. Arnold 1999:160, 2009; Lesure 2009) argue that certain vessel types such as tecomates are suited to a semisedentary lifestyle. On the basis of settlement and ceramic data from the Gulf Coast and Soconusco regions, these authors argue that the transition from Archaic period nomadism to Middle and Late Formative period sedentism was gradual. Given such varied opinions, a key question is thus whether ceramics and earthen architecture are evidence of sedentism, or if, instead, they encouraged transition *toward* it. I argue that La Consentida may exemplify the latter trend.

Data from La Consentida suggest a gradual shift toward sedentism during the Early Formative period. Although this evidence does not seem to fit well with the pattern found by Clark and colleagues (2007:35), wherein "village sedentism coincided with the appearance of pottery all across proto-Mesoamerica about 1900–1600 BC," another pattern discussed by these authors seems better supported by the La Consentida evidence. Specifically, manos and metates seem to gradually replace Archaic-style tools such as polisher/hammer stones over time at this site (figure 5.7.A). Whereas no obvious mortars have been recovered at La Consentida, many of the earliest tools bear shapes and grinding or polishing wear that suggest their possible use as pestles (e.g., figure 5.4.D). Although the wholesale replacement of these tool types by manos and metates is not clear at La Consentida (some possible pestles appear in later contexts and some possible manos appear in deep contexts), there is a general pattern wherein heavier ground stone tools specifically tailored to maize processing appear to have gradually replaced the earlier, multitool varieties. Rather than a stand-alone indicator of sedentism or absolute agricultural reliance, I see this pattern as one of several lines of evidence (in conjunction with an increasing emphasis on earthen architecture and the construction of more permanent and robust domestic structures) that indicates a gradual trend toward both of these hallmarks of later Mesoamerican history. Much of this argument hinges on evidence for La Consentida's subsistence economy, to which I turn my attention in the following chapter.

# 6

## Diet and Changing Culinary Tastes

In this chapter, I discuss results of faunal analysis, the study of dental pathologies, isotopic data (from human teeth, human long bones, and faunal bone), and patterns identified among food-processing tools to reconstruct La Consentida's subsistence practices. Although these diverse lines of evidence provide a good proxy for understanding diet at the site, ongoing and future investigations should help to bolster the conclusions presented here. Excavations and sediment sampling have recovered very few macrobotanical remains at La Consentida, for example. It is possible that this trend is a result of poor preservation of floral remains or simply that not enough (and not *large* enough) sediment samples have been collected during excavations. Future research will aim to correct for this absence of data through the collection of more sediment samples and improved flotation techniques.

Following discussion of various independent lines of evidence for diet, I conclude that subsistence at La Consentida incorporated a wide variety of resources and transitioned over time to greater reliance on maize. Stable isotope values suggest that the community consumed more maize than did contemporaneous communities in the Soconusco (Blake et al. 1992; Chisholm and Blake 2006) and Gulf Coast regions (Killion 2013). As isotope values and ground stone analysis demonstrate, however, the La Consentida community was not as fully reliant on agriculture as were later Mesoamerican groups, whose ground stone tools demonstrate a greater emphasis on maize-flour-grinding efficiency over tool portability and multifunctionality (see P. Arnold 2009; Clark et al. 2007:29). Changes in ground stone tool form over time also suggest a transition in the way maize was processed, perhaps indicating changing culinary practices such as the shift from consuming maize in liquid form to consuming it as a processed flour (see figure 5.7.A). Such shifts in cuisine have been identified in other Early Formative Mesoamerican contexts in which foods such as maize and cacao were employed as components of public events tied to the negotiation of status and intercommunal affiliations (Joyce and Henderson 2007).

## FAUNAL ANALYSIS RESULTS

To date, analysis of La Consentida's faunal remains has focused on the Ops. LC09 B, LC12 D, LC12 E, and LC12 H middens, along with the LC12 A-F15 ritual cache and a few miscellaneous finds such as a crocodile or caiman mandible fragment recovered with burial B2-I3.[1] Figure 6.1 provides a summary of the vertebrate faunal remains recovered from the four midden contexts. For this study, a biologist and I (Pérez Hernández and Hepp 2015) followed standard faunal analysis procedure of estimating NISP (number of identified specimens) and MNI (minimum number of individuals) for each identifiable taxon (e.g., Banning 2000:187–211). We did not extrapolate the quantity of meat provided by these animals (see, for example, Wing 1978). The faunal material analyzed consisted of a sample of 1,668 elements. Of those, 1,276 bones came from screened material and 392 came from the heavy fractions resulting from sediment sample flotation. Regarding subsistence practices suggested by faunal analysis of the middens, we identified a good deal of variability among contexts (tables 6.1–6.6). The Op. LC12 E midden (F16–F9) produced about 90 percent fish remains in screened sediments. Many of these remains came from Osteichthyes (bony fishes) and a species of catfish known as *Ariopsis guatemalensis* (see figure 6.2.A). These fish were also present in the other middens, though the relative frequencies of all fish were higher in LC12 E than in any other midden at the site (85 percent of NISP in Op. LC12 D, 66 percent in Op. LC12 H, and only 17 percent in Op. LC09 B). Large marine fish were also present and must have provided rare but valuable resource packages, as demonstrated by the sizable vertebra shown in figure 6.2.B. Although the stable isotope data discussed below indicate that fish were not necessarily a primary staple of the La Consentida diet, excavations in several contexts indicate that they represented an important source of animal protein consumed at the site (see A. Joyce 2010:53).

Mammal remains varied in relative frequency among screened deposits, with the Op. LC09 B midden (F17) containing by far the most, at 57 percent of all faunal remains. It is notable that many of the Op. LC09 B mammal bones likely came from just a few deer. In other words, the Number of Identified Specimens (NISP)

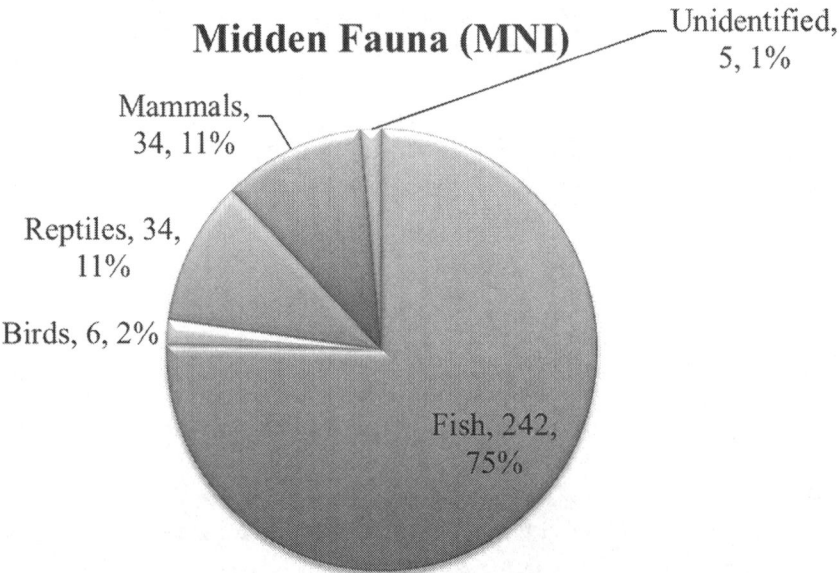

**FIGURE 6.1.** Pie chart summarizing the vertebrate faunal remains (analyzed by MNI) in screened deposits of four middens at La Consentida. Note that the MNI of fish tends to overrepresent their dietary importance relative to larger-bodied animals

of mammal remains was high in LC09 B-F17, while the Minimum Number of Individuals (MNI) was five. In general, mammal remains were rarer than expected in all excavated contexts. When they were recovered, deer bones were often burned (see figure 6.3.A), indicating that the processing of flesh from large terrestrial mammals (which were likely roasted over an open flame) differed from that of fish and shellfish, the remains of which often appear to have been boiled in soups or stews.[2] In some cases, the durable long bones of deer were valued for the production of tools (figure 6.3.B). When viewed under a microscope (e.g., figure 6.3.C), it is clear that these deer bone tools often bear use marks in the form of scratches in a perpendicular orientation to the tools themselves, suggesting that they were used on a hard, brittle material. One rare type of mammal remains recovered at La Consentida was that of canids. As demonstrated by the coyote or possible dog mandible fragment shown in figure 6.3.D, canid remains sometimes occurred in middens. This provenience perhaps indicates their use as food. Due to the lack of dental crowding on the mandible fragment shown in figure 6.3.D, this example likely comes from a coyote rather than from a domesticated dog (Banning 2000:202; Silvia Pérez Hernández, personal communication, 2013).

**TABLE 6.1.** List of the vertebrate faunal remains identified in screened samples from the Op. LC12 E midden

| | | Op. LC12 E | | | | |
|---|---|---|---|---|---|---|
| Class and order | Family | Species and common name | NISP | %NISP | MNI | %MNI |
| ACTINOPTERYGII (FISHES) | | | | | | |
| Osteichthyes | Various | Various | 774 | 84.1 | 159 | 69.73 |
| Siluriformes | Ariidae | *Ariopsis guatemalensis* Catfish | 51 | 5.55 | 24 | 10.52 |
| AVES | | | | | | |
| Ciconiiformes | Ardeidae | *Ardea* sp. (possible) Heron | 1 | 0.1 | 1 | 0.43 |
| REPTILIA | | | | | | |
| Crocodylia | Crocodylidae | *Crocodylus acutus* or *Crocodylus moreletii* Crocodile | 3 | 0.32 | 3 | 1.31 |
| Lacertilia | Iguanidae | *Iguana iguana* Green iguana | 3 | 0.32 | 3 | 1.31 |
| Testudines | Kinosternidae | *Kinosternon integrum* or *Kinosternon scorpioides* Mexican mud turtle or Scorpion mud turtle | 13 | 1.41 | 7 | 5.7 |
| Unidentified | | | 7 | 0.76 | 0 | 0 |
| MAMMALIA | | | | | | |
| Carnivora | Canidae | *Canis latrans* or *Canis familiaris* Coyote or dog | 1 | 0.1 | 1 | 0.43 |
| Lagomorpha | Leporidae | *Lepus* sp. or *Sylvilagus* sp. Hare or rabbit | 2 | 0.21 | 1 | 0.43 |
| Artiodactyla | Tayassuidae | *Pecari tajacu humeralis* Peccary | 2 | 0.21 | 2 | 0.87 |
| Artiodactyla | Cervidae | *Odocoileus virginianus* Deer | 34 | 3.7 | 12 | 5.26 |
| Artiodactyla | Tayassuidae or Cervidae | *Pecari tajacu humeralis* or *Odocoileus virginianus* Peccary or deer | 2 | 0.21 | 2 | 0.87 |
| Unidentified | | Various | 11 | | | 0% |
| | | TOTAL | 904 | 96.99 | 215 | 96.86 |

**TABLE 6.2.** List of the vertebrate faunal remains identified in screened samples from the Op. LC12 D middens

| | | Op. LC12 D | | | | |
|---|---|---|---|---|---|---|
| Class and order | Family | Species and common name | NISP | %NISP | MNI | %MNI |
| ACTINOPTERYGII (FISHES) | | | | | | |
| Osteichthyes | Various | Various | 115 | 44.23 | 24 | 40.67 |
| Siluriformes | Ariidae | *Ariopsis guatemalensis* Catfish | 105 | 40.38 | ·14 | 23.72 |
| Lepisosteiformes | Lepisosteidae | *Atractosteus tropicus* (possible) Tropical gar | 3 | 1.15 | 3 | 5.08 |
| Reptilia | | | | | | |
| Crocodylia | Crocodylidae | *Crocodylus acutus* or *Crocodylus moreletii* Crocodile | 8 | 2.99 | 4 | 6.77 |
| Lacertilia | Iguanidae | *Iguana iguana* Green iguana | 2 | 0.76 | 2 | 3.38 |
| | Iguanidae | *Ctenosaura* sp. Black iguana | 1 | 0.38 | 1 | 1.69 |
| Unidentified | | | 1 | 0.38 | 1 | 1.69 |
| Testudines | Kinosternidae | *Kinosternon integrum* or *Kinosternon scorpoioides* Mexican mud turtle or Scorpion mud turtle | 6 | 2.3 | 4 | 6.67 |
| Mammalia | | | | | | |
| Artiodactyla | Cervidae | *Odocoileus virginianus* Deer | 7 | 2.69 | 5 | 8.47 |
| Artiodactyla | Tayassuidae or Cervidae | *Pecari tajacu humeralis* or *Odocoileus virginianus* Peccary or deer | 1 | 0.38 | 1 | 1.69 |
| Unidentified | | | 4 | 1.53 | | 0 |
| Unidentified | | | | 7 | 2.69 | |
| | | TOTAL | 260 | 99.86 | 59 | 99.83 |

TABLE 6.3. List of the vertebrate faunal remains identified in screened samples from the Op. LC12 H midden

| Class and order | Family | Species and common name | NISP | %NISP | MNI | %MNI |
|---|---|---|---|---|---|---|
| ACTINOPTERYGII (FISHES) | | | | | | |
| Osteichthyes | Various | Various | 26 | 44.82 | 8 | 28.56 |
| Siluriformes | Ariidae | *Ariopsis guatemalensis* Catfish | 12 | 20.68 | 5 | 17.85 |
| REPTILIA | | | | | | |
| Crocodylia | Crocodylidae | *Crocodylus acutus* or *Crocodylus moreletii* Crocodile | 2 | 3.44 | 2 | 7.14 |
| Testudines | Kinosternidae | *Kinosternon integrum* or *Kinosternon scorpoioides* Mexican mud turtle or Scorpion mud turtle | 2 | 3.44 | 2 | 7.14 |
| MAMMALIA | | | | | | |
| Carnivora | Canidae | *Canis latrans* or *Canis familiaris* Coyote or dog | 1 | 1.72 | 1 | 3.57 |
| Lagomorpha | Leporidae | *Lepus* sp. Or *Sylvilagus* sp. Hare or rabbit | 1 | 1.72 | 1 | 3.57 |
| Artiodactyla | Cervidae | *Odocoileus virginianus* Deer | 1 | 1.72 | 1 | 3.57 |
| Artiodactyla | Tayassuidae or Cervidae | *Pecari tajacu humeralis* or *Odocoileus virginianus* Peccary or deer | 1 | 1.72 | 1 | 3.57 |
| Unidentified | | | 2 | 3.44 | 1 | 3.57 |
| Unidentified | | | 10 | | | |
| | | TOTAL | 58 | 82.7 | 22 | 78.54 |

Invertebrate marine animal remains were rarer than expected in screened contexts at La Consentida, which was likely about 4 km from an open-ocean bay at the time of its occupation (Goman et al. 2005). Other pacific coastal regions, including the Soconusco, are known for large Archaic period shell middens near rich estuarine areas (see Voorhies 1989, 2004; Voorhies and Kennett 2011). Despite

TABLE 6.4. List of the vertebrate faunal remains identified in LC12 A-F15 ritual cache

| Class and order | Family | Species and common name | NISP | %NISP | MNI | %MNI |
|---|---|---|---|---|---|---|
| ACTINOPTERYGII (FISHES) | | | | | | |
| Osteichthyes | Various | Various | 13 | N/A | 4 | 44.44 |
| Siluriformes | Ariidae | *Ariopsis guatemalensis* Catfish | 3 | N/A | 3 | 33.33 |
| REPTILIA | | | | | | |
| Lacertilia | Helodermatidae | *Heloderma horridum* Mexican beaded lizard | Nearly complete skeleton | N/A | 1 | 11.11 |
| Testudines | Kinosternidae | *Kinosternon integrum* or *Kinosternon scorpoioides* Mexican mud turtle or Scorpion mud turtle | 1 | N/A | 1 | 11.11 |
| | | TOTAL | N/A | N/A | 9 | 99.99 |

the relative lack of shellfish at La Consentida in comparison to other coastal sites, some contexts nonetheless produced shell, including those of mangrove Mytilid mussels (or *tichinda*), as well as oyster, clams, barnacles, and crustaceans such as crabs (Farías Sánchez 2006; figure 6.4.A). Many of the shellfish and fish remains were too small to have been collected individually, suggesting that they were likely collected in traps or nets and then boiled whole in soups or stews. The recovery of some very tiny fish and shellfish remains in sediment sample heavy fractions supports this interpretation. The presence of *tichinda* mussels suggests that the edges of the ancient bay reconstructed by Michelle Goman and colleagues (2005) likely contained a mangrove habitat. Aside from their relationship to subsistence, pieces of shell were also worked to produce jewelry and possibly other decorative artifacts (figures 6.4.B and 6.4.C). Shell remains from La Consentida do not seem to indicate the production of shell pendants on the same impressive scale as that demonstrated for Laguna Zope on the Isthmus of Tehuantepec (see Zeitlin 1979:ch. 5). Due to the time constraints of the 2014 faunal analysis project, a detailed study of the shell recovered at La Consentida was not possible. In order to discuss the relative quantities of shell from different deposits, however, I was able to quantify the shell recovered in sediment samples taken in the Op. LC12 D, LC12 E, and LC12 H middens. The results of flotation heavy fraction analysis are discussed below.

TABLE 6.5. List of the vertebrate faunal remains identified in screened samples from the Op. LC09 B midden

| Class and order | Family | Species and common name | NISP | %NISP | MNI | %MNI |
|---|---|---|---|---|---|---|
| ACTINOPTERYGII (FISHES) | | | | | | |
| Osteichthyes | Various | Various | 3 | 11.53 | 2 | 14.28 |
| Siluriformes | Ariidae | *Ariopsis guatemalensis* Catfish | 1 | 4.34 | 1 | 7.14 |
| Aves | | | | | | |
| Unidentified | Unidentified | Unidentified | 1 | 4.34 | 1 | 7.14 |
| REPTILIA | | | | | | |
| Crocodylia | Crocodylidae | *Crocodylus acutus* or *Crocodylus moreletii* Crocodile | 2 | 8.69 | 2 | 14.28 |
| Unidentified | | | 1 | 4.34 | 1 | 7.14 |
| MAMMALIA | | | | | | |
| Artiodactyla | Cervidae | *Odocoileus virginianus* Deer | 12 | 52.17 | 4 | 28.57 |
| Artiodactyla | Tayassuidae or Cervidae | *Pecari tajacu humeralis* or *Odocoileus virginianus* Peccary or deer | 1 | 4.34 | 1 | 7.14 |
| Unidentified | | | 2 | 8.69 | 2 | 14.28 |
| | | TOTAL | 23 | 98.44 | 14 | 99.97 |

Among screened contexts analyzed, remains of large reptiles (especially crocodiles or caimans) were surprisingly common. For example, the crocodilian MNI was 3 for the Op. LC12 E midden, 4 for the Op. LC12 D middens, 2 for the Op. LC12 H midden, and 2 for the Op. LC09 B midden. Many crocodile remains were burned (e.g., figure 6.5.A), suggesting that they were used for food rather than just occurring at the site naturally or through the scavenging of bone for tools. In some cases, the mandibles of crocodiles or caimans were used as tools, as that particular bone is extremely dense (almost entirely lacking spongy bone when viewed in cross-section) and resistant to wear. The bone fragment recovered with burial B2-I3, for example, came from the mandible of a small crocodile or caiman (figure 6.5.B). It is likely that this artifact is a broken tool fragment, as several similar bones were fashioned into punches

**TABLE 6.6.** List of the vertebrate faunal remains identified in heavy fractions of sediment samples analyzed from the Op. LC12 D, LC12 E, and LC12 H middens

|  |  |  | Heavy fraction samples | | | |
|---|---|---|---|---|---|---|
| Class and order | Family | Species and common name | NISP | %NISP | MNI | %MNI |
| ACTINOPTERYGII (FISHES) | | | | | | |
| Osteictios | Various | Various | 352 | 89.79% | 27 | 84.37% |
| Siluriformes | Ariidae | *Ariopsis guatemalensis* Catfish | 29 | 7.39% | 4 | 12.5% |
| REPTILIA | | | | | | |
| Lacertilia | Iguanidae | *Iguana iguana* Green iguana | 1 | 0.25% | 1 | 3.12% |
| Unidentified | | | 10 | 2.55% | | |
| | | TOTAL | 392 | 99.98% | 32 | 99.99% |

**FIGURE 6.2.** Examples of fish remains from La Consentida: (A) skull fragment from a large catfish recovered in LC12 E-F9-s1 midden context; (B) two views of a large marine fish vertebra from LC12 E-F9-s1 midden context

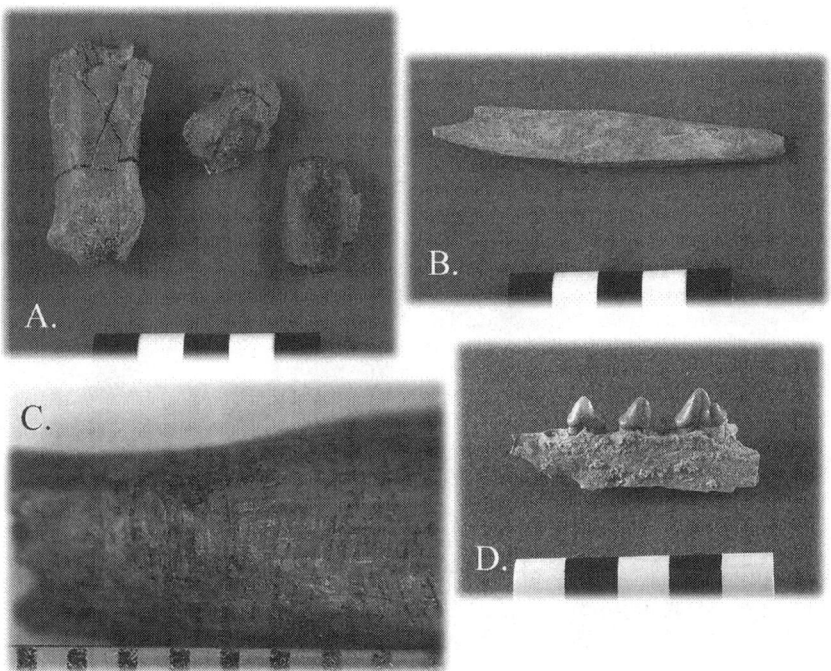

**FIGURE 6.3.** Examples of terrestrial faunal remains from La Consentida: (A) burned deer bones recovered from LC12 E-F9-s1 midden; (B) deer bone tool from near interface between LC12 E-F4 fill and LC12 E-F9-s1 midden; (C) deer bone tool (same as pictured in figure 6.3.B) viewed at 40× magnification; (D) probable coyote mandible fragment from the bottom of a midden (LC12 H-F4-s3)

or awls (figures 6.5.C and 6.5.D). When these tools are viewed under a microscope (figure 6.5.E) it is apparent that they were artificially shaped. Their dense consistency resisted obvious striations from use, however. Such rounded distal ends (see figures 6.5.C and 6.5.E) may result from the impressing of designs on decorated pottery (see appendix 2). Other reptile remains recovered included those of iguanas (which were most common in the Op. LC12 D and LC12 E middens) and the Mexican beaded lizard (*Heloderma horridum*) (Beck and Lowe 1991). While iguanas were certainly used for food, and are in fact still consumed in the region today, the beaded lizard (a venomous predator) was part of the LC12A-F15 ritual cache, and was likely interred whole, with no signs of processing for consumption (see figure 4.5).

In order to control for the bias toward bones from larger animals that the analysis of screened sediments presents, Pérez Hernández and I sorted and analyzed the

114  DIET AND CHANGING CULINARY TASTES

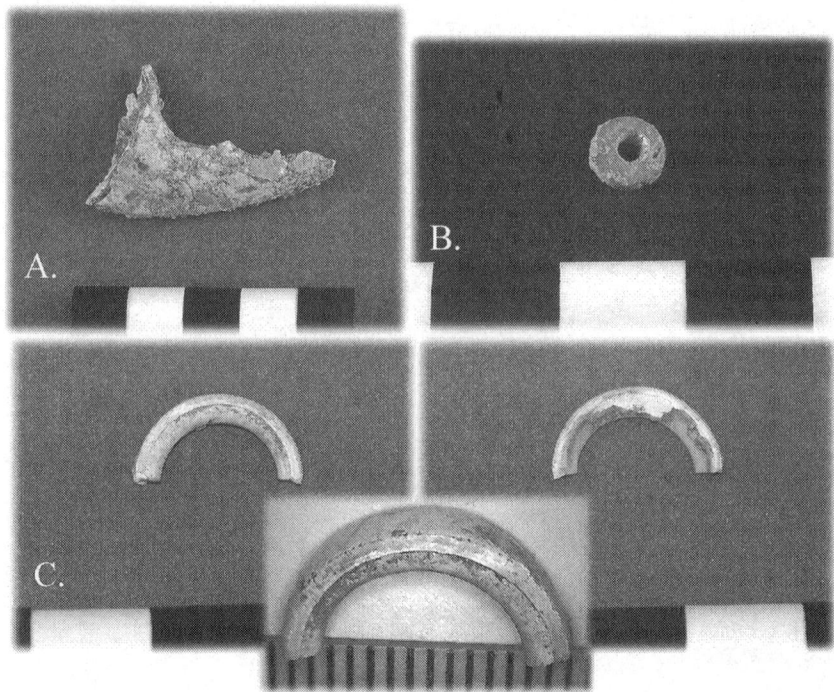

FIGURE 6.4. Examples of marine animal remains used for food and for adornment: (A) crab claw from LC12 E-F9-s2 midden context; (B) shell bead from LC12 D-F4 fill or LC12 D-F5 fill and shell dump (found in screen); (C) three views of a small, circular shell ornament from LC12 B-F6-s1 fill

heavy fractions of six flotation samples (see table 6.7). Two of these samples came from the Op. LC12 D midden, two from the Op. LC12 E midden, and two from the Op. LC12 H midden. This was done with the specific aims of identifying bones from small animals such as tiny fish and to provide an estimate of the relative frequencies of shell in these deposits. As discussed above, in-field observation demonstrated that the Op. LC12 E midden contained far more shell than any other excavated context, but that La Consentida generally contains less shell than expected for a circumcoastal occupation (see Voorhies 2004; Voorhies and Kennett 2011). The results of heavy fraction analysis indicated that many small Osteichthyes (bony fish) are represented in floated sediment samples. Some of the tiny bones, and especially vertebrae, recovered in the heavy fractions would surely be lost in screened sediments. These samples also demonstrated variation in the shell content of the middens analyzed, as indicated in table 6.7. Note that the shell quantities

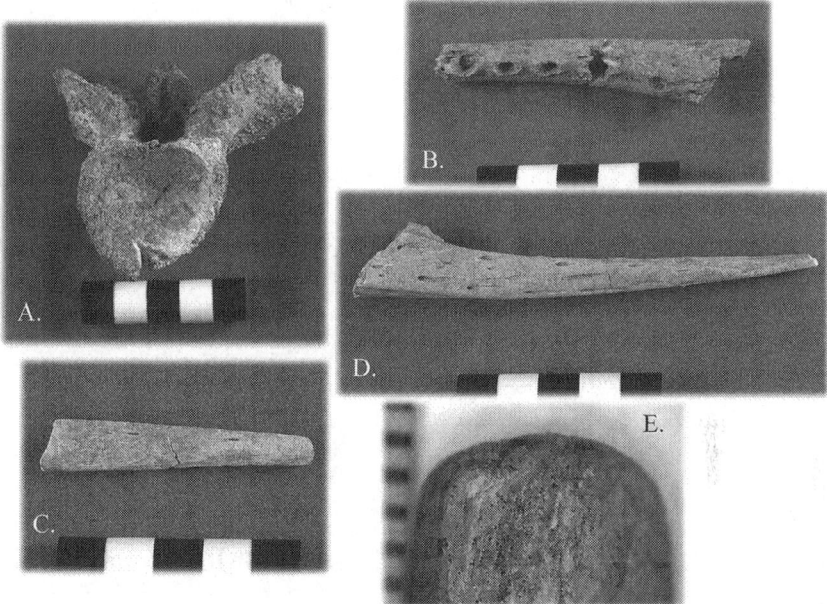

**FIGURE 6.5.** Crocodilian remains from La Consentida: (A) burned crocodile vertebra from interface between LC12 D-F4 and LC12 D-F6 fill layers; (B) crocodile or caiman mandible fragment from burial B2-I3; (C) crocodile or caiman mandible tool from near interface between LC12 E-F4 fill and LC12 E-F9-s1 midden; (D) crocodile or caiman mandible tool from near interface between LC12 E-F4 fill and LC12 E-F9-s1 midden; (E) distal end of crocodile or caiman mandible tool (same as pictured in figure 6.5.C) viewed at 40× magnification

varied drastically among middens, suggesting strong differences in the subsistence practices represented by the different deposits. One deposit in Op. LC12 D (F5) and both deposits sampled in Op. LC12 E were so shell-rich, in fact, that we were forced to sample 50 percent of the shell from the heavy fractions in order to extrapolate total shell quantities. The small fish bones and sometimes very tiny shells recovered in flotation heavy fractions appear to have been boiled whole rather than processed individually. This pattern suggests that the La Consentida community employed relatively advanced fishing technologies such as baskets, nets, traps, or weirs. When tiny marine animals were procured, evidence for boiling rather than individual processing indicates that the cooks of La Consentida may have combined diverse marine resources into a boiling pot to make soups or stews.

In general, the faunal remains demonstrate a broad diet exploiting marine, freshwater, and terrestrial resources. Surprises include the relatively low frequency of

TABLE 6.7. Relative percentages of shell recovered in flotation sample heavy fractions

| Sample number and provenience | Heavy fraction sample weight (2-liter sediment samples) | Shell recovered |
|---|---|---|
| D1 (LC12 D-F5 fill and shell dump) | 304.3 g (50% sampled) | 54.9% of heavy fraction (83.5 g) |
| D8 (LC12 D-F10-s1 fill with occ. surface) | 103.6 g (90% sampled) | 10.9% of heavy fraction (10.2 g) |
| E1 (LC12 E-F9-s1 midden) | 494.4 g (50% sampled) | 84.6% of heavy fraction (209.2 g) |
| E21 (LC12 E-F10 hearth or shell dump) | 520.6 g (50% sampled) | 94.2% of heavy fraction (245.3 g) |
| H2 (LC12 H-F4-s1 midden) | 25.1 g (100% sampled) | 4.0% of heavy fraction (1 g) |
| H4 (LC12 H-F4-s1/F4-s2 midden interface) | 24.7 g (100% sampled) | 0.4% of heavy fraction (.1 g) |

terrestrial mammal and avian remains and the relatively high frequency of animals that might be dangerous to hunt, such as crocodilians and large, open-ocean fish. Although many of the crocodilians that appear to have entered the La Consentida subsistence economy were young individuals, some were very large (e.g., figure 6.5.A). The importance of catfish in the La Consentida diet is also noteworthy. Catfish varied in relative frequency across middens, but their presence to some degree was ubiquitous among analyzed contexts. Given that the middens studied also varied in terms of the ceramic vessel types they contained (see appendix 2), the dietary variation identified suggests practices of resource selection that likely correspond to seasonal availability of different animal species and to the social events that produced the middens themselves.

### FAUNAL COMPARISONS WITH LATER SITES

The La Consentida faunal analysis offers the possibility of comparison with later sites in the lower Río Verde Valley (Fernández 2004). Before making such comparisons, it is worth noting that we did not attempt to extrapolate the quantity of meat represented by identified taxa, but instead restricted ourselves to identification of NISP and MNI (Fernández 2004; see also Wing 1978; chapter 3). In general, patterns of animal resource use at La Consentida are similar to those of Middle and Late Formative period sites in the region. Middle Formative deposits at the site of Corozo demonstrated minimal consumption of mollusks, which appears to be consistent with the pattern identified at La Consentida (Fernández 2004:128). Similarly, Late Formative Cerro de la Cruz and Río Viejo faunal assemblages suggest that mollusks contributed little to the diet in comparison to fish and terrestrial fauna (Fernández 2004:125). Occupants of Río Viejo in particular apparently did not consume much shellfish, though what shells were present indicated the exploitation of diverse coastal settings including lagoons (Fernández 2004:127). These patterns of shellfish exploitation may be complicated by the possibility that people shelled some mollusks at the shore before returning home with the meat. The lack of identified coastal shell middens is not supportive of this hypothesis, however (Fernández 2004:125, 127). By the Terminal Formative period, occupants of the site of Yugüe were obtaining more of their diet from shellfish than had previous communities. This circumstance, coupled with an increased consumption of coastal fish compared to previous sites, suggests overexploitation of faunal resources in the immediate area (Fernández 2004:129; see also Barber 2005).

Fernández (2004:124–28) found that Middle and Late Formative period occupants of the lower Río Verde Valley consumed both aquatic and terrestrial resources, but that the ratios varied by time period and location. She concluded that Late Formative occupants of Cerro de la Cruz probably gained more meat from terrestrial resources than from aquatic ones, despite higher total numbers of remains from the latter (Fernández 2004:124). Río Viejo, which is located along the Río Verde and closer to the Pacific than is Cerro de la Cruz, appears to have obtained more of its resources during the Late Formative from fish than did Cerro de la Cruz (Fernández 2004:126; A. Joyce 1991a). Although remains from La Consentida also demonstrate the exploitation of both terrestrial and aquatic resources, its location closer to the coast than the later sites likely influenced the very high percentage of aquatic remains in some deposits. As A. Joyce (1991b:137) demonstrated, proximity to the coast significantly influenced human bone chemistry within the lower Río Verde region. Also, the formation of coastal estuaries by the Middle Formative period (see Goman et al. 2005, 2013) likely influenced regional dietary change over time, particularly regarding increased estuarine resource consumption. In general, the importance of catfish, cichlids, and jacks appears to be consistent between La Consentida and later sites in the region (Fernández 2004:124–28).

Animal remains from La Consentida demonstrate other similarities and differences when compared to Middle and Late Formative sites in the region. The presence of probable coyote and deer remains at La Consentida is consistent with piedmont fauna identified at Cerro de la Cruz (Fernández 2004:131; see figures 6.3.A and 6.3.D). The recovery of iguana remains at La Consentida is generally similar to finds at later sites in the region, though the relatively high quantity of iguanas (14 percent of MNI) at Río Viejo (Fernández 2004:132, 185) differs from the La Consentida finds, which never were higher than 5 percent MNI in any context analyzed. Fernández (2004:180, 181) identified 1 percent MNI of crocodilian remains at Cerro de la Cruz and about 3 percent MNI of crocodilians at Río Viejo. In contrast, we identified as high as 7 percent and 14 percent MNI in screened contexts at La Consentida. Although our smaller overall sample may skew these results, the presence of large and even cooked crocodilian remains (e.g., figure 6.5) suggests that these animals were an important resource for the La Consentida community. Another point of worthwhile comparison relates to the order Siluriformes (catfish). We identified as high as 24 percent MNI of catfish (especially *Ariopsis* [or *Sciades*] *guatemalensis*) in screened deposits at La Consentida (see Marceniuk and Menezes 2007). The use of catfish is also indicated at Middle Formative Corozo and Late Formative Río Viejo (Fernández 2004:134). The largest amount of catfish remains identified in screened samples at later sites was 12 percent MNI at Río Viejo (Fernández 2004:184). It therefore seems that the La Consentida community placed more emphasis on catfish than did later people in the region. Although catfish are often found in fresh water (Fernández 2004:25), they are also remarkably adaptable to salt water and brackish conditions. Their recovery at La Consentida may thus relate either to the site's location close to coastal lagoons or to use of the river that must have run near the site (see chapter 4).

One interesting point of contrast between the La Consentida faunal results and those from the later sites is that Fernández (2004:127) found little to no evidence of deep-sea fishing in Middle, Late, and Terminal Formative period contexts. As evidenced by some very large fish vertebra (e.g., figure 6.2.B), the La Consentida community likely exploited potentially dangerous open-ocean environments to a greater degree than did later communities. Fernández (2004:132) proposed a tentative model of dietary change over time for the lower Río Verde Valley, which included a shift from a Middle Formative diet emphasizing fish to a broader Late Formative diet of fish, shellfish, and terrestrial mammals, and a Terminal Formative and Classic period diet focused on shellfish and coastal fish (Fernández 2004:133–34). Although La Consentida's location being closer to the Pacific than any of the later sites considered by Fernández may complicate such comparisons, the high percentage of fish remains (up to 90 percent NISP) in some contexts at La Consentida appears consistent with Fernández's model.

## ANALYSIS OF HUMAN TEETH AND LONG BONES

The analysis of human remains, and particularly of teeth, can aid in the reconstruction of ancient diets (M. Pearson 1999:80–82; Schwarcz and Schoeninger 2011). At La Consentida, both pathological and isotopic indicators are useful for understanding changes in the community's subsistence strategies. While adults in the earlier burials at the site do sometimes have dental caries, overall dental health seems to have initially been good and thereafter to have declined. For example, while the earliest burials are of robust people with a comparatively large stature and relatively healthy teeth (figure 6.6), later burials show an increased presence of caries, dental attrition, and even one example of possibly fatal mandibular abscesses (figure 6.7). Among later burials (e.g., B5-I6, a probable female aged 20–35 years), even relatively younger individuals bore dental indicators of an agricultural diet such as caries. Although such dental pathologies are a secondary indicator of specific dietary practices, they nonetheless suggest change over time in the community's diet. The health implications of adopting agriculture and particularly of the impacts of a starchy diet and of the abrasive quality of maize flour ground on stone metates have been identified in a wide variety of archaeological contexts (e.g., Cohen and Armelagos 2013; e.g., Hodges 1987; Larsen 1987; Mayes 2016). I discussed literature regarding such skeletal and dental markers in chapter 2, but a few are most significant for understanding ancient diet. Porotic hyperostosis may suggest malnutrition, linear enamel hypoplasia might indicate juvenile growth interruption (Cook 1981; Goodman 1991; Skinner and Goodman 1992), and cribra orbitalia is suggestive of anemia. Significant consumption of starchy foods, particularly those prepared using stone mills that leave grit in the processed flour, may result in tooth loss, dental wear or attrition, and caries (Larsen 1995:187–89).

As discussed in chapter 3, Paul Sandberg and I sampled human dental enamel from nine adult individuals and dentin from eight of those individuals. To this analysis was added the study of long bone collagen from six of those same adults, as well as from thirty faunal bones (fifteen terrestrial and fifteen aquatic) (Hepp et al. 2017). We prepared the enamel and extracted the collagen in labs at the University of Colorado. The enamel samples were then processed at the University of California Santa Cruz Stable Isotope Laboratory, and the collagen samples were processed at the University of California Davis Stable Isotope Facility. All nine human individuals had adequate enamel preservation for analysis. We identified a general pattern wherein collagen preservation, for both human and animal bone, is poor. This seems to be a problem throughout the region, though the data we produced are sufficient for inclusion in a forthcoming regional diachronic study of diet and land use throughout the Formative period (Joyce et al. 2017).

The $\delta^{13}C$ values in the enamel apatite data presented in table 6.8 indicate a diet that included C4 plants. It should be noted that isotopic indicators of diet in Mesoamerica are complicated by the presence of crassulacean acid metabolism (CAM) plants, which include cacti and agave and metabolize carbon in a way that produces a signature between those of C3 and C4 plants. The consumption of CAM plants could produce elevated $\delta^{13}C$ levels and confuse interpretations of maize

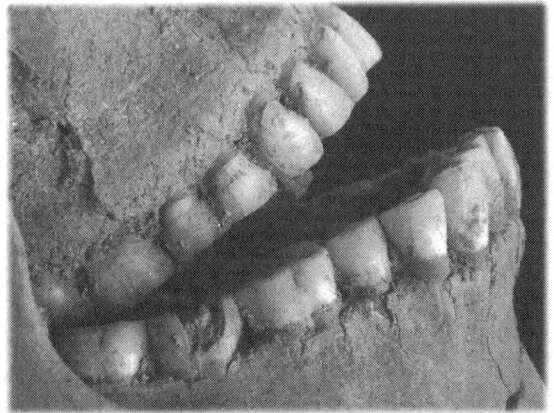

**FIGURE 6.6.** Two views of relatively healthy dentition of B12-I14 adult female (aged 45–50 years) (Images courtesy of José Aguilar)

consumption (Blake 2006; Blake et al. 1992; Llano and Ugan 2014; Tykot 2006). Although relatively little evidence for CAM plant exploitation has been found in pre-Hispanic contexts in the lower Río Verde Valley, Stacie Marie King (2003:265) did identify a few probably imported cactus seeds in Early Postclassic deposits at Río Viejo. The presence of ground stone tools known for use in maize processing, however, further indicates that maize was a component of the La Consentida diet. The $\delta^{18}O$ levels presented in table 6.8 denote qualities of the water sources used by the La Consentida community and demonstrate a lack of evidence for significant outliers that might suggest immigration from another region. Only burial B12-I14 produced enough preserved dentin collagen for analysis (table 6.9). This individual was directly dated via AMS with XAD purification to 1690–1530 cal BC (Cummings 2017; table 1.1, chapter 3). She was a healthy and relatively robust adult female with good dentition (see figure 6.6). She had a relatively low $\delta^{15}N$ value ($\delta^{15}N$ [‰] = 7.0), suggesting little marine input and a relatively high $\delta^{13}C$ value ($\delta^{13}C$

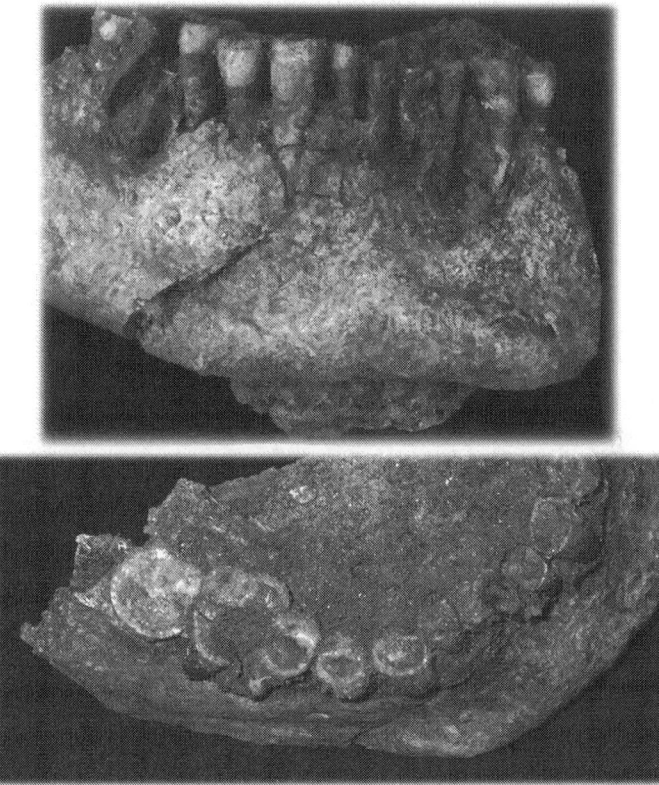

**FIGURE 6.7.** Two views of diseased mandible of B1-I1 adult male (aged 35–50 years) with caries and possibly fatal mandibular abscesses (Images courtesy of José Aguilar)

[‰] = −12.8), which indicates roughly 50 percent of the carbon in her diet was derived from $^{13}$C-enriched foods such as maize. The relative lack of dental attrition she experienced in comparison to some others (e.g., figure 6.7) suggests that her diet was not gritty despite its relatively high maize component. In order to address the issue of poor dentin collagen preservation, we analyzed the long bones of several of the same adult individuals (table 6.9). This effort produced viable results from the same B12-I14 individual, as well from B6-I7, which apparently postdated B12-I14 in the burial sequence. B6-I7 was an adult male between twenty and thirty-five years of age with minor tooth wear and several dental and skeletal pathologies (Aguilar and Hepp 2015; Hepp et al. 2017). During this additional stable isotope study, we also included the analysis of bone collagen from terrestrial and aquatic fauna and

TABLE 6.8. Stable carbon and oxygen isotope values from human enamel ($\delta^{13}C$ [‰] range = −3.8–7.2; average [‰] (±SD) = −5.7 (± 1))

| Burial number | $\delta^{13}C$ (‰) | $\delta^{18}O$ (‰) |
|---|---|---|
| (B1-I1) | −3.8 | −5.6 |
| (B3-I4) | −4.9 | −6.0 |
| (B4-I5) | −6.1 | −6.1 |
| (B5-I6) | −6.2 | −6.3 |
| (B6-I7) | −7.2 | −6.6 |
| (B6-I8) | −5.1 | −7.9 |
| (B8-I10) | −5.1 | −7.7 |
| (B10-I12) | −6.3 | −6.2 |
| (B12-I14) | −6.2 | −7.6 |
| Average | −5.7 | −6.7 |

TABLE 6.9. Stable carbon and nitrogen isotope values from human and faunal collagen

| Burial or sample number | $\delta^{13}C$ (‰) | $\delta^{15}N$ (‰) | C:N |
|---|---|---|---|
| Human | | | |
| B12-I14 (dentin) | −12.8 | 7.0 | 2.8 |
| B12-I14 (long bone) | −11.6 | 6.6 | 2.8 |
| B6-I7 (long bone) | −14.4 | 7.5 | 2.8 |
| Deer | | | |
| F27 | −19.4 | 5.5 | 2.8 |
| F29 | −22.1 | 6.1 | 2.9 |
| Fish | | | |
| F1 | −13.0 | 8.3 | 2.8 |
| F2 | −18.1 | 7.3 | 3.0 |
| F6 | −12.8 | 11.0 | 2.8 |
| F8 | −14.4 | 8.6 | 2.8 |
| F13 | −19.1 | N/A | 2.9 |
| F15 | −18.6 | N/A | 3.2 |

received viable results from two deer and six fish (table 6.9). While these data indicate Early Formative maize consumption, they also demonstrate (when compared to later evidence from the region) that maize reliance increased significantly during the Formative period in coastal Oaxaca. The faunal results presented in table 6.9 help to contextualize the human isotope data and suggest that those values fall within expected ranges for the region.

When bioarchaeological data on dental pathologies are combined with the $\delta^{13}C$ values from enamel apatite, the general pattern that results is that dental attrition from a gritty diet appears to have increased over time in human remains at La Consentida. The remains of nine adult individuals were complete enough for comparisons between chronology, enamel carbon isotope composition, and dental attrition. The relatively low variation in $\delta^{13}C$ values among these burials suggests that maize consumption stayed reasonably consistent over time. This phenomenon may indicate that the nine adult individuals for whom burial chronology, dental attrition, and enamel apatite $\delta^{13}C$ values are available were all interred over a relatively short period of the site's occupation. The apparent increase in dental attrition over time, however, suggests that some other aspect of subsistence was changing even as maize consumption remained fairly constant. The increase over time in the use of manos and metates made of grainy stone for the probable processing of maize flour may account for this change. The three worst cases of dental wear (in burials B1-I1, B5-I6, and B10-I12) occurred in the individuals in the fifth, eighth, and ninth chronological positions, respectively (table 6.10). This pattern tends to support the interpretation that dental attrition likely caused by a gritty diet was increasing over time. A possible complicating factor for this argument is the age estimates for the individuals studied. As demonstrated in table 6.10, however, the oldest individuals were not always those with the most severe dental attrition. A linear regression analysis performed on burial chronology, enamel apatite $\delta^{13}C$ values, and dental attrition demonstrates a lack of significant relationships between any variables except dental attrition and chronology, which are positively correlated ($p = .0355$). In other words, dental attrition at La Consentida increased over time.[3] This linear regression analysis is based on the approximate chronological relationships of the burials, which may be stratigraphically clear within an excavation area (e.g., in Op. LC09 B) but less obvious between distant excavation areas (e.g., Ops. LC09 B and LC12 A). Dental attrition is coded simply as 1 = minor; 2 = moderate; and 3 = extreme. The teeth pictured in figure 6.7 exemplify "extreme" dental attrition. In the following section, I will discuss chronological patterns identified among ground stone tools as the key factor that I believe is responsible for this trend in dental pathology.

## GROUND STONE TOOLS

Numerous types of ground stone tools have been recovered at La Consentida. As mentioned in chapter 5, many of these artifacts bear use-wear indicating their

TABLE 6.10. Burial chronology, assessments of dental attrition, and age estimates

| Burial number | Chronological position | Dental attrition | Age estimate |
|---|---|---|---|
| B12-I14 | 1 (oldest) | Minor | 45–50 years |
| B8-I10 | 2 | Moderate | 15–18 years |
| B6-I8 | 3 | Moderate | Unknown adult |
| B6-I7 | 4 | Minor | 20–35 years |
| B1-I1 | 5 | Extreme | 35–50 years |
| B3-I4 | 6 | Moderate | Unknown adult |
| B4-I5 | 7 | Moderate | Unknown adult |
| B5-I6 | 8 | Extreme | 20–35 years |
| B10-I12 | 9 (most recent) | Extreme | 20–35 years |

<sup>a</sup> The burial chronology discussed here pertains just to the nine burials for which all relevant data were available. Chronological positions are thus listed differently than those for the *total* sample of La Consentida burials (see Table 7.2).

employment in a variety of food-processing and/or crafting activities. Often, as with the examples demonstrated in figures 6.8.A and 6.8.B, these tools bear impact marks from their use as hammer stones. It is not clear if this hammering was all from food processing or if other pounding activities, such as woodworking or construction of domestic structures, produced some of the use-wear. The artifact shown in figure 6.8.C may explain some of the battering damage, as its slightly concave surface indicates its use as an anvil, perhaps for grinding or pounding activities employing small hammer stones. Other fragments of ground stone are more enigmatic in their use. The curving fragment demonstrated in figure 6.8.D is made of a fine-grained dark material (likely basalt) and may have been some kind of polisher. This artifact is narrow and almost bladed or lozenge-like in cross-section. Its crescent form appears to have widened at the point where it is broken, challenging interpretations of its original shape. Although it is likely that not all such tools were directly tied to subsistence, their complex wear patterns and portable size suggest that at least the earlier ground stone at La Consentida was well suited to the portability needs of a semimobile people (see P. Arnold 2009).

Although excavations uncovered portable and multipurpose tools in a wide variety of contexts at La Consentida, heavier ground stone artifacts tended to occur in later deposits at the site (see chapter 5 and figure 5.7.A). Where tools can be clearly linked to food processing, as is the case with manos and especially with metates, they suggest that La Consentida's later occupants placed greater emphasis on the processing of flour from domesticates such as maize than did their ancestors who initially founded the site (see Clark et al. 2007). Nonetheless, all manos so far

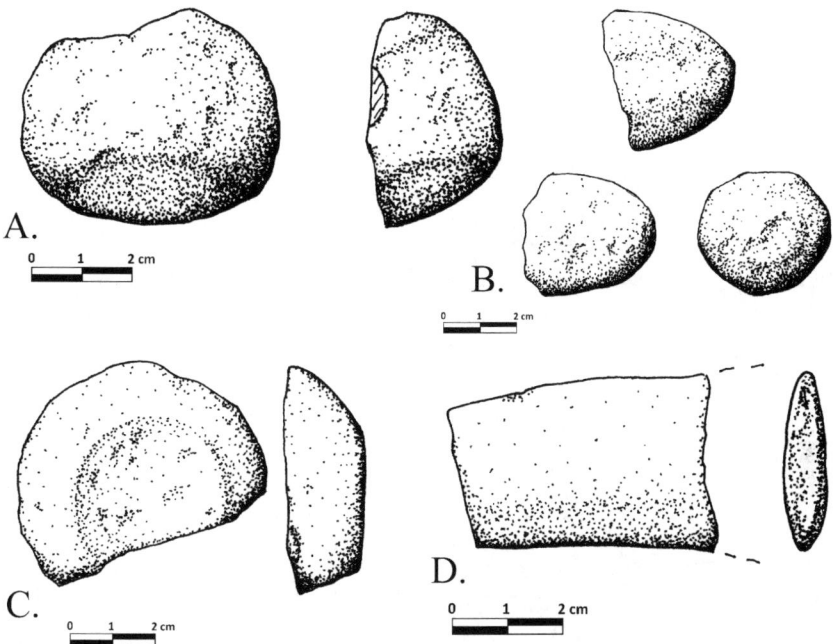

**FIGURE 6.8.** Examples of ground stone tools from La Consentida: (A) hammer stone or pestle from fill with occupation layer (LC12 C-F8); (B) polisher / hammer stone from ancient soil formed in fill (LC09 B-F12); (C) hammer stone or anvil from fill with architectural debris (LC12 E-F4); (D) possible polisher from LC12 E-F9-s1 midden

recovered at La Consentida are of the earliest "one-handed" variety, which is a style consistent with ground stone tools of the Archaic period (P. Arnold 2009; Clark et al. 2007; Hard et al. 1996; Winter and Mateos 2010; Winter and Sánchez Santiago 2014a:10–11). In fact, one of the best examples of the small size and portability of the manos at the site was actually recovered at the surface (figure 5.5.C).

Even the larger manos at La Consentida (e.g., figures 6.9.A and 6.9.B) are relatively small in comparison to later Mesoamerican examples (Clark et al. 2007:29; Winter and Mateos 2010). La Consentida's "larger" manos measure up to 140 × 103 × 44 mm in dimensions, but are still well within the realm of "one-handed" tools (Winter and Mateos 2010). Furthermore, most manos at La Consentida bear use-wear not only from grinding (probably from processing flour on a metate) but also from some additional activities. These patterns include impact marks on one or both ends from use as a hammer stone (e.g., figure 6.9.C) or pitting and concavity on at least one flat side from use as an anvil (e.g., figure 6.9.D). P. Arnold (2009:404) has described

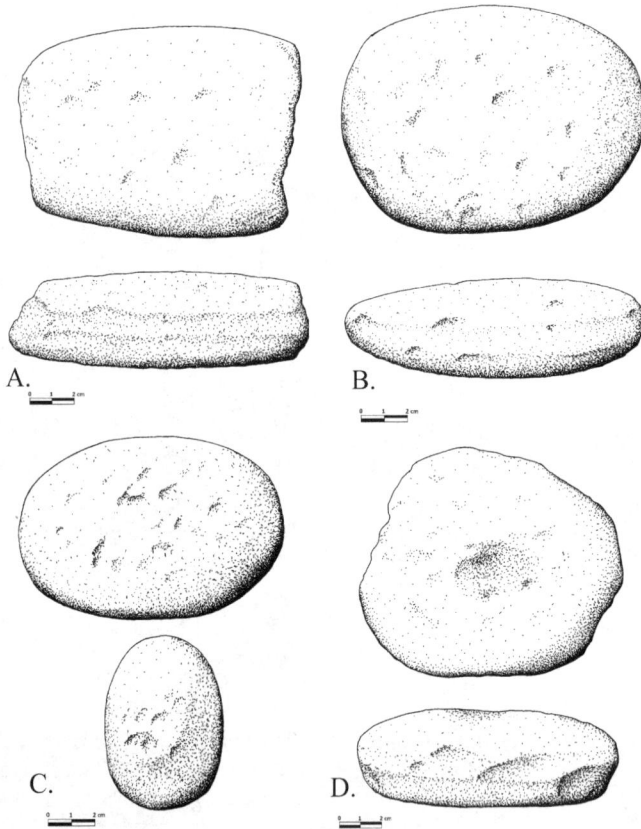

**FIGURE 6.9.** Manos from La Consentida: (A) mano and possible anvil from occupational debris/surface (LC12 C-F2); (B) mano from Structure 1 domestic architecture (LC12 C-F4); (C) mano and probable hammer stone from burial B2-I3; (D) mano and anvil from occupational debris/surface (LC12 C-F2)

one-handed manos as "associated with a multiplicity of tasks" (perhaps *not* focused on maize processing) in contrast to the "targeted grinding" (especially for maize flour) performed with later, two-handed manos. Combinations of diverse types of use-wear on most of La Consentida's ground stone may indicate an interest in conserving stone. As they pertain more specifically to diet, such palimpsests of use-wear also suggest that the community—even in the later years of site occupation, when diet appears to have become increasingly based on maize flour—required of its tools a wide variety of tasks, likely to complement a diet that was still in a state of transformation from the

DIET AND CHANGING CULINARY TASTES    **127**

diverse foraging on wild resources that was a hallmark of the Archaic to the more dedicated agriculture of the later Formative and Classic periods. As discussed in appendix 2, a possible decline in ceramic bottles over time at the site corroborates the ground stone data in suggesting a shift in maize processing from liquid to ground flour.

Metates have been recovered at La Consentida only in fragmentary form, though in some instances (e.g., figure 5.6) they are complete enough to give a good impression of the original shape and dimensions of the artifact. Some metates (e.g., figure 6.10.A) have heavily concaved upper surfaces indicating relatively significant grinding. Other tools bear grinding wear like a metate but lack the concave shape typical of most metates. Such artifacts may more appropriately be termed grinding platforms (e.g., figure 6.10.B). Although their intended use is not as apparent as that of the more obvious metates, they may have been employed for maize flour processing and/or other grinding activities. As discussed below, their recovery in domestic contexts (sometimes in the same deposits where obvious metates were recovered) further supports their use in household activities such as food preparation. Arnold (2009:404–5) has discussed a similar discrepancy between "multi-purpose metates (grinding slabs)" and "single-purpose metates," which were more exclusively used for producing flour by grinding.

The majority (75 percent) of La Consentida's metates are made of fairly coarse-grained granite (e.g., figures 5.6, 6.10.A, 6.10.C, and 6.10.D). This pattern differs from the overall sample of ground stone, which is composed of 29 percent coarse, 30 percent fine, and 41 percent medium-texture material. A few examples of large grinding implements (e.g., figure 6.10.B) are made from a softer, more fine-grained material, at least some of which is also granite. This slight difference in material preference may further indicate the subtly different uses to which true metates versus grinding platforms were put. Also, while most of the metate fragments came from large artifacts unsuitable to transportation across great distances, at least a few (e.g., figure 6.10.D) indicate that the community also used smaller grinding implements. Finally, it is worth noting that most of the metates do not appear to have been particularly heavily used prior to breakage or discard. This pattern contradicts that among some other ground or polished stone implements at the site, which bear use-wear indicative of employment for a variety of tasks, sometimes to the point of exhaustion (e.g., figure 5.5.D). Whether this pattern indicates change over time in material availability or patterns of use is not clear based on the existing evidence. As mentioned previously, ongoing paleoethnobotanical studies should help clarify how these implements were used.

Both manos and metates recovered at La Consentida are most common in domestic contexts (see chapters 4 and 5), and this pattern is especially strong with the metates. Of eight obvious metate or grinding platform fragments, six (75 percent) come from the areas of the Structure 1 and 2 domestic buildings.

**FIGURE 6.10.** Metate and grinding platform fragments from La Consentida: (A) metate fragment from Structure 1 domestic architecture (LC12 C-F4); (B) refitting grinding platform fragments from Structure 1 (LC12 C-F4); (C) metate fragment from Structure 1 (LC12 C-F4); (D) metate or grinding platform fragment from Structure 2 domestic architecture (LC12 G-F2)

Among thirty-two manos, fragmentary manos, or possible manos, fifteen (47 percent) came from within or near Structures 1 and 2. The domestic nature of manos and metates at the site is not surprising and underscores the importance of the household scale of social interaction and economic production in the community's subsistence practices. Notable exceptions to this pattern include the mano (figure 6.9.C) and the large grinding platform or grave marker stone (figure 6.11) recovered with burial B2-I3. The latter bears little use-wear but does have a circular, ground depression (approximately 2 cm in diameter) on one side. The artifact was placed atop the torso of the B2-I3 individual, probably at the time of interment (see figure 7.22). Although it is not clear for what sort of activities the tabular stone might have been intended, the mano at least indicates that probable food-processing tools held meanings beyond the household and could even be included as mortuary offerings.

FIGURE 6.11. Grinding platform and/or marker stone from burial B2-I3

## CHIPPED STONE TOOLS AND OTHER SUBSISTENCE/ CRAFTING TECHNOLOGY

The majority of chipped stone recovered at La Consentida occurs as informal flakes of clear gray and gray obsidian (D. Williams 2012:64–68; chapter 8). Black obsidian, chert, chalcedony, and quartzite are less common material types present at the site (D. Williams 2012). In general, the informal bipolar flaking technology (and the relatively few formalized tools or prismatic obsidian blades) is similar to what archaeologists have found in other Early Formative contexts (Clark 1987; Clark and Lee 2007; G. Lowe 1967, 1975, 1977) and fits the pattern wherein prismatic blade technology did not become common in much of Mesoamerica until later in the Early Formative and Middle Formative periods (Jackson and Love 1991; though see Hirth et al. 2013). As I discussed in chapter 2, the ostensibly "random" nature of Early Formative obsidian reduction has interested researchers for many years, with some suggesting subsistence-related activities such as manioc processing as possible explanations (e.g., Davis 1975; G. Lowe 1967, 1975, 1977; but see DeBoer [1975:431] and Lewenstein and Walker [1984] for cautionary remarks). Recent plant microfossil analyses of chipped stone artifacts (e.g., Morell-Hart et al. 2014) are beginning to provide more concrete evidence for the use of obsidian in root crop processing. In the case of the La Consentida obsidian, macroscopic analysis has encountered little use-wear (D. Williams 2012). Future microscopic use-wear study may help to identify the crafting and food production practices for which these artifacts were used.

The analysis of starch grains on unwashed obsidian samples may indicate whether some of these tools were used to process manioc.

Based on his analysis of the obsidian recovered during the 2009 excavations at La Consentida, Williams (2012:68) reported that no prismatic blades have been found at the site. This is no longer the case. The more extensive 2012 excavations uncovered a few of these artifacts in shallower deposits, particularly in the area of Substructure 2, where the Structure 1 and 2 domestic contexts were excavated. These strata, such as LC12 C-F1 and C-F2, include occupational refuse and artifacts from after site abandonment. A handful of prismatic blades or blade fragments may date to Early Formative period site occupation, however. Several authors (Hirth et al. 2013; MacNeish et al. 1967:22; Niederberger 1976; Zeitlin 1978, 1979) have noted that prismatic blades do sometimes occur in Early Formative and possibly even Archaic period deposits. In the Mixteca Alta, for instance, prismatic obsidian blades are present by the Cruz B phase (Blomster and Glascock 2010:192). The few blades at La Consentida may either help date the site's Formative period abandonment to the later Early Formative or indicate the early production (or more likely, the importation) of these technologically innovative tools (Hirth et al. 2013; Jackson and Love 1991). The overwhelming majority of La Consentida's obsidian falls into the category of "expedient" lithic reduction. Elsewhere in the lower Río Verde Valley, Andrew Workinger (2002:313–29) identified Pachuca obsidian in Late Formative, Terminal Formative, and Classic period contexts at San Francisco de Arriba. To date, no Pachuca obsidian has been recovered at La Consentida. This lack of this green obsidian emphasizes the site's early date, as Terminal Formative and Early Classic period sites in the region often contain high quantities of this material (Joyce et al. 1995; D. Williams 2012:76, 77). This observation is consistent with a general pattern identified by R. Cobean (2002:41), wherein Pachuca obsidian was not widely traded in Mesoamerica until 1200 BCE or about 1450 cal BC.

Although a few obsidian artifacts from La Consentida may exhibit light use-wear, most appear to be flakes and shatter with only a few formal unifacial or bifacial tools (D. Williams 2012:65; figure 6.12.A). In general, the lithic assemblage suggests the use of relatively informal flake tools for a variety of cutting purposes. Despite this pattern, formal tools do occur. The chert blade or knife pictured in figure 6.12.B represents one such tool type. This artifact occurred as an offering with B2-I3, the same burial that was accompanied with the mano (figure 6.9.C) and tabular stone (figure 6.11) discussed above, as well as a ceramic bottle (figure 8.11) and partial figurines (figures 7.2.A and 7.2.B). Another special type of artifact, chert drills, indicates the use of specific material and tool forms for some type of crafting activity (figures 6.12.C–6.12.E). One such drill (figure 6.12.E) was recovered as a probable offering with burial B2-I3. Such drills have been identified in other archaeological contexts as a component of "an elaborate economic system that was based on maritime exchange" (Gamble 2002:301). Archaeological research on maritime economies has found evidence for the more generalized use of "chert macrodrills" in activities including the drilling of shell for the production of fishhooks (J. Arnold and Bernard 2005). There is precedent for similar use of stone tools in coastal Oaxaca, as Robert Zeitlin (1979:ch. 5) discussed flaked quartz tools used as shell working implements at Formative period Laguna Zope. Coastal populations elsewhere in the Americas have also used similar stone drills (Gamble 2002; Jones and Klar 2005). For example, "trifacial stone drills" made of chert have been associated with sewn-plank canoe production among the Chumash of California (Gamble 2002:306). Although not directly tied to food processing, the production of boats, nets, traps, weirs, and other technologies for exploiting the mangrove, bay, and open-ocean habitats would nonetheless be intricately involved with subsistence in this coastal environment. Other specialized crafting tools identified at La Consentida include worked sherd discs (figure 6.13). While it is not yet clear what purpose these discs served, similar examples occur in Tierras Largas phase deposits in the Valley of Oaxaca (Ramírez Urrea 1993:fig. 72).

## SUMMARY

A key distinction for discussions of Early Formative diet is that between horticulturalists, for whom domesticates are a minor component of a diet based on wild resources, and agriculturalists, who rely on their crops (P. Arnold 2009; Clark et al. 2007; Kennett et al. 2010; Killion 2013; VanDerwarker 2006). For that reason, dietary reconstructions for Early Formative period communities should not seek to identify the absolute presence or absence of domesticated maize as sufficient evidence for subsistence strategies. The date of 6700 cal BC for maize domestication in the Río Balsas region (Piperno 2011; Piperno et al. 2009; Ranere et al. 2009:5015) indicates that early communities such as La Consentida had access to maize, at least as a supplemental or feasting food (Blake et al. 1992; Clark and Blake 1994; Goman et al. 2005, 2013; Smalley and Blake 2003).

Dental isotopic data suggest more maize consumption at La Consentida than in other coastal regions of Early Formative Mesoamerica (e.g., Blake et al. 1992; Chisholm and Blake 2006). The human enamel bone apatite and limited collagen results, along with some faunal isotope data, represent valuable additions to subsistence reconstructions for the site. Taken together, the enamel and collagen data available to date indicate a greater degree of C4 and/or CAM plant consumption than that recorded for the Early Formative period in other coastal regions of Mesoamerica such as the Soconusco. As indicated by the general lack of evidence

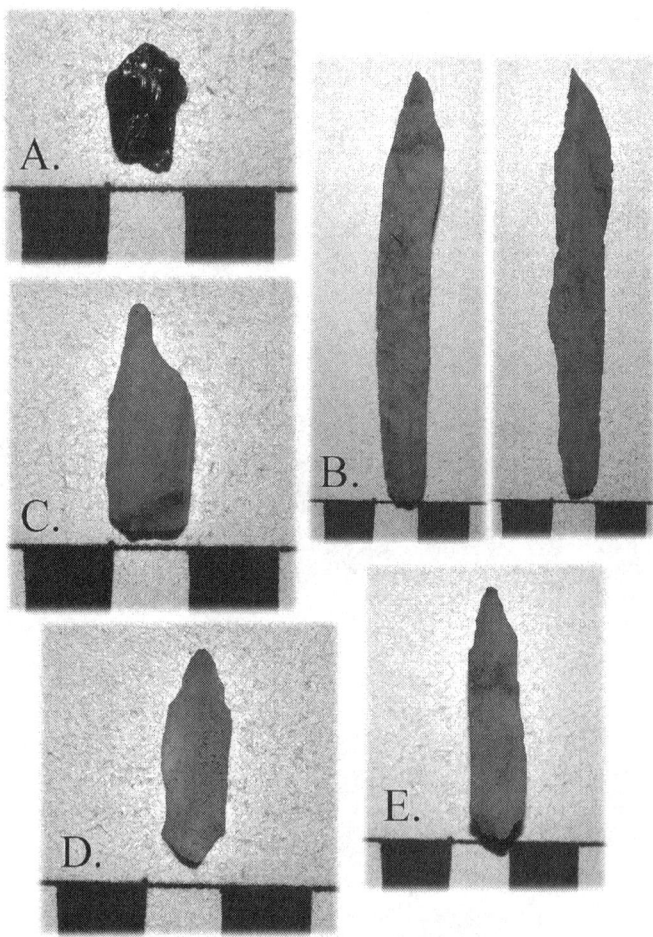

**FIGURE 6.12.** Chipped stone tools from La Consentida: (A) probable bifacially flaked tool from initial Platform 1 fill (LC09 A-F5). (B) two views of chert blade or knife with probable retouch flaking and use-wear. From burial B2-I3; (C) drill from early fill (LC09 B-F14); (D) drill from early fill (LC09 B-F14 or F-16); (E) drill from burial B2-I3 (scale in centimeters; individual images courtesy of David T. Williams)

for CAM plant consumption in the region in pre-Hispanic times (see King 2003:265), C4 plants such as maize are likely the main contributor to this pattern. Comparisons with later isotopic results from coastal Oaxaca also demonstrate a shift over time toward greater agricultural reliance in the region (Joyce et al. 2017).

DIET AND CHANGING CULINARY TASTES  133

**FIGURE 6.13.** Sherd disc artifacts from Structure 2 and surrounding domestic context

The analysis of faunal remains from various contexts at La Consentida, and especially from middens, bolsters the isotope data with evidence for animal resources used at the site. The prevalence of *Ariopsis guatemalensis* in the Op. LC12 E midden is intriguing. These adaptable fish are capable of thriving in marine, brackish, and freshwater conditions, and may have migrated seasonally up freshwater channels such as the Río Verde and/or frequented mangrove habitats, where the La Consentida community could capture them in large numbers during certain times of the year (Kailola and Bussing 1995). The presence of these fish remains in Op. LC12 E corresponds to that of decorated ceramics in an ashy midden context indicative of large cooking and food consumption events (see table A2.28). This correlation suggests intermittent feasting, perhaps timed to correspond with natural catfish behavior or with mangrove-harvesting events, and during which community members could negotiate aspects of the novel social identities of the Early Formative period through practices such as displaying fancy, decorated pottery (see Clark and Blake 1994 regarding the role of decorated ceramics in feasting).

The transition over time from small, multipurpose, portable ground stone tools in earlier occupation levels to the much larger metates used shortly before La Consentida's abandonment appears significant (see figure 5.7.A). These results suggest that the site's earliest food-processing tools may have met the needs of a semisedentary people, though isotopic data indicate some maize consumption (see P. Arnold 2009; Clark et al. 2007; Killion 2013; VanDerwarker 2006). The increase over time in the size and grinding or flour-processing efficiency of the site's ground stone suggests a transition towards a more grain-based diet, probably focusing on maize flour. Based on the small size of artifacts such as manos from throughout the site's occupation, however, it is possible that the community was not fully reliant on maize until closer to site abandonment, if even then. What is clear, based on the ground stone evidence, is that the La Consentida diet (or at least culinary practices at the site) changed between initial occupation and site abandonment. A probable decline in the use of ceramic bottles (see appendix 2), concurrent with an increase in ground stone manos and metates (and in the paleopathologies associated with their use), further implies a culinary change from liquid maize to ground flour.

Faunal remains, human dental pathologies, isotopic values, and food-processing tools suggest that La Consentida's diet was based a wide variety of resources including maize. Maize flour consumption likely increased during occupation. From early in the site's history, however, maize (perhaps initially in liquids such as gruel or beer) appears to have been a larger component of the diet than it was for contemporaneous Soconusco and Gulf Coast communities (Blake et al. 1992; Killion 2013). The $\delta^{13}C$ (‰) range identified in the La Consentida enamel sample is $-3.8--7.2$, (average = $-5.7$). The dentin and long bone collagen $\delta^{13}C$ (‰) range is $-14.4--11.6$. Excluding the one tooth dentin sample to avoid using data from the same individual twice, the bone collagen $\delta^{13}C$ (‰) average at La Consentida is $-13.0$. Blake and colleagues (1992:87) reported Barra phase bone collagen $\delta^{13}C$ levels of $-20.5$ and $-18.7$, suggesting relatively little maize consumption. Similarly, Brian Chisholm and Blake (2006:166) reported a Locona phase bone collagen $\delta^{13}C$ value of $-19.3$ (reconstructed diet = $-23.8$). The possible culinary change at La Consentida (from liquid maize to maize flour) demonstrates how socially constructed "cuisine" (see R. Joyce and Henderson 2007) may impact subsistence and ultimately a community's health and is relevant to the study of changing Early Formative dietary regimes.

# 7

## Social Organization

*Diverse Identities at La Consentida*

As discussed in chapters 1 and 2, the timing and nature of initial social complexity have been major concerns of Early Formative period archaeology (Blake and Clark 1999; Clark 2004a; Clark and Blake 1994; Flannery 1968b; Flannery and Marcus 2003; Hill and Clark 2001; R. Joyce 2004a; Lesure and Blake 2002; Love 2007). In this chapter, I will discuss various lines of evidence for social organization at La Consentida. Layers of earthen architecture demonstrate the community's practices of communal labor. Anthropomorphic imagery, particularly in the form of ceramic figurines, provides clues about how La Consentida's occupants perhaps saw themselves or wished to show themselves. Evidence of ancient jewelry and clothing supports the figurine data by suggesting that there was variety in the dress of community members. I will argue that these dissimilar costumes represent people with distinct social roles. Indications of ceremonial practices, particularly regarding mortuary ritual and community events—such as music, dancing, and feasting—can help us to better grasp the sensory experience of life at ancient La Consentida, and perhaps understand how the community's diverse members interacted (Howes 2003, 2006; Seeger 1987). Calling on these lines of evidence, I will argue that the La Consentida community was heterarchically diverse, though it bears little evidence of the hereditary hierarchies of later pre-Colombian history (R. Adams 1966; Drennan 2009; Feinman and Nicholas 1989; A. Joyce 2000, 2010; Kowalewski 1990; Lesure and Blake 2002; J. Parsons 1974; Sanders and Nichols 1988; Spores 1997).

DOI: 10.5876/9781607328537.c007

## EARTHEN ARCHITECTURE AND COMMUNAL LABOR

As discussed in chapter 5, a population of approximately 80 people, of whom perhaps half or slightly fewer were suited for heavy labor at a given time, could have constructed all of La Consentida's earthen architecture in 250 years or less. As demonstrated in chapter 4 and appendix 1 (e.g., figures A1.6–A1.8), building phases composing Platform 1 and its seven substructures varied in volume and in the amount of labor required for construction. When analyzing excavation profiles for information about practices of architectural construction, a few general patterns emerge. First, initial fill layers tend to be thinner than some later fill episodes, an observation that is consistent with the interpretation that the first occupants may have only seasonally occupied the site (see chapter 5). Second, at least some of the substructural mounds were under construction from early in the site's occupational history, indicating that these probable domestic zones were planned from an early date (see LC12 A-F11-s1, for example). Finally, later construction layers (represented by strata such as LC12 A-F4-s1) became relatively uniform in their thickness (see figure A1.8). One potential explanation for the relative uniformity of the uppermost fill deposits (e.g., LC12 A-F10-s1, A-F4-s1, and A-F2) is that construction of Platform 1 and its substructures may have become a routine occurrence, perhaps taking place in distinct but regular intervals when subsistence interests permitted, as suggested by the "working season" component of most labor estimate models (Erasmus 1965; A. Joyce et al. 2013; see tables 5.1–5.4). These explanations for variation in earthen fill strata are speculative but may suggest increasing organization of labor. From early fits and starts of varied and intermittent construction, the community developed more regular labor practices indicated by increasingly uniform deposits. As indicated by the early planning of substructural mounds (note the shape and extent of stratum LC12 A-F11-s1, for example) certain aspects of community planning appear to have increased over the course of site occupation (see figure A1.8).

## ANTHROPOMORPHIC ICONOGRAPHY

The analysis of ceramic figural artifacts has proven useful for helping to understand social dynamics in ancient Mesoamerica (e.g., Blomster 2009; Cyphers Guillén 1993; Faust and Halperin 2009; Hepp 2007; Hepp and Joyce 2013; Hepp and Rieger 2014; R. Joyce 2000b; Lesure 1997a, 1999a, 2011c; Marcus 1998, 2009). Patterns apparent among anthropomorphic figurines in particular can shed light on social roles pertaining to status, gender, age, and occupation. Ceramic figurines are common at La Consentida, with portions of over 250 different artifacts recovered during the 2009 and 2012 excavations. Figure 7.1 demonstrates basic patterns among figurines, musical instruments, and other iconographic ceramic artifacts. Findings

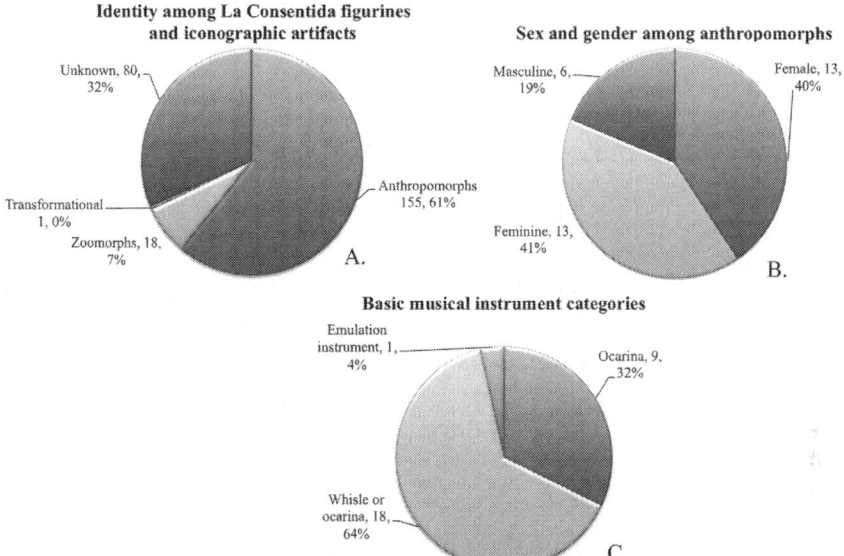

**FIGURE 7.1.** Descriptive statistics for La Consentida's figurines and related artifacts: (A) basic identity categories among figurines and iconographic artifacts; (B) sex and gender among anthropomorphs; (C) basic musical instrument categories among ceramic aerophones.

of note include the predominance of anthropomorphic imagery and the emphasis on feminine and female imagery among anthropomorphs. The musical instruments demonstrate the prominence of ocarinas (ceramic aerophones with finger stops for playing multiple notes) rather than simple whistles. This pattern indicates a relatively sophisticated instrumentality present at this early site. Similar ceramic musical instruments have been identified in highland Oaxaca during the Tierras Largas (Ramírez Urrea 1993:143) and San José (Sánchez Santiago 2014:248) phases. As discussed elsewhere (Hepp et al. 2014), these early musical instrument traditions share the attribute of top-oriented apertures. They also share an emphasis on avian imagery, as discussed in greater detail later in this chapter.

Many of La Consentida's figurines are anthropomorphic, and they are often appear female or feminine in identity. As discussed elsewhere (Hepp and Joyce 2013), when studying anthropomorphic artifacts, it is useful to distinguish between *sex* (i.e., primary or secondary indications of biological sex) and *gender* (i.e., clothing, jewelry, and other accoutrements and attributes not directly indicating sex but which are correlated with females or males and which appear to have carried culturally coded meanings relevant to constructions of femininity and masculinity).[1] Even in the earliest fill layers, where redeposited artifacts must date to very early in site occupation, these figurines bear a variety of head garments, hairstyles, and iconographic elements. Some figurines have generalized or schematic features, while others have more individualized faces. Other research in the region (e.g., Barber and Hepp 2012; Hepp et al. 2014) has suggested that such variations in facial styles indicate that some musical instruments and figurines represent individual people (either living or dead), while others may suggest more generic categories of "ancestor," "deity," "spirit," and the like. Another intriguing interpretation for figurine use was suggested by Elsie Parsons (1936:71), who noted that wax anthropomorphic figurines were used during the early twentieth century in the Zapotec towns of Mitla and Zaachila to pray to the saints for children. In these cases, male figurines were used to pray for a boy and female figurines for a baby girl.

One of the common figurine types at La Consentida is that of nude or nearly nude anthropomorphs representing women. The figurines shown in figures 7.2.A and 7.2.B, for example, are nearly identical broken female torsos recovered with Burial 2. The B2-I3 individual was an adult male, indicating that figurines left as burial offerings need not have represented interred individuals in a strictly anatomical sense. Figure 7.2.C depicts a similar nude female torso. As with most of the figurines, these examples appear intentionally broken, often at their thickest points. This pattern suggests the intentional breakage of figurines as part of a termination ritual (see R. Joyce 2009:416; Shafer and Taylor 1986:51; Smith 1932).

When anthropomorphs are more complete, as pictured in figures 7.3.A and 7.3.B, it is clear that the original artifacts sometimes depicted people in a seated position. As these two artifacts also indicate, figurine provenience is diverse, with examples occurring in mortuary, midden, fill, and domestic contexts. Other artifacts, such as the example represented in figure 7.3.C, demonstrate that some figurines were constructed to show standing figures. This object (see also figure 7.3.B) also exemplifies the interest of figurine makers in depicting the human form in a full-figured, even corpulent style. As Julia Guernsey (2012:121–22) and others (e.g., Lesure 1999a) have discussed, seated figures with big bellies may suggest high social status in some Formative period Mesoamerican contexts. Standing figurines from La Consentida (e.g., figure 7.3.D) also demonstrate that limbs were often represented in a simplified manner.

The attention of La Consentida's figurine makers was often focused on overall body postures and on accoutrements, heads, faces, and headgear. Limbs seem to have served to demonstrate posture but (with at least one notable exception that I will discuss later in this chapter) were not typically a locus for the display of obvious social indicators.

SOCIAL ORGANIZATION 139

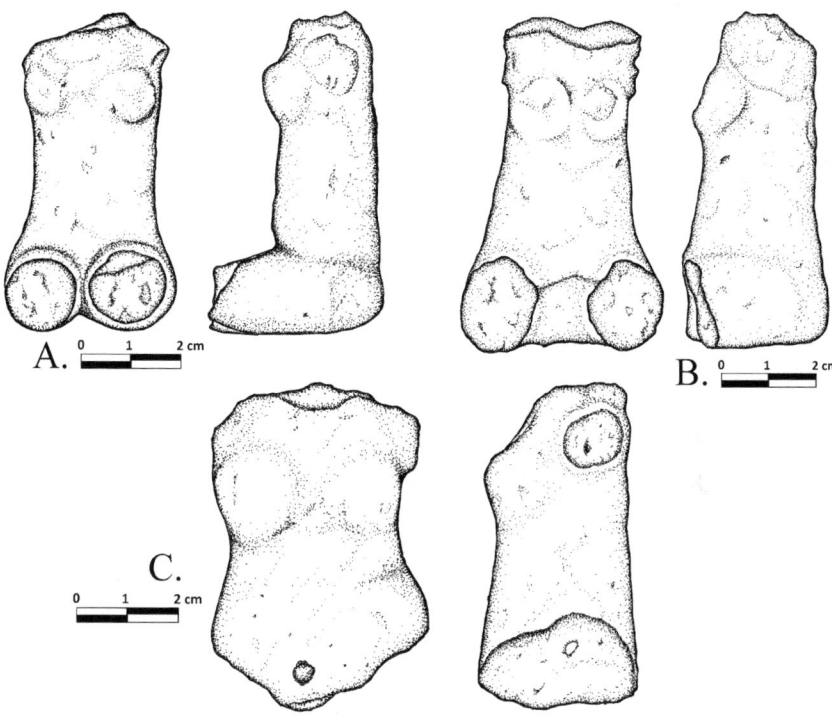

**FIGURE 7.2.** Nude female anthropomorphs from La Consentida: (A) possible burial offering from B2-I3; (B) partial figurine from LC09 B-F12 fill near burial B2-I3; (C) partial figurine from LC09 B-F16 early fill near hearth and burials

Figurines that emphasize some parts of the human body at the expense of others are common in Formative period Mesoamerica, as discussed by authors such as Rosemary Joyce (2009:410). The figurines discussed in Joyce's example tended to accentuate the human torso at the expense of the head and face. Figurines demonstrating a similar trend occur at La Consentida, as indicated by two artifacts found close to one another at the edge (perhaps against the wall) of the Structure 1 domestic building (figures 7.4.A and 7.4.B). These artifacts bear only the vaguest resemblance to real human bodies and have simplified facial characteristics akin to some of the earliest Archaic and Early Formative period figurines known for regions such as Texas and Central Mexico, respectively (Niederberger 1976, 1987, 2000:176; Shafer 1975 [cited in R. Joyce 2009:409–10]). The figurine head pictured in figure 7.4.C demonstrates that even figurines from shallow deposits at the site bear these early-style, simplified features. Furthermore, this artifact (in comparison with others such as the examples in

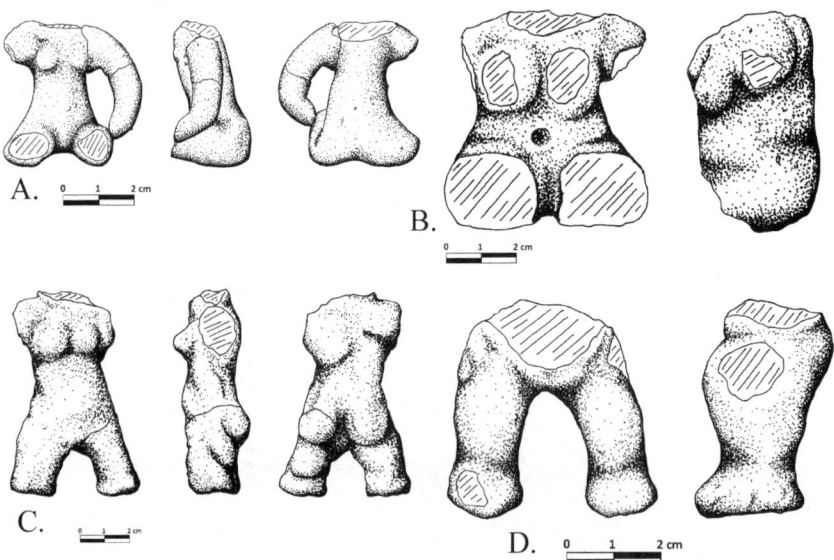

**FIGURE 7.3.** Basic anthropomorphs from La Consentida: (A) partial figurine from floor of Structure 2 house; (B) partial figurine from LC12 G-F1 surface and domestic context; (C) partial figurine from LC12 A-F4-s1/F7 interface (fill with burials); (D) partial figurine from LC12 C-F2 domestic occupational debris

figures 7.4.A and 7.4.B) demonstrates the considerable variability at the site in terms of artistic style and conventions influencing the depiction of the human face.

Although many of the La Consentida anthropomorphic figurines appear to be nude, some show indications of minimal clothing. The figurines displayed in figure 7.5, for example, demonstrate the depiction of simple skirts or loincloths. Both of these figures appear to represent females, though the artifact pictured in figure 7.5.B has fewer obvious indications of sex due to the locations of its breakage in antiquity. It is not possible to be completely sure that the minimal garments of the La Consentida figurines represent the actual clothing worn by members of the ancient community, but archaeological, ethnohistoric, and ethnographic research in various regions of Mesoamerica (e.g., Altman and West 1992; Brumfiel 2006:866, 868; King 2003:81–82, 217–22, 301–3; Klein 1997; Stephen 1991:107–8; Urcid and Joyce 2001) has demonstrated the use of similar outfits by indigenous peoples dating from ancient through modern times. Systematic study of various forms of anthropomorphic imagery in coastal Oaxaca has found evidence suggesting correspondence between representations of attire on figurines, carved stelae, and other media, and the human groups they represented (Hepp and Rieger 2014; Jennings 2010). For

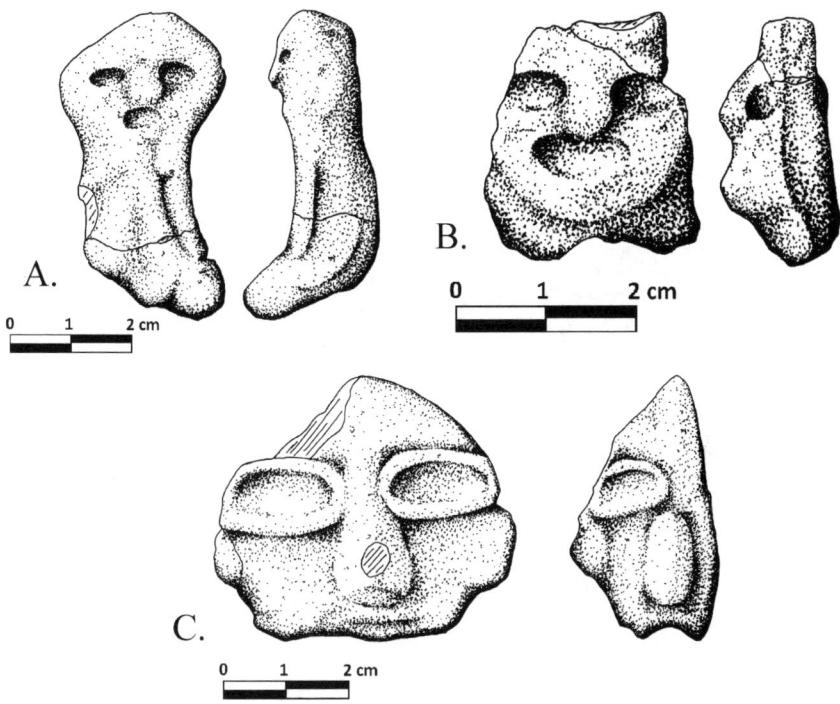

**FIGURE 7.4.** Simplified anthropomorphs from La Consentida: (A) figurine from edge of Structure 1 domestic building; (B) figurine fragment from edge of Structure 1; (C) figurine fragment from LC12 A-F2 occupational debris and fill

example, fancy hairstyles and headgear were more frequently depicted with female or feminine characters than with masculine, male, or gender-neutral imagery in the region during the Formative period. Also, while figurines and stelae suggest that men wore pendants more often than women, some accoutrements, such as ear spools, crosscut gender categories. In general, ubiquitous recovery contexts of figural artifacts indicate that portraying the human body in Formative period coastal Oaxaca was not restricted by social status (Hepp and Rieger 2014:123, 133–34).

Figurine heads were often the venue for appliques and other details that apparently served to personalize anthropomorphs at the site. The artifact pictured in figure 7.6.A, for example, shows a good deal of attention paid to the hair. As argued by Marcus (1998:4, 158), the depiction of elaborate hairstyles in Formative period Oaxaca was mostly associated with representations of women. Another interesting aspect of this figurine is its early date. Recovered from the interface between two deep fill strata (LC12 A-F17-s2 and A-F18-s2), it was likely redeposited and therefore

142  SOCIAL ORGANIZATION

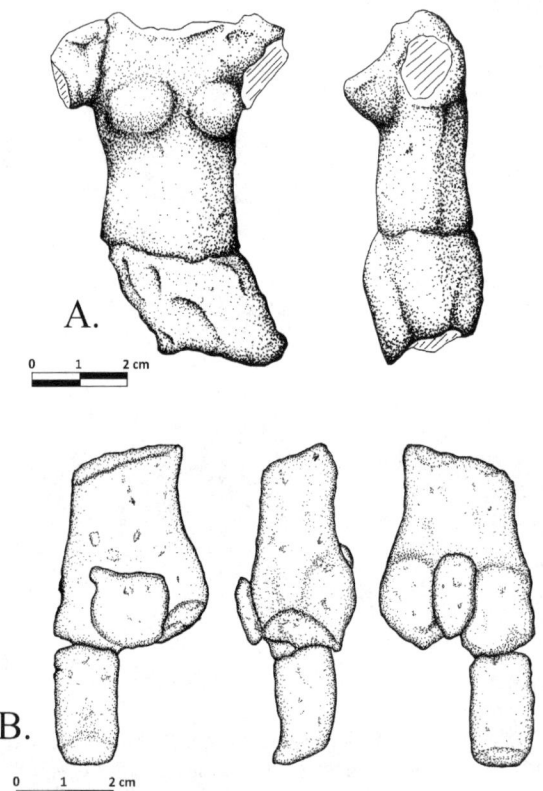

FIGURE 7.5. Anthropomorphs bearing representations of clothing: (A) partial figurine from LC12 A-F4-s1, A-F7, A-F8, and A-F9 fill and intrusive pit—likely associated with burial B12-I14; (B) partial figurine from LC09 B-F12 fill near burials

dates to very early in site occupation. The depiction of individualized identity or at least of femininity was thus a long-standing tradition at La Consentida. The artifact pictured in figure 7.6.B also bears an elaborate hairstyle and is one of numerous similar artifacts (see also figures 7.2, 7.6.C, and 7.6.D) to occur in close association with a human burial. In fact, the artifacts in figures 7.6.B and 7.6.C were found face down near the remains of interred individuals, some of whom were also in a prone position (see Hepp et al. 2017). The artifact shown in figure 7.6.B was found with burial B5-I6, and the figurine depicted in figure 7.6.C was recovered 5 cm above burials B6 and B7. This pattern of artifact recovery may further emphasize the links between figurines and mortuary practice at La Consentida.

One of the earliest figurines recovered at La Consentida is also among the most enigmatic. Complete with what appears to be a bald head and masculine face with aquiline features, this figurine (figure 7.7.A) bears interesting raised markings on the chest that may represent an arm and a breast, some kind of iconography, or even

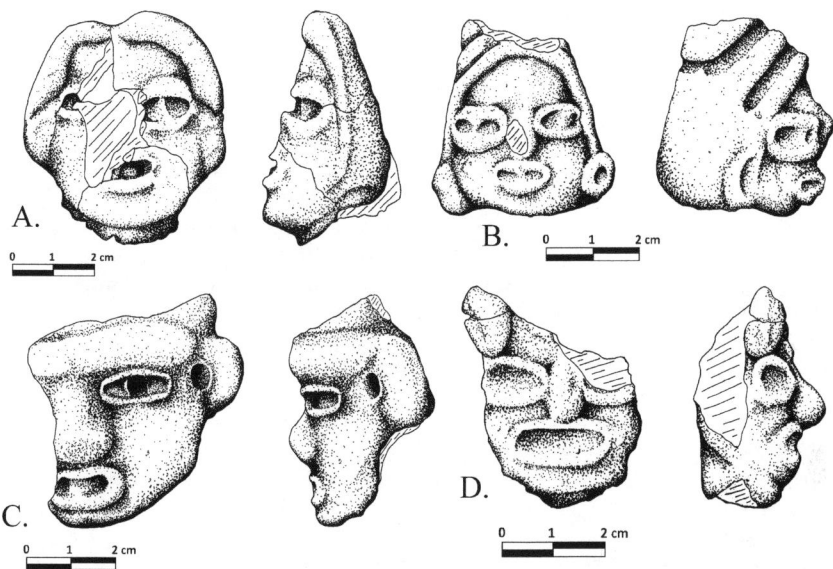

**FIGURE 7.6.** Anthropomorphic heads from La Consentida: (A) figurine fragment from LC12 A-F17-s2 and A-F18-s2 interface; (B) figurine fragment from burial B5-I6; (C) figurine fragment from the area of burials B6 and B7; (D) figurine fragment from burial B8-I10

scarification (see Hepp and Rieger 2014). Guernsey has commented that this figurine has jowly cheeks that link it to a series of figurines she has identified in the Soconusco and elsewhere, including in the lower Río Verde Valley, whose traits anticipate key features of Late Preclassic "potbelly" sculptures (see Guernsey 2012:107–8, ch. 7 [note 3]; Hepp 2007:176 [photo 81], 180 [photo 181]; Hepp et al. 2014:fig. 4.d). While this artifact lacks the puffy eyes diagnostic of those figurines presaging the potbelly tradition, the position of the figure's arms (which appear to wrap around the body) relates not only to widespread Middle Preclassic figurine traditions but also to later "potbelly" sculptures themselves (Julia Guernsey, personal communication, 2015). Guernsey (2012:143) has interpreted such figurines as representing "vital ancestors" imbued with references to communication and intimately involved with domestic ritual. Such imagery began with figurines during the Early Formative, and some of their iconographic elements were later co-opted for larger-scale sculptural art (likely with different symbolic value than the preceding figurines) by the Middle Formative and Middle-Late Formative period transition (Guernsey 2010). Also of note are the La Consentida figurine's rounded, hollow shape and the hole through its neck, which suggests it was worn as a pendant or was otherwise suspended. Jeffrey Blomster (personal communication, 2015) has suggested that this figurine may be

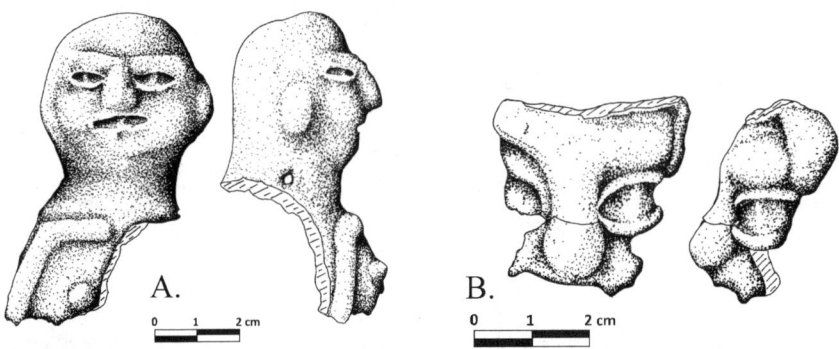

**FIGURE 7.7.** Figurine fragments from a midden: (A) partial Anthropomorph from LC12 E-F10 hearth or burning feature; (B) anthropomorphic fragment from LC12 E-F9-s1

iconographically similar to examples from the Cruz A phase in the Mixteca Alta. Significantly, the figurine comes directly from a hearth or burning feature that has been dated to 1885–1635 cal BC (see table 1.1). This means that the object was already used, broken, and discarded by early in La Consentida's occupation. Another early figurine (figure 7.7.B) had a large and skillfully executed face with what appears to have been colored pigment, slip, or paint. Unfortunately, the small fragment that remains from this artifact gives few impressions of its initial form. What is clear is that some of La Consentida's earliest figurines were well made. Other general attributes of the figurine collection include that all appear to have been hand formed and most are solid bodied. About 76 percent are of medium brown ware paste, and about 24 percent are coarse brown ware artifacts. Pendants are rare but do occur. Sizes vary considerably, with one large, hollow figurine or statue leg measuring nearly 7 cm in height and 5 cm in width. Figural artifacts have been recovered from the full range of depositional contexts at La Consentida.

Although many of La Consentida's figurines are fragmentary, several examples are complete enough to show how accoutrements likely signified key aspects of status and identity. The figurine pictured in figure 7.8.A demonstrates how some of La Consentida's "simple" anthropomorphic torsos were likely connected to detailed heads bearing elaborate hairstyles, headgear, and jewelry. This artifact also wears minimal clothing similar to that portrayed in figure 7.5.B. While not necessarily directly linked to social status, the knotted headband, hairstyle, and ear spools of this figurine have later Formative period counterparts in the lower Río Verde Valley that correlate with the display of social identity (Hepp and Joyce 2013; Hepp and Rieger 2014). Artifacts from other contexts at La Consentida (e.g., figures 7.8.B–7.8.E) also include diverse forms of headgear, headdresses, and jewelry. Given

SOCIAL ORGANIZATION   **145**

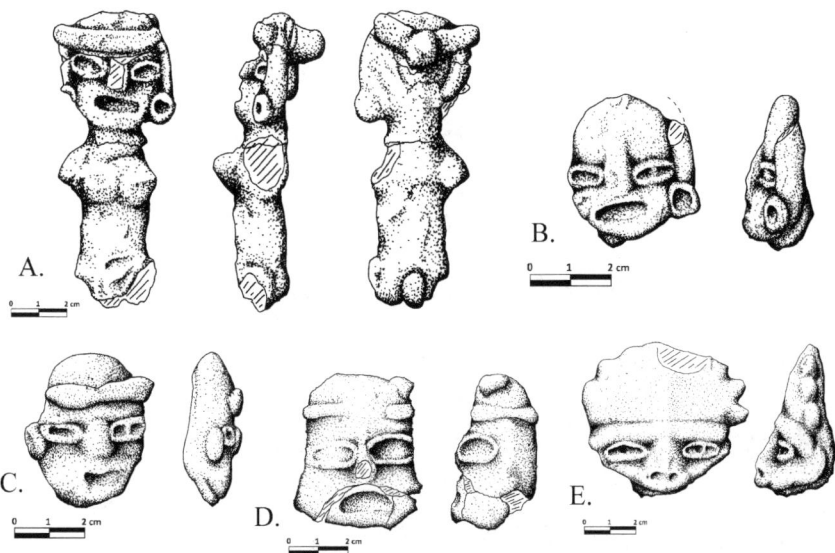

**FIGURE 7.8.** Anthropomorphs bearing indications of clothing or accoutrements: (A) partial figurine from LC12 A-F7/A-F10-s1 interface (fill with burials); (B) figurine fragment from LC12 F-F5 fill with occupation surface; (C) figurine fragment from LC12 A-F7 fill or burial B8-I10 intrusion; (D) figurine fragment from LC12 G-F1/G-F16 fill and domestic context. (Likely associated with Structure 2; 9E) Figurine fragment from Structure 2 fill

the site's early date and the wide variety of contexts in which such artifacts occur, it seems clear that the display of accoutrements representing aspects of social identity was a significant concern from a very early date on the western Oaxaca coast.

Two artifacts from very different contexts (the figurine shown figure 7.9.A from the LC12 E-F9-s1 midden and the possible effigy vessel represented in figure 7.9.B from just outside Structure 2) further demonstrate the use of ceramic iconography to display individual identity at La Consentida. The artifact shown in figure 7.9.A comes from an early deposit, and the artifact pictured in figure 7.9.B was probably discarded shortly before Formative period site abandonment. Although likely representing different artifact classes, found in different parts of the site, occurring in different sorts of deposits, and probably coming from different times in the site's occupation, these artifacts nonetheless share a striking resemblance. The half-lidded eye apparent in the figure 7.9.A artifact, according to Mesoamerican conventions identified in other Preclassic contexts, may represent death (e.g., Guernsey 2010). As a matter of speculation, the exposed teeth might also represent death, as the lips of the decaying human body may pull away from the teeth and thus give

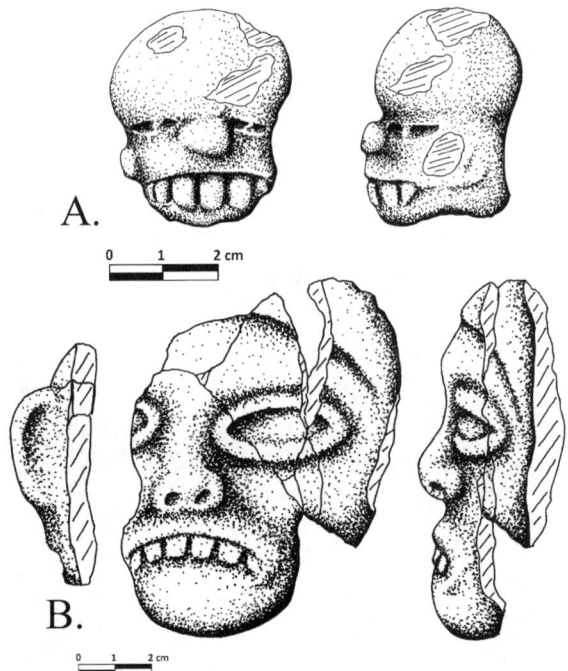

FIGURE 7.9. Anthropomorphs with similar grimaces: (A) figurine fragment from LC12 E-F9-s1 midden; (B) fragments of probable effigy vessel from LC12 G-F16 fill likely associated with Structure 2 domestic building

the impression of a smile or grimace. Notably, the "frowning teeth" depicted on the figure 7.9.B artifact convey an expression that is impossible for the human face to produce, further suggesting that very deliberate attention has been paid to giving the face a particular appearance. It is not possible, given available data, to determine whether these artifacts represent a deceased and revered ancestor, a powerful god or spirit, a ritual practitioner wearing a mask, or perhaps a specific community member known for a fearsome, intimidating gaze. Whatever the case, this seems to have been an artistic trope with some longevity at the site.

Although many figural artifacts permit only conjecture about the types of social roles they represented, a couple of examples from La Consentida appear to more specifically refer to activities or roles carried out by community members. The figurine shown in figure 7.10.A comes from an early midden context (LC12 H-F4-s1) and appears (by virtue of its thick "belt") to represent a ballplayer. Some authors (e.g., Hill and Clark 2001) have proposed that the ball game may have been one of the primary catalysts for sparking initial social complexity in Mesoamerica. As discussed briefly in chapter 2, the colossal heads of the later Early Formative Olmec may represent ballplayers, further suggesting the important role of these

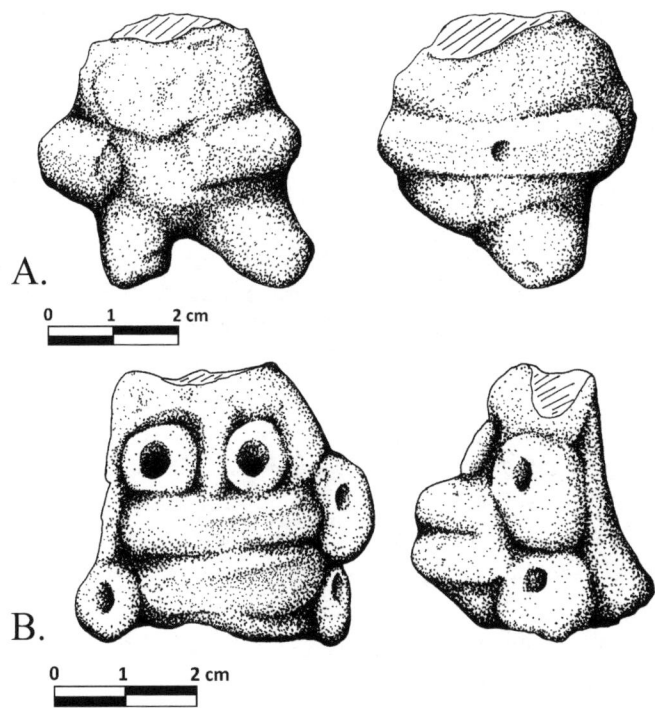

FIGURE 7.10. Artifacts suggesting specialized social roles: (A) possible ballplayer figurine from LC12 H-F4-s1 midden; (B) transformational figurine fragment from surface at Op. LC12 F

community members in the early establishment of chiefly power (Clark 2007:30). Although a single figurine is not sufficient evidence to demonstrate that some elite class of athletes or ritual practitioners reigned at La Consentida, the discovery of this artifact nonetheless demonstrates the site's connections to broader traditions of symbolism and representation in Mesoamerica.

Other figurines also provide tantalizing clues about the social roles fulfilled by La Consentida's occupants. By the time of the site's abandonment, a figurine combining human and animal features was discarded at or near the surface at Op. LC12 F (figure 7.10.B). While apparently discarded late in the site's occupational history, this artifact appears to be executed in an Early Formative period style. Ethnographic, ethnohistoric, and archaeological data from many regions of Mesoamerica (e.g., de Sahagún 1974; Foster 1944; Gutiérrez and Pye 2010; Hepp and Joyce 2013; Kaplan 1956; Rojas 1947) demonstrate the long-held tradition of ritual practitioners who can transform into animals. Along with later Formative period transformational figurines from the lower Río Verde Valley, this artifact suggests the deep historical significance of nagualism and shamanism in the region (Hepp and Joyce 2013).

Although at times speculative, the analysis of figurines has proven to be useful for helping to understand social dynamics in ancient Mesoamerica (Lesure 2011c). Resulting interpretations should be cautious ones, however, as the prevalence of ancestor veneration and the depiction of deities in ancient Mesoamerica may mean that some figurines do not represent actual living people (e.g., Hepp and Joyce 2013; Marcus 1998). Furthermore, the identities of humans and divinities may not always have been completely distinct, and aspects of identity and social status represented on figurines are not necessarily self-evident. Also, figurine attributes may at times indicate the prescription of social ideals rather than reflecting actual dress and comportment (Hepp and Rieger 2014; R. Joyce 2000a; Lesure 1997b, 1999a). Nonetheless, it is in the medium of the figurine—along with rarer artifacts such as carved stones, decorated pottery, and musical instruments—that Formative period Mesoamericans left a record of their communities, ancestors, and deities, according to how they saw themselves.

The patterns identified in La Consentida's figurine sample may appear simple but are nonetheless important. The key emphases of the site's figurine makers included the human, and particularly the *female*, form. This pattern is not unique to La Consentida and is, in fact, a general trend throughout much of Formative and Classic period Mesoamerica (R. Joyce 2000b, 2002, 2003; Marcus 1998; Wolf 1959:57). Whereas the nude or nearly nude bodies often represented in these figurines convey the artists' concern for basic concepts of humanity, sex, and gender, other figurines tell a different story. Diverse headdresses and various types of headgear, hairstyles, and jewelry, for example, imply that the expression of distinct identities was also a significant concern. Rare artifacts that may represent a specific deity, ancestor, or community member further emphasize this pattern. Finally, some figurines even appear to portray individuals carrying out important and uncommon social roles, including that of ballplayer and perhaps that of ritual practitioner or shaman. Although hardly solid evidence of formalized hierarchy, the figurines thus represent the interests of a community of people who saw themselves as socially diverse and appear to have differentiated themselves according to sex and/or gender, age, accoutrements and dress, personal comportment, and occupation.

## FURTHER EVIDENCE FOR BODILY ADORNMENT

Few elements of Early Formative attire preserve beyond what was portrayed on ceramic figurines, but some tantalizing clues to the ways in which La Consentida's occupants dressed and distinguished themselves do remain. Table 7.1 summarizes the recovery context and typology of remnants of jewelry recovered at the site. Beads, often made of ceramic (figure 7.11.A), but sometimes made of shell (figure 6.4.B), black stone (figures 7.11.B and 7.11.C), or greenstone (figures 7.11.D–7.11.F) are fairly common. Although often recovered in fill or in domestic contexts, they also occur as burial offerings (see figure 7.11.A). The recovery of more ceramic and greenstone beads (figures 7.11.D–7.11.F) from primary deposits would be necessary to support any claims about their possible role as markers of inequality. At the very least, the diversity of beads recovered at the site, in both earlier and later deposits, serves to confirm what the figurines also suggest, namely, that community members distinguished themselves with their attire. Furthermore, specialized goods such as delicate shell beads and imported, often difficult to work greenstone (see Tremain 2014) suggest either that considerable crafting took place at the site, or at least that La Consentida had access to such goods through its interaction networks (Carballo 2009; see chapter 8).

Other artifacts from La Consentida provide secondary evidence of clothing. Probable bone needles, such as the one pictured in figure 7.12, imply sewn fabrics or skins that have not survived the millennia in the site's tropical climate. Tiny pieces of other jewelry and accoutrements, such as the possible polished hematite[2] mirror fragment pictured in figure 7.13, may be more directly related to status distinction. As several archaeologists (e.g., Ashmore 2004:184–85; Blomster 2004:85, 186; Clark 1994:126; Heyden 1991:195; Saunders 2001) have argued, accoutrements such as mirrors and headdresses were sometimes associated with elevated social status in Mesoamerica. Such connections between elements of dress and social differentiation have been suggested for coastal Oaxaca during the later Formative period (Barber and Olvera Sánchez 2012; Hepp and Rieger 2014). Even when the evidence suggests some social differences, demonstrating that inequality was hereditary is another matter.

## RITUAL PRACTICE AT LA CONSENTIDA

Archaeologists have long used mortuary evidence to infer aspects of social organization (Binford 1971; Gillespie 2001; Higelin Ponce de León and Hepp 2017a; R. Joyce 1999; Saxe 1971; Spencer and Redmond 2004; Whalen 1983:30–33; Winter 2002:68). As I discussed in chapter 2, some scholars (e.g., Carr 1995; Love 2007) have attempted to refine mortuary archaeology by problematizing the degree to which mortuary data are used to infer aspects of social organization, particularly in instances where social value placed on grave offerings in the past is not well understood. I agree that assuming ancient social dynamics are somehow "fossilized" in

TABLE 7.1. Typology and recovery context of jewelry from La Consentida

| Op. | Unit | Lot | Context | Artifact type |
|---|---|---|---|---|
| LC09 B | 0Z | 4 | LC09 B-F10 ancient surface formed in fill | Bead, ceramic |
| LC12 A | -1P | 4 | Burial B5-I6 | Bead, ceramic |
| LC12 A | -2Q | 4 | Burial B6-I7 intrusion | Bead, ceramic |
| LC12 A | -2Q | 8 | Burial B6-I7 | Bead, ceramic |
| LC12 A | -1R | 3 | LC12 A-F4-s1 fill / resurfacing context | Bead, ceramic |
| LC12 A | -1R | 7 | LC12 A-F4-s1/F7 interface | Bead, ceramic |
| LC12 C | 4D | 3 | LC12 C-F1 occ. debris | Bead, ceramic |
| LC12 C | 3Z | 5 | LC12 C-F2 occ. debris/fill | Bead, ceramic |
| LC12 E | 2B | 10(2) | LC12 E-F9-s1 midden | Bead, ceramic |
| LC12 E | -6Z | 6 | LC12 E-F4 fill with architectural debris | Bead, ceramic |
| LC12 F | 1A | 9 | LC12 F-F3 fill | Bead, ceramic |
| LC12 F | 1A | 8 | LC12 F-F3 fill | Bead, ceramic |
| LC12 A | 0J | 3 | LC12 A-F4-s1 fill / resurfacing layer | Bead, greenstone |
| LC12 A | -1Q | 6 | Burial pit near burials B6 and B7 | Bead, greenstone |
| LC12 G | 1D | 3 | Occ. debris above Structure 2 floor | Bead, greenstone |
| LC12 D | 0A | 10 | LC12 D-F4 or F5 fill or shell dump | Bead, shell |
| LC12 A | -3Q | B.8 | Burial B12-I14 | Bead, stone |
| LC12 E | 1C | 6 | LC12 E-F9-s1 midden | Bead, stone |
| LC12 B | 5D | 13 | LC12 B-F5 fill | Bead or ring, shell |
| LC12 C | 3B | 3 | LC12 C-F1 occ. debris near Structure 1 | Mirror fragment (possible) |
| LC12 A | 0C | 3 | LC12 A-F2 occ. debris/fill | Mirror or pectoral fragment (possible) |

the patterns identified in mortuary contexts, and that certain burial offerings have universal meaning, is problematic. Honoring a burial with offerings of textiles or other perishables may have been a sign of great respect or social preeminence in Early Formative Oaxaca, but such a grave might appear barren when exposed by the archaeologist's trowel after thousands of years of organic deterioration. Despite these complications and the caution they rightly instill in an archaeologist hoping to understand past social dynamics, it would be misguided to overlook patterns of mortuary variation as one of several lines of inquiry for piecing together ancient social organization. In the following paragraphs, I will summarize variations in mortuary treatment and grave offerings at La Consentida. I will also discuss evidence for other ceremonial deposits that may relate to mortuary practice.

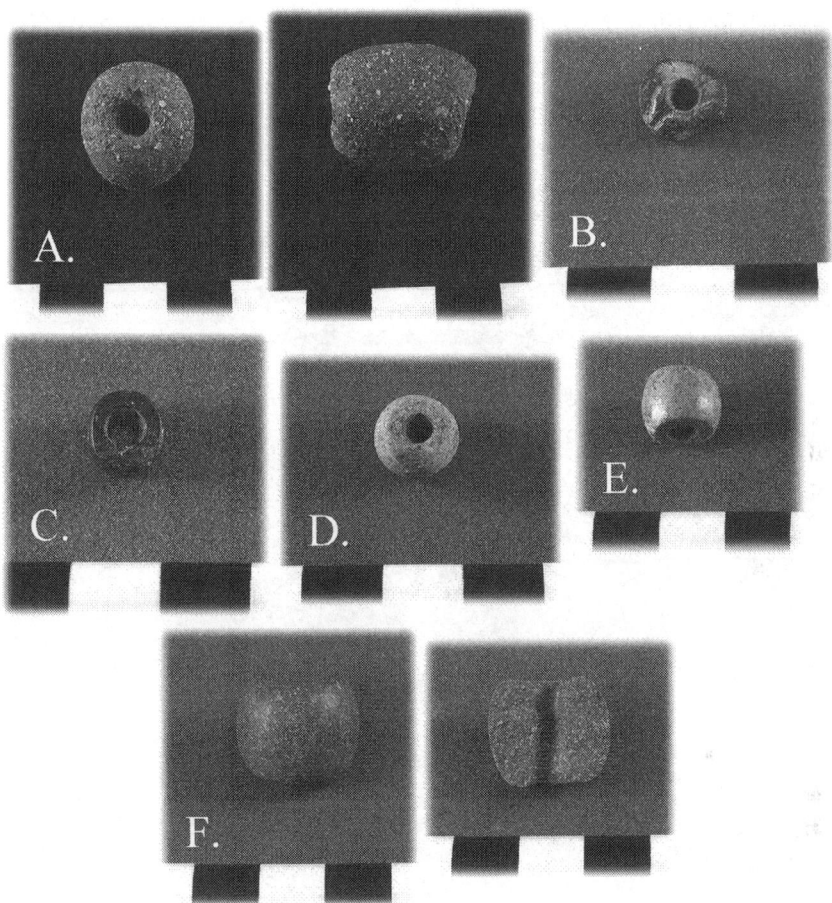

**FIGURE 7.11.** Beads from La Consentida. (A) Two views of a ceramic bead recovered as a probable offering with burial B5-I6; (B) Black stone bead from LC12 E-F9-s1 midden; (C) Black stone bead from burial B12-I14; (D) Greenstone bead from LC12 A-F4-s1 fill / resurfacing context; (E) Greenstone bead from occupational debris above Structure 2 floor; (F) Two views of a partial, biconically drilled (drilled from both sides) greenstone bead from burial pit near burials B6 and B7.

Burials at La Consentida tend to be accompanied by relatively few offerings in comparison to later examples in coastal Oaxaca and other regions of Mesoamerica (e.g., Barber 2005:382–406; Hepp et al. 2017; Higelin Ponce de León and Hepp 2017a; A. Joyce 1991a:718–87; Marcus and Flannery 1996:97–106; Whalen 1981). Burials at the site appear to be relatively simple primary inhumations, rather than

152  SOCIAL ORGANIZATION

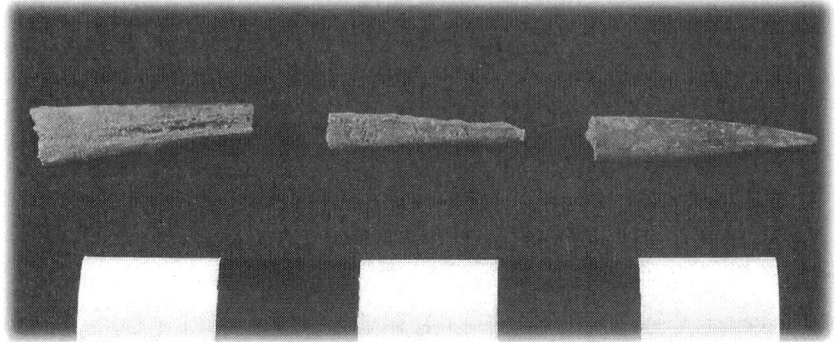

FIGURE 7.12. A probable bone needle from fill near burial B12-I14

FIGURE 7.13. Two views of possible hematite mirror fragment from LC12 C-F1 occupational debris near Structure 1. Left: no light reflecting off surface; right: light reflecting off surface

the secondary interments of mummy bundles seen later in Oaxaca (e.g., Levine 2007:192–94) or the elaborate tombs found in later contexts in such regions as highland Oaxaca, Colima, and Jalisco (Caso et al. 1967:447; Feinman et al. 2002:261; Marcus 2008; Mountjoy 2006, 2012:212). Some fragmentary burned human bones occur in fill and midden contexts at La Consentida, but it is not clear if these indicate accidental burning or very early cremation.

Although several burials are accompanied by offerings of ceramic vessels, figurines, or stone tools, there appears to be little differentiation among them in terms of offerings. The burial B2-I3 individual (a male aged 40–50 years) was buried with more numerous and diverse offerings than others (see figures 6.5.B, 6.9.C, 6.10, 6.12.B, 6.12.E, 7.2.A, 7.2.B, 8.10.A, 8.11). Accomplishments during life, rather than the inheritance of status by birthright, may easily explain that circumstance, however. The burial B12-I14 adult female (aged 45–50 years) appears to have been interred with only a single stone bead, perhaps from a bracelet (figure 7.11.C). Currently the only La Consentida burial with a direct AMS date (1690–1530 cal BC, see table 1.1), this burial appears stratigraphically earlier than other adult burials in the Op. LC12 A area. Her relatively early date of burial, her age or gender, or some aspect of her social identity may explain the relative lack of offerings that accompanied her. The evidence is equivocal. Interestingly, some of La Consentida's only complete or fully reconstructable ceramic vessels occur as burial offerings with children. Two individuals in particular (B11-I13, aged 2–4 years and B9-I11, aged 3–4 years) were buried early in the site's history, each with a grater bowl and one also accompanied by a small jar (see figures 8.3.A, 8.3.B, and A2.6). While it is unclear whether most children at the site were interred with some sort of offering, at least one other child (burial B1-I2, aged 1–2 years) apparently was not. Interestingly, Marcus and Flannery (1996:96, fig. 87) identify the "mat motif" as incised on San José phase ceramic vessels as an early indication of political authority. It is possible that the geometric designs found in the La Consentida grater bowls (see chapter 8 and appendix 2) were inspired by woven basketry or even *petate* mats. They may thus represent an early example of woven mat imagery referenced on later pottery. I do not argue that these designs are evidence of a hierarchical social organization, though they may refer to the sorts of textiles and mats that later became emblematic of Mesoamerican nobility (see Carrasco and Englehardt 2015; Cheetham 2010:180; Marcus and Flannery 1996:96).

Additional evidence for ritual practice at La Consentida comes in the form of a ceremonial cache (LC12 A-F15) that was deposited at the base of Platform 1, at the foot of Substructure 1, and near the burials identified in Op. LC12 A (see table A1.3, figures 4.5, 7.14–7.16.A). This cache incorporated the complete skeletal remains of a large, predatory reptile (*Heloderma horridum*), remains of terrestrial turtles, shell, a playable ceramic ocarina in the form of a bird (see discussion below), and a fossil bull shark tooth. Marine biologist Dr. Gordon Hubbell (personal communication, 2016) identifies this as the fossilized "lower lateral tooth from a Bull Shark, *Carcharhinus leucas.*" Bull sharks have an extremely broad natural range that includes the Pacific Ocean along the Oaxaca coast (Fernández 2004:178). Based on the preservation of this example, Hubbell (personal communication, 2016) believes it to be a several-million-year-old fossil, perhaps dating to the Miocene epoch. This would mean that ancient coastal Oaxacans found the fossilized tooth, curated it, and then

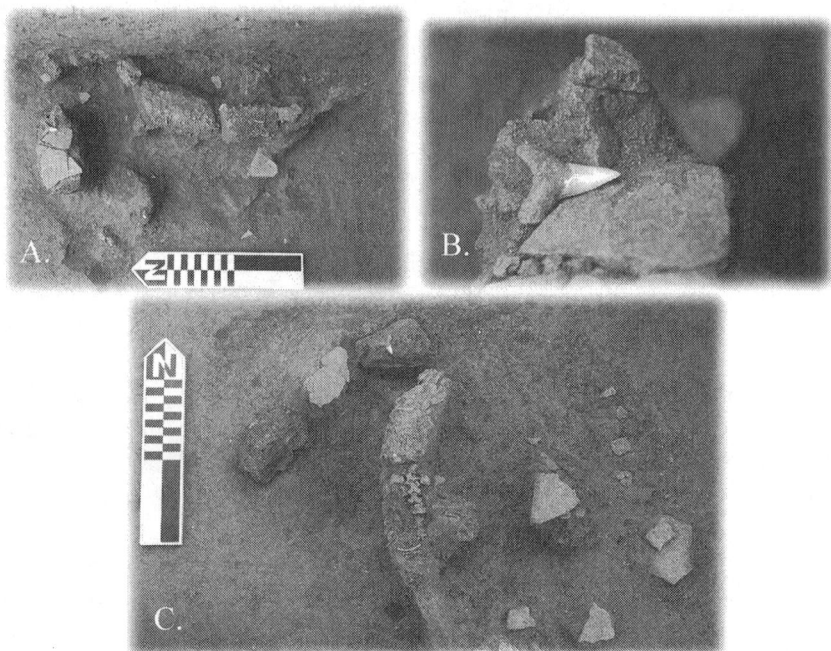

**FIGURE 7.14.** Photographs of the LC12 A-F15 ritual cache during excavation: (A) plan view of partially excavated feature; (B) close-up of shark tooth in situ; (C) plan view with feature mostly excavated

interred it as part of a special ritual deposit. It is not immediately clear what the tooth may have been used for, but Kent Flannery (2009b:344) has suggested that shark teeth were sometimes used in bloodletting ceremonies. Karl Taube (2010) has discussed the significance of symbolic unions between the earth, the sky, and the sea in Maya cosmology. Although many of Taube's examples are Classic or Postclassic in date, similar beliefs in areas as distant as western Mexico and the American Southwest suggest the great geographic breadth (and likely historical depth) of concepts such as reptilian sea monsters, earth turtles, and sky symbolism as connected with everything from the movements of the sun, to flooding, to the watery Underworld of the afterlife. While the La Consentida cache would be an extremely early example of these symbolic motifs, it bears many of the elements of this "primordial sea" concept, itself linked to broader notions of human creation and the Underworld (Reilly 1994; Taube 1986, 1992, 1993, 2000). Many Mesoamerican groups saw the world as apportioned into four cardinal directions, each associated with a sacred color and, in some cases, a patron deity. The Aztecs, for instance, related the north and west with

SOCIAL ORGANIZATION 155

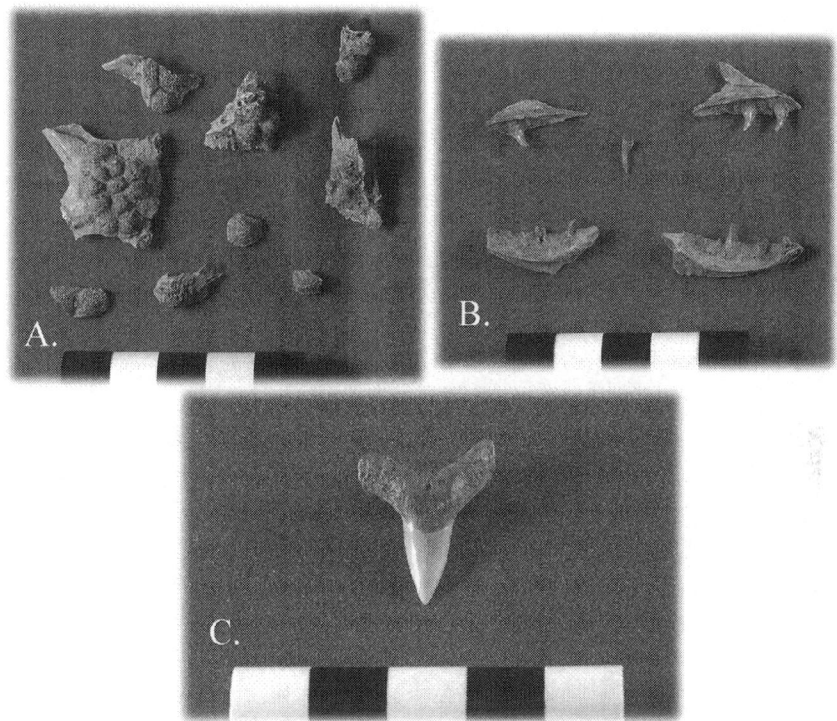

FIGURE 7.15. Faunal remains from LC12 A-F15 ritual cache: (A) diagnostic cranial bones of Heloderma horridum; (B) mandibular and dental fragments of Heloderma horridum; (C) close-up of buccal side of fossil bull shark (Carcharhinus leucas) tooth

the Underworld and with Tezcatlipoca and Quetzalcoatl, respectively. The east was the realm of fertility and of the macabre Xipe Totec. The south was the purview of the warrior bird-god Huitzilopochtli (Carmack et al. 2016:99–101). Although they are separated by thousands of years and hundreds of miles, I see compelling parallels between this Early Formative ritual deposit at La Consentida and those later and better documented cosmological themes. The La Consentida cache's recipe of symbolism suggests, I argue, the work of a ritual practitioner with specialized knowledge and thus likely a discrete social identity, at least in relation to special religious events. The cache may thus refer to the individuals buried in the Op. LC12 A area entering the watery Underworld, the movements of the sun, cardinal directions, the primordial power of the sea, and/or calendrical ritual.

Music also appears to have been an important aspect of ritual life at La Consentida and to have been incorporated into the LC12 A-F15 cache. The site's ceramic

156  SOCIAL ORGANIZATION

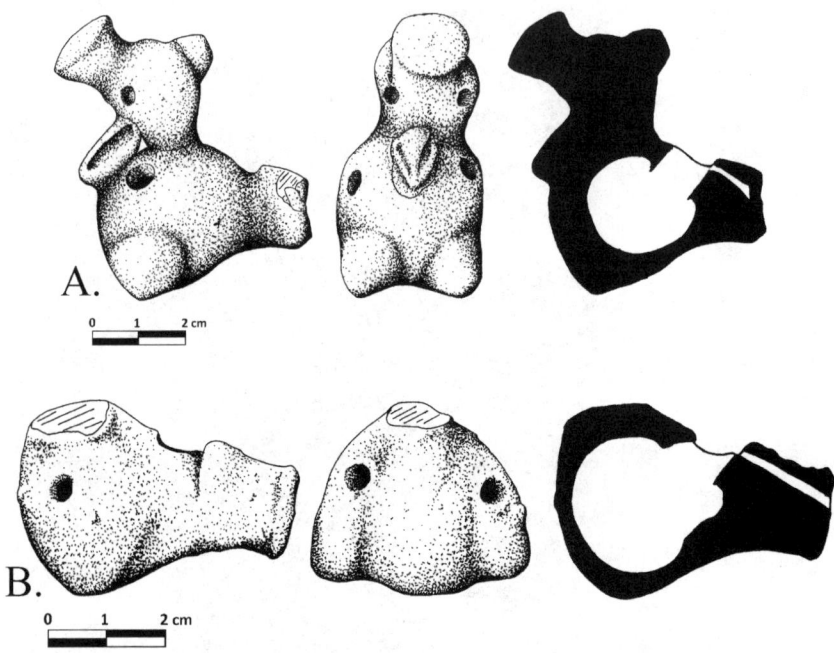

**FIGURE 7.16.** Playable bird instruments from La Consentida: (A) ceramic ocarina from LC12 A-F15 ritual cache; (B) ceramic ocarina from fill near burial B6-I7

whistles and ocarinas may be some of Oaxaca's earliest examples, as they appear to predate similar instruments of the Tierras Largas and San José phases (Hepp et al. 2014; Ramírez Urrea 1993:143; Sánchez Santiago 2014:248). I mentioned earlier in this chapter that most of La Consentida's figural iconography is anthropomorphic. The few zoomorphic artifacts often represent birds (figures 7.16–7.18; though also see figure 8.14). These artifacts are frequently musical instruments (specifically aerophones), two of which still play. The instrument pictured in figure 7.16.A was recovered at the edge of the LC12 A-F15 cache (figure 4.5). Its association with that ritual offering further emphasizes the cardinal directions and a cosmic union of sea, earth, and sky as meaningful for interpreting this feature. Whistles and ocarinas like those pictured in figure 7.16 play consistent groups of notes, suggesting their use in unison performances and perhaps for public events such as dedicatory rituals, contacting ancestors, and feasts (Hepp et al. 2014). Although the La Consentida examples are the oldest-known musical instruments in the region, their sophisticated instrumentality suggests that they come from a long tradition of Archaic period instruments constructed of perishable materials. The ocarina pictured in figure 7.16.A is capable

SOCIAL ORGANIZATION 157

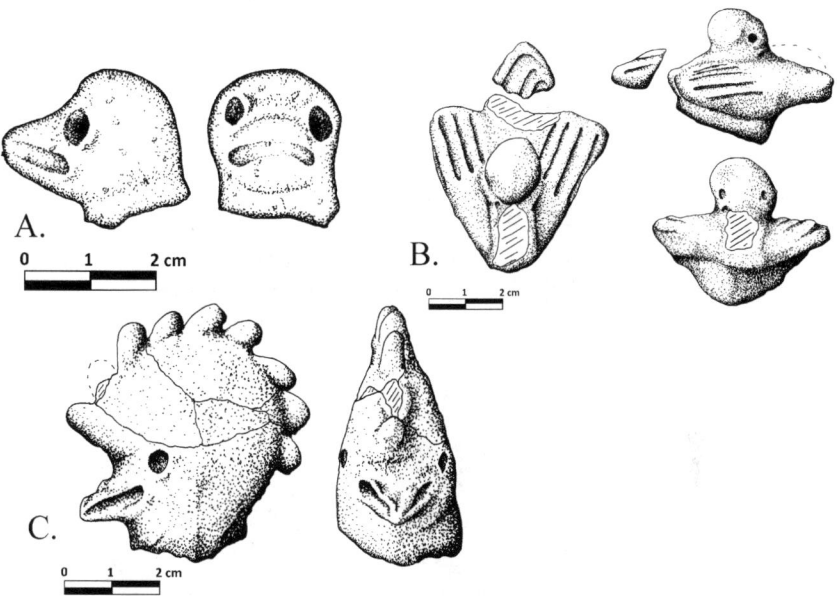

**FIGURE 7.17.** Partial ceramic bird instruments: (A) probable whistle or ocarina fragment from LC09 B-F14 fill or B1-I1 burial intrusion; (B) possible instrument fragment from fill likely associated with burial B8-I10; (C) partial whistle or ocarina from LC12 A-F10-s1 early fill near burial B11-I13

of playing notes from B5 to G6. It was likely originally covered with a red slip or paint, as indicated by a few remaining traces of pigment. The ocarina pictured in figure 7.16.B plays notes from D6 to G6. It has a coarser paste than the instrument represented in figure 7.16.A. Other artifacts from La Consentida (e.g., figure 7.17) are from definite or probable ceramic aerophones, though they have been broken and no longer play (see Barber and Hepp 2012; Barber and Olvera Sánchez 2012; Hepp et al. 2014; King and Sánchez Santiago 2011).

A few rare artifacts (figure 7.18) are likely bird representations that were not musical instruments. The artifact pictured in figure 7.18.B, for example, appears to be an unplayable and solid-bodied emulation of a ceramic bird ocarina, perhaps made by a child (Hepp et al. 2014). Such artifacts are a useful reminder of the importance of children, who often remain overlooked in archaeological interpretations (Baxter 2008; Kamp 2001; Lopiparo 2006). As active members of the community, La Consentida's children would have been integral to social organization at the site, a fact perhaps reflected by the offerings carefully placed with child burials (see discussion below). Although bird instruments and similar artifacts may not immediately appear to be

158  SOCIAL ORGANIZATION

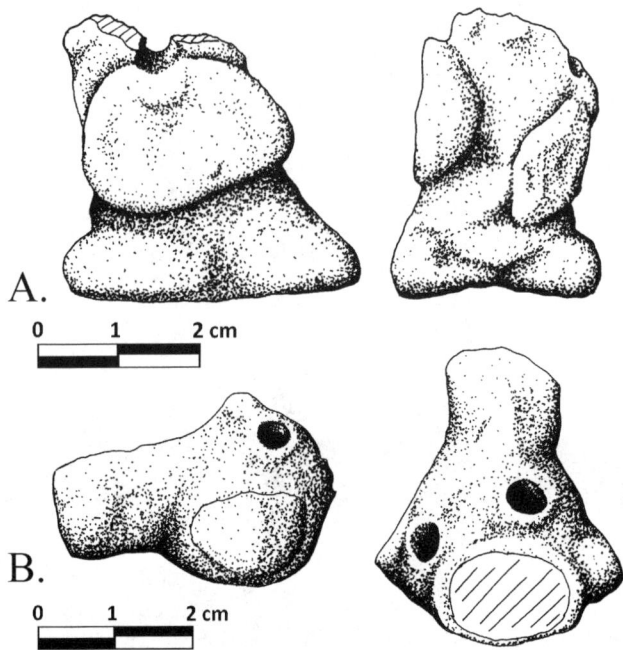

**FIGURE 7.18.** Other ceramic representations of birds: (A) possible bird figurine fragment from fill near burials B6 and B7 and above B9-I111; (B) possible emulation instrument, perhaps made by a child. From LC12 E-F3 fill

related to social organization, I argue that the opposite is true. As evidence of probable public events in which ritual practitioners utilized well-established tropes of Mesoamerican bird symbolism to communicate with an audience, such instruments suggest cosmological and social sophistication. As birds were considered messengers between the worlds of the living and the dead in ancient Mesoamerica (Marcus 1998; Urcid 2005:41–42, 62–63), these artifacts may even suggest that some community members served as ritual specialists, as discussed above with respect to figurines (see Hepp et al. 2014; Hepp and Joyce 2013; figure 7.10.B).

Further evidence of public ceremony at La Consentida comes in the form of ceramic figurine and mask fragments (figures 7.19 and 7.20) that suggest dancing and costumes likely used in public performance. The figurine fragment pictured in figure 7.19 may represent a human leg festooned with bells, probably for making noise during dances or other performances. Its recovery in close association with burial B8-I10, a feature that also included the crushed remains of at least one

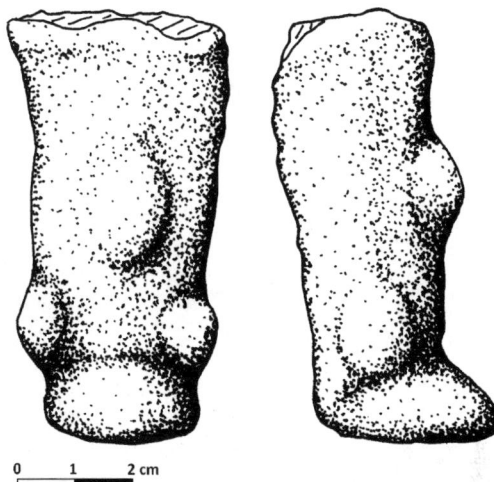

FIGURE 7.19.
Anthropomorphic leg fragment, possibly depicting a dancer wearing bells. From burial B8-I10

ceramic musical instrument, may even suggest that certain community members fulfilled specialized roles as musicians and dancers for public events. The recovery of female figurines with the adult male interred in Burial 2 serves as a cautionary reminder not to assume that figurines represent interred individuals, but the recovery of both an instrument and the "dancer" figurine fragment with B8-I10 slightly strengthens the case that the individual was a musician and/or dancer. At the very least, the presence of numerous ceramic aerophones demonstrates the community's emphasis on sound.

Life-size fragments of anthropomorphic ceramic faces (figure 7.20) represent another class of artifacts suggesting public performances likely related to music and dancing. Although some large ceramic faces may come from statues or effigy vessels, the holes at the top and side of the example pictured in figure 7.20.B suggest that some of these artifacts were tied to the head of a dancer or performer, as is often done with modern ritual masks in Oaxaca. The figure 7.20.B artifact even appears to have had eyeholes for allowing a dancer or performer to see while wearing the mask. Only eyes and noses from these probable masks have been recovered, so it is not possible to know if the mouths may have resembled the toothy frown of the aforementioned figurine and probable effigy vessel pictured in figure 7.9, or if the mouths were perhaps left uncovered to allow for clearer speaking or singing. The red paint or slip on the example pictured in figure 7.20.A (see also Hepp 2015:fig. 7.52) and the traces of a similar pigment on the nose represented in figure 7.20.D provide another clue as to the original appearance of these artifacts. Much like the diverse figurines, musical instruments, and elaborate ritual

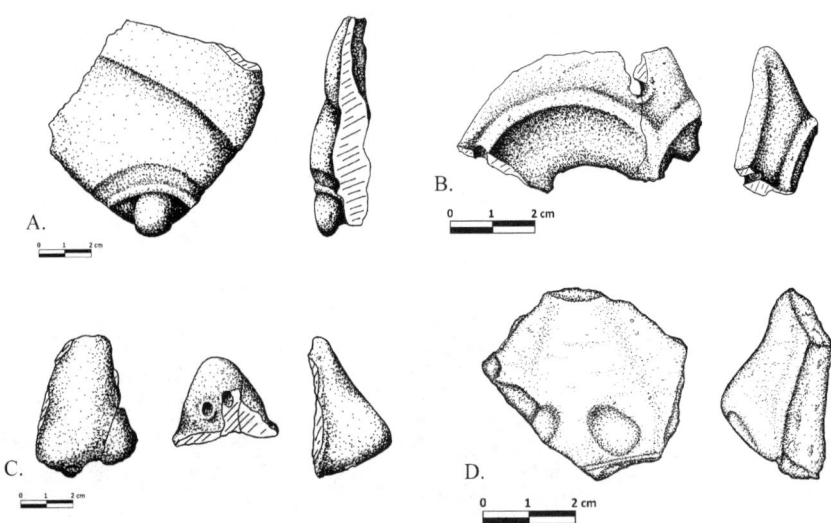

**FIGURE 7.20.** Probable ceramic mask fragments: (A) mask or statue fragment from LC09B-F17-s1 midden. Has a red paint or slip; (B) probable mask fragment from LC12 A-F4-S1 fill near burials B6 and B7. Note holes for tying to headgear; (C) possible mask fragment from fill likely associated with burial B12-I14; (D) possible mask fragment from LC09 A-F5 initial fill

offerings discussed above, these probable masks suggest diverse occupational and ritual roles enacted by La Consentida's community members. Masks remained a rare but important artifact type in coastal Oaxaca throughout the Formative period (Brzezinski et al. 2017)

## ANALYSIS OF MORTUARY CONTEXTS

Fourteen sets of human remains were identified in twelve burial deposits in two loci at La Consentida during the 2009 and 2012 excavations (Hepp et al. 2017). For detailed discussion of each burial, including lists of pathologies and offerings, as well as plan view drawings, refer to Aguilar and Hepp (2015). Also, see chapter 6 of this book for a discussion of the human remains as they relate to ancient diet. In this chapter, I will provide limited discussion of the burials in light of evidence for social organization at La Consentida.

To date, the burial of the dead at La Consentida has been identified in two discrete areas (uncovered in Ops. LC09 B and LC12 A) at the northern edge of Platform 1. Both of these locations are away from known domestic structures (see figure 3.4). As discussed below, many of the burials share basic elements of body orientation and positioning. Such practices suggest the early stages of establishing communal cemeteries away from houses. The dating of the B12-I14 individual to 1690–1530 cal BC (and, most probably, to the mid-seventeenth century cal BC), indicates that the LC12 A mortuary area was produced quite early in the history of cemeteries in Oaxaca and in Mesoamerica more generally (Higelin Ponce de León and Hepp 2017b; table 1.1). The closest domestic context possibly associated with any of the burials at La Consentida was the LC09 B-F15 hearth (figure 4.4), which was only a few centimeters from burial B1-I1. The burials located near this feature (especially B1) were likely later interments only coincidentally associated with the hearth, however. Early Formative cemeteries are also known in highland Oaxaca (Cervantes Pérez et al. 2017; Whalen 1981, 2009:78). Unlike in some highland contexts, though, children and infants may not have been interred under the floors of houses at La Consentida (Whalen 2009:78). In later Formative period coastal Oaxaca, cemeteries were fairly common (e.g., Barber et al. 2013; A. Joyce 1991a; Mayes and Joyce 2017). As demonstrated by A. Joyce (1991a, 2010:181–85), for example, later Formative sites such as Cerro de la Cruz contain formal cemeteries and burials associated with domestic structures.

Although most burials at La Consentida were interred in sediments containing numerous artifacts such as broken pottery and obsidian, objects clearly intended as grave offerings were relatively rare. For example, one early adult burial (B12-I14; figure 7.21) was apparently interred only with a single black stone bead near the left wrist, which may have been part of a bracelet (see figure 7.11.C). While the stratigraphic position of this burial suggests that it dates to early in the sequence of mortuary contexts uncovered at La Consentida, the remains of the woman (an adult of 45–50 years) were in excellent condition. She had healthy teeth and a robust build, as I discussed in chapter 6 as evidence for dietary change. The relative lack of offerings in her grave may indicate a trend over time wherein earlier burials had fewer offerings than did later ones, or this condition may relate instead to some aspect of her social role in relation to other members of the community. Signs of occupational stress from squatting facets on both tibiae suggest that she spent much of her working hours in a squatting position (Aguilar and Hepp 2015:550). Although early in the burial sequence, this woman's remains were placed in the southwest/northeast orientation that would become a tradition at La Consentida, probably for several generations.

Some burials did include numerous offerings, as demonstrated by B2-I3 (figure 7.22). Like several other burials at La Consentida, the individual interred here (a male aged 40–50 years) was placed in a southwest/northeast orientation. He had the most numerous burial offerings thus far identified at La Consentida, including a tabular stone, a mano, a ceramic bottle, and chert tools (figures 6.9.C, 6.11,

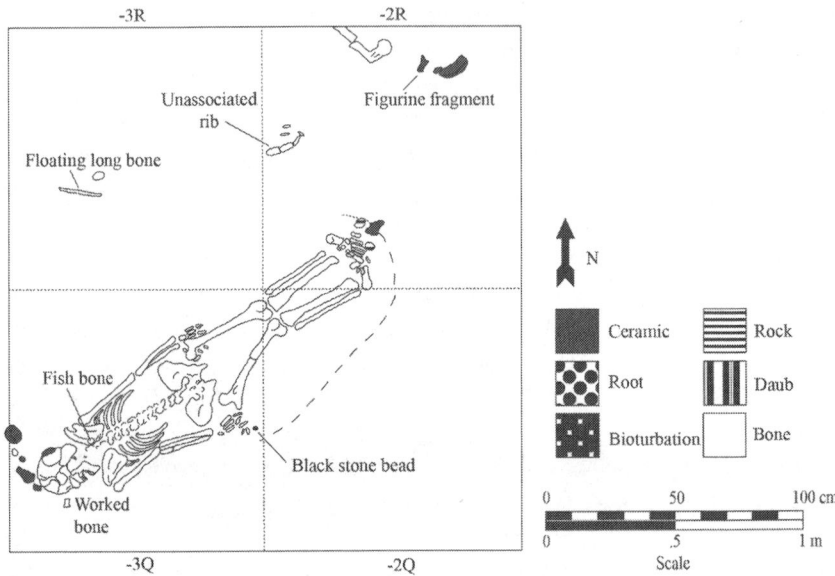

**FIGURE 7.21.** Plan view of B12-I14

6.12.B, 6.12.E, 8.11). He also appears to have been buried with two partial ceramic figurines (figures 7.2.A and 7.2.B) and a crocodile mandible fragment (figure 6.5.B). The fact that B2-I3 was subsequently disturbed by the interment of B3-I4 indicates the importance of burying the dead in consistent areas of the site through time. This trend is also exemplified by several burials jumbled together in Op. LC12 A. For instance, B6-I7 and B6-I8 overlapped one another and were located near several other burials in a small space (Aguilar and Hepp 2015:fig. A.5.4). The B6-I7 individual was an adult male of 20–35 years. Like other burials at the site, he was in a southwest/northeast orientation, with his head to the southwest. The B6-I8 individual was another probable male of undetermined age. Although his head was oriented to the northeast, his overall body positioning is consistent with numerous other burials at the site that follow the general southwest/northeast orientation of Platform 1 itself (see figure 3.3). The Burial 6 event was accompanied by numerous offerings and possible offerings, including figurine fragments, a complete jar, a ceramic bead, most of another broken jar, a possible crocodile bone, and a partial ocarina. This multi-individual burial was apparently later disturbed by the interment of B7-I9. The resulting commingling of human remains made for difficult interpretation (both in the field and in the laboratory), and it is possible that small fragments from additional individuals were also present among these burials.

SOCIAL ORGANIZATION **163**

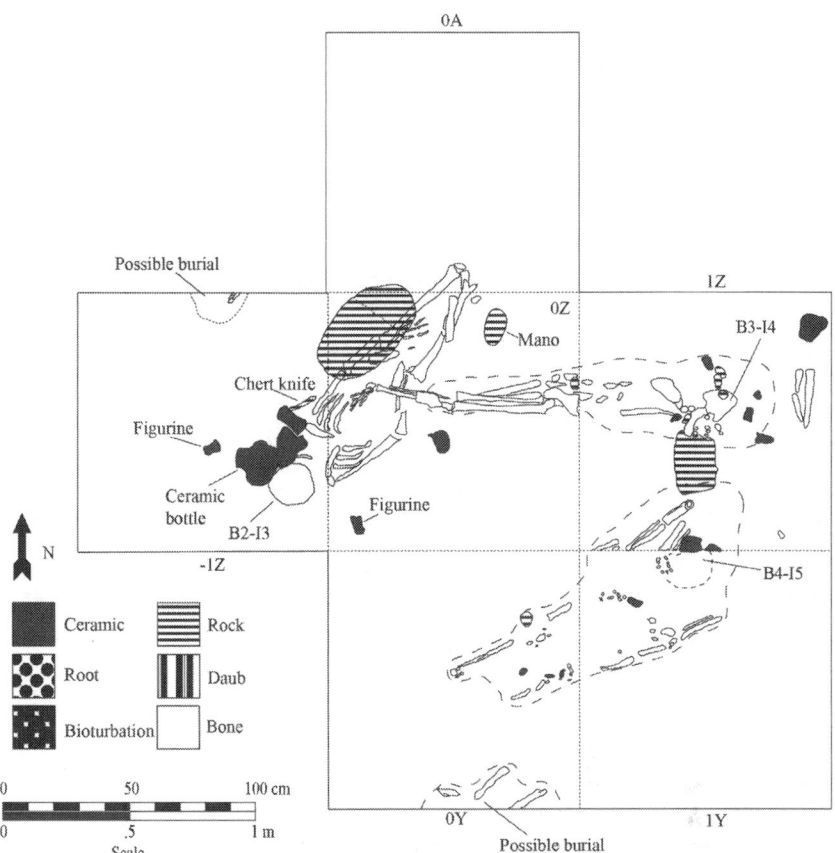

**FIGURE 7.22.** Plan view of B2-I3, B3-I4, and B4-I5

Another burial (B8-I10) provides an example of the types of items likely left as offerings with the deceased at La Consentida. The young adult male (aged 15–18 years) in this burial was accompanied by the probable grave offerings of figurine fragments, a partial ocarina, large ceramic vessel fragments, and a carbonized seed. One figurine fragment (figure 7.19) was recovered just above the chest of the individual, while the ocarina fragment was found near his pelvis (figure 7.23). Although this interpretation is speculative, the combination of a figurine suggesting music and dance (by virtue of its possible depiction of bells worn on the leg of a dancer) with the remnants of an actual musical instrument potentially underscores the importance of music and public performance in the burial of the dead, or perhaps even the specialized social roles of the deceased himself. While this individual was oriented approximately east/

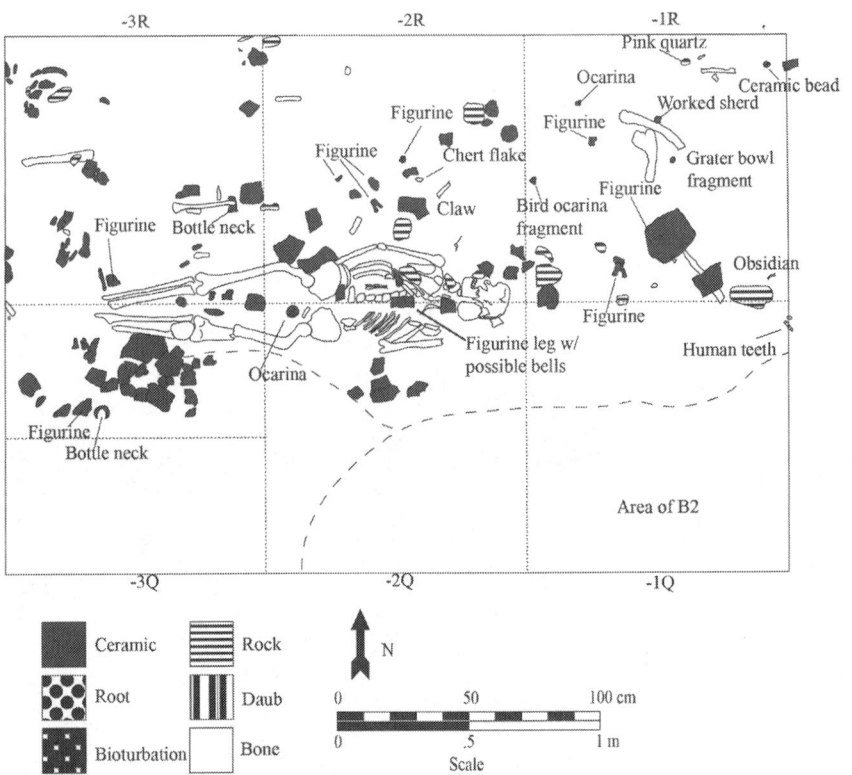

**FIGURE 7.23.** Plan view of B8-I10

west, he nonetheless follows the general trend of body orientation among burials at La Consentida, where no burials have yet been found in a north/south orientation.

Not all burials with offerings at La Consentida are those of adults. Two small children were interred in separate burial events in the Op. LC12 A mortuary area (figure 7.24). These burials (B9-I11, a child of 3–4 years, and B11-I13, child of 2–4 years) were accompanied by offerings including a complete collared jar, green and black minerals, ceramic vessel fragments, lithics, and two complete or nearly complete grater bowls (figures 8.3.A and 8.3.B). Both of these burials also follow the general trend of body orientation at the site. It was the association of the grater bowls with these small children's burials that led to the interpretation that grater bowls may have been somehow related to children at La Consentida, and perhaps to the weaning process (see Hepp et al. 2017; chapter 8). One grater bowl in particular (figure 8.3.B) has a spout and may have been used to pour some liquid processed in the grater bowl.

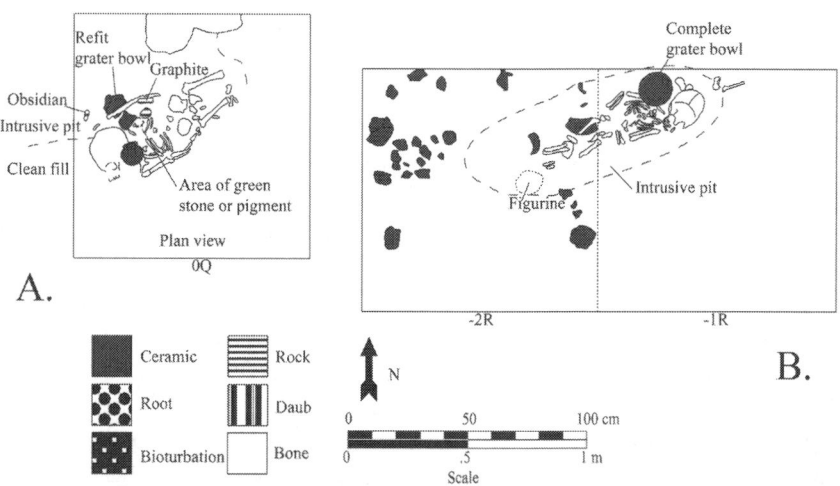

**FIGURE 7.24.** Child burials at La Consentida—not depicted in original relative position: (A) plan view of B9-I11; (B) plan view of B11-I13

The numerous burials consistently placed in two discrete areas (at least nine in Op. LC12 A) suggest an early stage in the development of formalized, extrahousehold cemeteries beginning, according to direct dating of B12-I14, by about the mid-seventeenth century cal BC (table 1.1). The Op. LC12 A burials, in particular, seem to have been placed in a consistent area over time, with later burials frequently disturbing earlier ones. That the northwest edge of Platform 1 was considered an appropriate mortuary area for those burials found both in Op. LC09 B and in Op. LC12 A is also suggested by the lack of human remains in other excavated contexts at the site. Only a few redeposited human bone fragments have been identified anywhere at the site outside the two areas discussed here. This pattern of burial placement may have implications for the importance of memory and shifting concepts of landscape and place in the gradual adoption of sedentism at La Consentida (see Ashmore 2002, 2004; Brady and Ashmore 1999; A. Joyce and Goman 2012).

It is also interesting that many of La Consentida's burials were oriented southwest/northeast and that several were interred in a prone position. The northwestern portion of the site itself (where the burials were located) is oriented in a southwest/northeast fashion, and it is possible that the burials follow this orientation intentionally, leaving many of the human bodies aligned with the layout of the site (see figure 3.3). Another possibility is that the burials are aligned to face the direction of the rising and setting sun as measured on some particular day in the solar cycle, such as the summer solstice. Attempts to reach the site on the date of the summer solstice in 2012 (June 21) to measure the angle of the rising sun were not successful, due to flooding from the recent Hurricane Carlota. The angle of the rising sun from the closest point that my friend and colleague Pepe Aguilar and I were able to reach on that day was 61 degrees east of north, which is roughly equivalent to the orientation of some interred individuals (either of the head or of the feet). Measurements of the angle of the rising sun in the days immediately preceding the solstice suggest that it approximately coincides with both the orientation of platform's northwestern edge and with the burials themselves. Future research may indicate whether other features at La Consentida align with celestial bodies during specific times of the year.[3] The prone positioning of the bodies themselves is rare in the lower Río Verde region (see Barber 2005:app. F; A. Joyce 1991a:app. 1). Future excavations at La Consentida and at other sites in the region (such as Álvar Núñez Cabeza de Vaca) with possible Early Formative period occupations may help to determine whether the prone positioning of human remains is diagnostic of the Early Formative period in the region. As discussed earlier in this chapter and in chapter 4, the LC12 A-F15 ritual cache is located just a few centimeters from some of the Op. LC12 A burials. The combination of marine and terrestrial faunal remains in this deposit, along with the bird ocarina that suggests more celestial ideological connotations, may represent significant aspects of mortuary ritual. As with the burials themselves, this cache may have astronomical significance. It appears to have combined marine and reptile symbolism with the movement of the sun as a reference to the "watery Underworld" (e.g., Taube 2010).

As discussed previously, one direct AMS date from human remains at La Consentida (B12-I14) is currently available (table 1.1). Although the burials within the LC12 A mortuary context can be relatively dated according to their stratigraphic relationships to this burial, those burials in LC09 B cannot yet be directly dated. Deciphering the sequence of mortuary events thus involves comparisons of strata both within and among different areas of the site. Stratigraphic sequences within the LC12 A and LC09 B areas, as well as identification of early platform fill layers (e.g., LC09 B-F14 and LC12 A-F17-s2) versus later fill deposits (e.g., LC09 B-F10 and LC12 A-F4-s1) permit an approximate chronological sequencing of the burials excavated thus far (table 7.2).

## FEASTING AND ITS IMPLICATIONS FOR SOCIAL ORGANIZATION

Feasting is suggested by the presence of faunal remains (chapter 6), decorated ceramics (chapter 8 and table A2.28), and rapid deposition of pottery identified in various middens at La Consentida. As discussed in chapter 4 and appendix 2, two main midden deposits, excavated in Ops. LC12 E and LC12 H, provide evidence of feasting. As demonstrated in appendix 2, the Op. LC12 H midden (LC12 H-F-4) contained about 94 percent utilitarian jars (table A2.13). These vessels were deposited quickly, with cross-fitting sherds occurring throughout the midden. This pattern suggests that the ceramics came from just one or very few events, likely tied to the preparation for some large community gathering. The Op. LC12 E midden (F-16 through F-9) was also deposited quickly but contained a wider variety of vessel types and included a number of decorated vessel fragments (tables A2.10 and A2.28) in addition to high percentages of marine animal remains among the fauna (see chapter 6). As discussed in appendix 2 (see table A2.10), many of the vessel fragments recovered in the LC12 E midden were serving wares such as bowls (26 percent of all diagnostic vessel fragments) and bottles (12 percent of all diagnostic vessel fragments). Contrast this with the pattern identified in the LC12 H midden, where almost all diagnostic remnants came from jars likely used for storage or cooking, and only about 6 percent of diagnostic sherds came from bowls and bottles, combined (table A2.13).

The LC12 H midden produced some of the most complete and well-preserved ceramic vessel fragments thus far identified at La Consentida. As exemplified by the vessels pictured in figure 8.2, these vessels were mostly undecorated, globular jars with out-curving or out-leaning necks. The large hemispherical bowl pictured in figure 8.2.E indicates some variability within the midden, though it is made of the same paste as the jars and also lacks decoration. I interpret these ceramic remains as evidence of a large, rapidly produced deposit of cooking vessels, probably discarded after the preparations for one or very few communal feasting events. Such rapid deposition is demonstrated by the fact that several refitting vessel fragments (e.g., figure 8.2.D) come from both the LC12 H-F4-s1 and H-F4-s2 substrata, and the fact that the deposit's largest and most complete hemispherical bowl fragment (figure 8.2.E) is composed of refitting sherds from up to five lots apart. These five lots (three of which contained identified fragments of the bowl) represent approximately 60 cm of excavated depth within Unit H.0A. The interpretation that these vessels come from feasting preparation, rather than from domestic consumption, is supported by the presence of very large jars in the sample, with one reaching 53 cm in rim diameter.

As discussed in chapter 2, archaeologists working in various regions have considered public events such as feasts to be among the primary loci for social maneuvering in ancient communities (e.g., Blake and Clark 1999; Clark and Blake 1994; Hayden 1990, 2009; Hill and Clark 2001). "Aggrandizers" displaying prestige goods such as exotic imports or decorated pottery at such occasions and establishing lasting relationships of indebtedness with other community members, may have been one of the main catalysts for shifting social organization in transegalitarian communities (Blake and Clark 1999; Hayden 1995, 2011). Similarly, feasts could have been

TABLE 7.2. Approximate chronology of burials excavated at La Consentida in 2009 and 2012. Refer to table 1.1 and chapter 3 regarding AMS radiocarbon dating of B12-I14.

| Burial number | AMS radiocarbon date | Chronological position | Sex | Age estimate | Orientation |
|---|---|---|---|---|---|
| B11-I13 | N/A | 1 (oldest) | Undetermined | 2–4 years | Head to NE |
| B12-I14 | 1690–1600 cal BC (p = .76) | 2 | Female | 45–50 years | Head to SW |
|  | 1585–1530 cal BC (p = .20) |  |  |  |  |
| B9-I11 | N/A | 3 | Undetermined | 3–4 years | Head to SW |
| B8-I10 | N/A | 4 | Male | 15–18 years | Head due E |
| B6-I8 | N/A | 5 | Probable male | Unknown adult | Head to NE |
| B6-I7 | N/A | 6 | Male | 20–35 years | Head to SW |
| B7-I9 | N/A | 7 | Probable male | 20–35 years | Head to SW (possible) |
| B1-I1 | N/A | 8 | Male | 35–50 years | Head to NE |
| B1-I2 | N/A | 8 | Undetermined | 1–2 years | Head to S (possible) |
| B2-I3 | N/A | 9 | Male | 40–50 years | Head to SW |
| B3-I4 | N/A | 10 | Probable female | Over 18 years | Head due E |
| B4-I5 | N/A | 11 | Undetermined | Unknown adult | Head to NE |
| B5-I6 | N/A | 12 | Probable female | 20–35 years | Head to SW |
| B10-I12 | N/A | 13 (most recent) | Probable female | 20–35 years | Head to NE, SW (if two individuals present) |

one of several venues for the solidarity-promoting activities of social collectives as a precursor to later development of more exclusionary forms of hierarchical social organization (R. Joyce 2004b; Joyce and Henderson 2001, 2007). I do not claim that evidence of feasting at La Consentida necessarily indicates an early establishment of hereditary hierarchical inequalities. As argued by Ian Kuijt (2009:643; see also Twiss 2008), archaeological and ethnographic evidence demonstrates that "feasting occurs in the social context of coexisting integrative and competitive processes, not just competition." Public events such as feasts at La Consentida certainly could have provided appropriate venues for competition and transitions to increasing social complexity, however. The presence of decorated ceramics in some of these midden deposits (see table A2.28) suggests that community members did have an interest in the display of prestige items that may indicate increasing economic/craft specialization.

## SUMMARY

Rather than strictly focusing on degrees of hierarchical social inequality, I have striven in this chapter to use multiple lines of evidence in order to paint a picture of the rich social life led by members of La Consentida's ancient community. Regarding the social organization/social hierarchy component of the central LCAP research question, however, a few comments are worthwhile here. The case for social complexity at La Consentida seems to hinge upon how one defines the term (J. Arnold 1996; McGuire 1983; Paynter and McGuire 1991). If we stay wedded to identifying it as inherited hierarchical inequality, the evidence appears relatively scant. As implied by increasing standardization of communal labor evident in later deposits of earthen fill discussed earlier in this chapter (see also chapters 4 and 5 and appendix 1), some community members may have come to organize group labor efforts at the site toward the end of its occupation in the Formative period. Perhaps La Consentida's best evidence for both heterarchical and hierarchical social complexity comes from small-scale ceramic iconography. The presence of seated, big-bellied, and headdress-wearing figurines is consistent with the depiction of elevated social status, perhaps held by community elders represented in Early and Middle Formative period contexts elsewhere in Mesoamerica and in later contexts in the lower Río Verde Valley (Guernsey 2012:108, 121–22; Hepp and Joyce 2013; Hepp and Rieger 2014; Lesure 1999a:121). Anthropomorphic imagery and remnants of actual jewelry from the site suggest diverse social roles and personal adornment (such as headdresses and jewelry) that are associated with elevated status in later Mesoamerican contexts (e.g., Ashmore 2004:184–85; Blomster 2004:85, 186; Clark 1994:126; Heyden 1991:195; Saunders 2001). Architectural, mortuary, public ceremony, and feasting evidence also suggests heterarchical social differences.

As Lesure and Blake (2002) discussed, identifying the initial stages of social complexity is difficult because it is to be expected in discontinuous and regionally variable patterns. Due to the nature of the artifacts recovered and contexts excavated during the 2009 and 2012 fieldwork at La Consentida, I have relied heavily in this chapter on iconography as one of several lines of evidence for understanding social organization at the site. Inferences promoted by analysis of figurines and similar artifacts—when bolstered by evidence provided by architectural, mortuary, personal adornment, public ceremony, and feasting data—can support productive discussions of ancient social organization. What these various lines of evidence suggest for the La Consentida community, I argue, is that it was heterarchically complex, with diverse social roles including communal labor organizers, dancers, musicians, ballplayers, feasting facilitators, and ritual specialists. Given evidence for the significance of these roles in later Mesoamerican hierarchical social complexity (e.g., the Olmec "ballplayer chiefs" discussed by Clark [2007]), I argue that La Consentida bore the seeds of hereditary social hierarchy. As discussed in chapter 2, most research regarding Mesoamerican social complexity has focused on identifying formalized social hierarchies rather than understanding the role of heterarchical complexity (though see J. Arnold 1996; Crumley 1995; McGuire 1983; Paynter and McGuire 1991). Key social distinctions at sites such as La Consentida, I suggest, were fundamental in the eventual establishment of hereditary social hierarchy in Mesoamerica.

# 8

## No Village Is an Island

*Interregional Interaction and Exchange*

In the preceding three chapters I have focused on the main components of my research question regarding the interrelatedness of domestic mobility, subsistence, and social organization at La Consentida. In this chapter, I turn my attention to evidence for the site's role in networks of interregional interaction. Understanding La Consentida's relationships beyond the lower Río Verde region is important for addressing questions about settlement, subsistence, and social organization for multiple reasons. First, in order to argue that the social and economic changes identified at La Consentida are related to those in better-documented Mesoamerican regions and time periods, it is necessary to demonstrate the site's participation in that larger cultural sphere. Second, understanding the nuances of La Consentida's interregional relationships may help to explain its uniquely early date and the particulars of its ceramic assemblage. In this chapter, I discuss styles of ceramic vessel form and decoration that suggest contact with regions such as West Mexico, the Valley of Oaxaca, and Central Mexico. I propose that the Tlacuache ceramic complex represents one of the earliest known examples of the Red-on-Buff horizon (see Clark 1991; Winter 1992; Winter and Sánchez Santiago 2014b). Furthermore, Tlacuache ceramics may represent part of a western Pacific coastal interaction sphere evinced by decorative styles held in common with those from the Capacha phase (see I. Kelly 1980). I also consider obsidian sourcing data, the presence of metamorphic greenstone, and figurine iconography to support my argument that La Consentida was not isolated or an anachronistic anomaly but rather an important early locus for many of the traditions, beliefs, and practices that came to define Mesoamerica in subsequent centuries (Clark 2004b; R. Joyce 2004b; Kirchhoff 1943).

## COMPARING BARRA AND TLACUACHE CERAMICS

The Barra phase of the Soconusco region is widely recognized as the earliest well-dated ceramic tradition in Pacific coastal Mesoamerica. Barra ceramics are "remarkably sophisticated," often bear decoration, and are composed largely of tecomates. Specifically, Clark and Blake (1994:25) reported that the assemblage is composed of 89.4 percent tecomates and 10.6 percent bowls. Many Barra phase vessels are phytomorphic or designed to resemble plants (Clark and Blake 1994). Numerous authors have argued that these ceramics were instrumental in competitive feasting that helped spark the development of social complexity in the Soconusco (Clark 2004b; Clark et al. 2007; Clark and Blake 1994; see also Hayden 1990, 1995, 2009, 2011). Although Tlacuache ceramics were contemporaneous with the Barra phase (see tables 1.1 and 1.2), they are dissimilar in form because they include far fewer tecomates and highly decorated phytomorphic vessels and instead consist of a majority of jars, followed in relative emphasis by bowls, bottles, and more specific variants of these basic vessel types (see table 8.1, figure 8.1, and appendix 2). While tecomates and a few probable phytomorphic vessels are present at La Consentida (see figures A2.8.C and A2.9), they are rare. The early dates for the Tlacuache phase, when considered in conjunction with basic differences between Barra and Tlacuache ceramic complexes, indicate that we should question the argument that Pacific coastal Mesoamerica's earliest ceramics were all introduced from Central America through Chiapas and Guatemala (Clark and Blake 1994; G. Lowe 2007). Certainly, early Central American sites with ceramics are candidates for interaction with some of Mesoamerica's first potters (see Bradley 1994; Hoopes 1994). In addition to those southeastward connections, however, a very early ceramic tradition seems to have appeared west of the Isthmus of Tehuantepec by as early as the nineteenth- or twentieth-century cal BC. As I will discuss below, Tlacuache wares exemplify the Red-on-Buff ceramic horizon, rather than the Locona horizon to which the Barra ceramics arguably belong (see Clark 1991; Winter 1992). These ceramic macro horizons (which I will here call "Locona" and "Red-on-Buff") may relate to broad patterns of cultural and linguistic distribution (see Clark 1991; Josserand et al. 1984; Winter 1992; Winter and Sánchez Santiago 2014b).

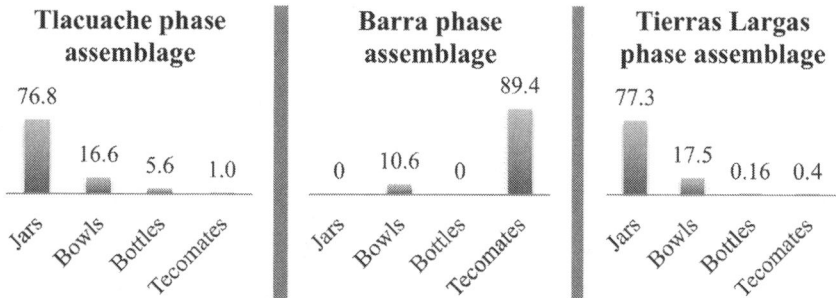

**FIGURE 8.1.** A comparison of Tlacuache, Barra, and Tierras Largas vessel types by percentage

### THE TLACUACHE COMPLEX AS AN EARLY EXEMPLAR OF THE RED-ON-BUFF HORIZON

In Oaxaca, the Espiridión, and Tierras Largas phases—the former of which is now in question as distinct from Tierras Largas—have previously been recognized as producing the earliest ceramic complexes and the first examples of Red-on-Buff horizon wares (see Winter 1992; table 1.2). These vessel assemblages appear to be similar to another early highland pottery tradition, that of the Purrón phase (1900–1680 cal BC) of the Tehuacán region (Clark and Gosser 1995; García Cook and Merino Carrión 2005). In fact, ceramics of both the Purrón and early Ajalpan (approx. 1680–1350 cal BC) phases are similar to Tlacuache pottery in the presence of globular jars, conical bowls, and a few tecomates (MacNeish et al. 1970:21–40). One notable difference between the Tlacuache and Purrón assemblages, besides the greater diversity of the former, is that the latter (at least as reconstructed) appears to consist mostly of vessels with rounded bottoms (MacNeish et al. 1970:fig. 6). As demonstrated in appendix 2, most vessels of the Tlacuache assemblage appear to have had flat or at least flattish bases.

In this section, I compare Tlacuache ceramics with these other early pottery traditions, placing special emphasis on Tierras Largas and Barra ceramics (see table 8.1 and figure 8.1). As I will demonstrate, despite some similarities, these assemblages differ significantly and must be assigned to discrete phases (refer to appendix 2 for a full description of Tlacuache ceramics). table 8.1 and figure 8.1 demonstrate the relative emphases on different vessel forms in the Tlacuache, Barra, and Tierras Largas assemblages. Of note are the general similarities in vessel ratios between Tierras Largas and Tlacuache, both of which are very different from the Barra phase, which lacks jars and bottles and is mostly composed of tecomates.[1] Between Tlacuache and Tierras Largas phases, important differences include the higher percentages

**TABLE 8.1.** A comparison of Tlacuache, Barra, and Tierras Largas vessel types by percentage

| Ceramic phase | Vessel type | Percentage |
| --- | --- | --- |
| Tlacuache | Jars | 76.8 |
| Tlacuache | Bowls | 16.6 |
| Tlacuache | Bottles | 5.6 |
| Tlacuache | Tecomates | 1.0 |
| Barra | Jars | 0.0 |
| Barra | Bowls | 10.6 |
| Barra | Bottles | 0.0 |
| Barra | Tecomates | 89.4 |
| Tierras Largas | Jars | 77.3 |
| Tierras Largas | Bowls | 17.5 |
| Tierras Largas | Bottles | 0.16 |
| Tierras Largas | Tecomates | 0.4 |

of bottles and tecomates in the former. For these comparisons, vessel varieties were collapsed into simple designations of "bowls," "jars," "bottles," and "tecomates." As the higher percentage of bottles and the presence of special vessel types such as grater bowls suggests, the Tlacuache assemblage is more diverse than Tierras Largas. Although the nonexistence of jars and bottles in the Barra phase and the very low percentages of tecomates in the other phases make a Chi-square statistical comparison useless, a Fisher's exact test demonstrates that these phases differ in a statistically significant way. When just the Tlacuache and Tierras Largas assemblages (which have similar percentages of jars and bowls) are compared using a Fisher's exact test, the differences between them are statistically significant ($p < .0001$).[2]

As I briefly discussed in chapter 4, ceramics from the LC12 H-F4 midden consisted almost entirely of globular jars (see appendix 2). These vessels (e.g., figures 8.2.A–8.2.D) appear formally similar to Tierras Largas phase jars (Flannery and Marcus 1994:frontispiece A, 45–101, fig. 8.37; Ramírez Urrea 1993:figs. 48, 57–59). Some Tierras Largas phase jars have rounded bases, while bases among all styles of Tlacuache vessels are usually flatter (see Ramírez Urrea 1993:figs. 36, 38). Undecorated semispherical bowls (e.g., figure 8.2.E) are also similar between the two phases. Four small fragments of a finely burnished and slipped "kidney-shaped bowl" (figure 8.2.F) from the LC12 E midden also appear similar to Tierras Largas phase hemispherical kidney-shaped bowls (Flannery and Marcus 1994:fig. 7.2). Such kidney-shaped bowls are not a particularly diagnostic vessel type, however, as they also appear at Zohapilco (Niederberger 1976:lám. LII.16, 25, lám. LIV.16, Foto

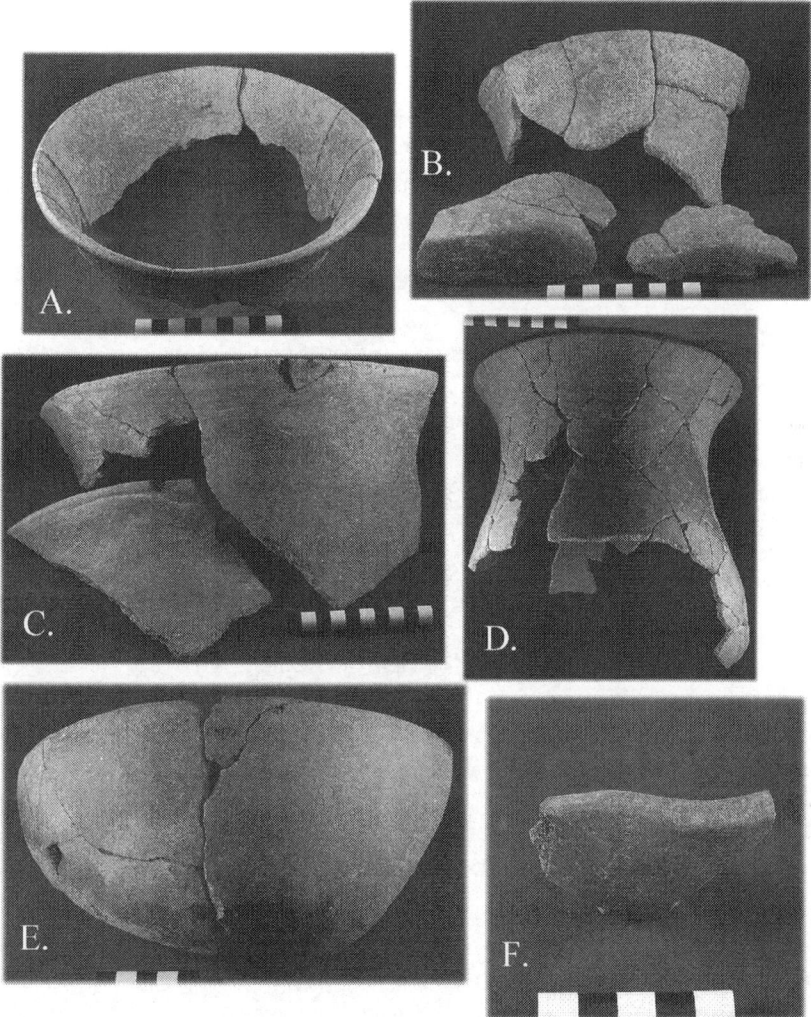

**FIGURE 8.2.** Ceramic remnants from the Op. LC12 H midden: (A) jar rim; (B) jar rim, neck, and base; (C) partial jar; (D) partial jar; (E) partial hemispherical bowl; (F) rim fragment of hemispherical kidney-shaped bowl

37), in the Tehuacán Valley (MacNeish et al. 1970:fig. 6), and at Tlatilco (Piña Chan 1958:fig. 40.j, lám. 21). Similarly, globular jars were not really a diagnostic form, regionally speaking, during the Early Formative period. For example, Tlacuache jars are generally similar to some from Tlatilco (e.g., Piña Chan 1958:fig. 36.c; Fig 41.b) and Zohapilco (Niederberger 1976:lám. LIX), as well as the earliest reconstructed forms of the Tehuacán Valley's Purrón phase (MacNeish et al. 1970:figs. 8, 9).

Tierras Largas phase ceramics are less formally diverse than those of the Tlacuache phase. Specifically, Tierras Largas pottery lacks the types of bottles and grater bowls identified at La Consentida, and what few bottles are present in Tierras Largas are rarer than Tlacuache bottles (see Ramírez Urrea 1993; appendix 2). Although Tierras Largas and Tlacuache ceramics have relatively similar vessel form ratios (table 8.1), the assemblages are markedly different in terms of both plastic and painted decoration styles. The "rocker stamping" found on some Tierras Largas ceramics (e.g., Flannery and Marcus 1994:fig. 8.18) and in the late Ajalpan phase of the Tehuacán Valley (MacNeish et al. 1970:fig. 21) appears to be completely absent from the Tlacuache ceramics. Furthermore, bands of red paint on Tierras Largas bowls and jars (e.g., Flannery and Marcus 1994:figs. 8.22–26) are dissimilar to Tlacuache painted decorations. Both assemblages share the use of red paint and/or slip for exterior vessel decoration, however. Generally, vessel forms of greatest similarity between the Tlacuache and Tierras Largas phase assemblages are generic types that are also shared in common with other early ceramics of the Red-on-Buff horizon (see discussion below) rather than particularly diagnostic forms. Tlacuache ceramics predate the Tierras Largas phase, at least as it is currently categorized. As demonstrated in table 1.1 and table A1.10, carbonized food remains from the interior of a jar fragment found in the LC12 H midden returned an AMS radiocarbon date of 1880–1625 cal BC. Both the Tierras Largas and Tlacuache ceramics differ markedly from the more tecomate-emphasizing and highly decorated Barra tradition (Clark and Blake 1994).

Relatively little is known about Purrón phase (1900–1680 cal BC) ceramics from the Tehuacán Valley, as the assemblage is based on very few known artifacts (MacNeish et al. 1970:21–25). The timing of the onset of several central and southern Mexican Early Formative period pottery traditions (including Tlacuache, Barra, Purrón, and Espiridión) is remarkable, however. This trend seems to imply a broad shift toward village farming and the use of ceramic technologies in Mesoamerica. More specifically, similarities between coastal and highland undecorated, utilitarian cooking jars and hemispherical bowls, despite marked dissimilarity among the decorated vessels of these traditions, deserve some explanation. For example, contrast the decorated Tlacuache wares with those from the Tierras Largas phase (Flannery and Marcus 1994:figs. 8.22–8.27, 8.30, 8.31, and 8.34; Ramírez Urrea 1993:figs. 62–65). A tentative interpretation of this discrepancy is that utilitarian wares at La Consentida adhered to traditions of jar manufacture shared with communities in highland Oaxaca and Central Mexico. A Pacific coastal interaction sphere of the kind proposed by Isabel Kelly (1980; see also Anawalt 1998), and involving navigation along the Pacific coast in boats, could explain why the decorated wares share more in common with coastal traditions far to the west than they do with highland Oaxacan ceramics (see discussion below). Perhaps the occurrence of decorated serving vessels such as bowls and bottles in feasting middens (LC12 E-F16 through E-F9), rather than in cooking middens (LC12 H-F4), suggests that they were meant for public display employing motifs meaningful to visitors from nearby and distant coastal zones sharing a decorative tradition (see chapter 7 and appendix 2; see also Clark and Blake 1994).

While some of La Consentida's ceramics suggest interaction with the highlands and West Mexico, other vessels are of a style whose interregional affiliations are more difficult to trace. Small grater bowls from the site come in various forms, including as rounded conical bowls with flat bottoms and (more rarely), as conical bowls with pouring spouts, as square bowls, and as semispherical bowls (figure 8.3). Some examples exhibit considerable use-wear. The two most complete of these ashtray-sized vessels were recovered as offerings with children's burials. These grater bowls are not totally without counterparts in other regions; vessels bearing similar weaving-inspired incisions are found in highland Oaxaca (Flannery and Marcus 1994:figs. 12.142 and 12.143; Marcus and Flannery 1996:96) and at Cantón Corralito (Cheetham 2010:180). Robert Rosenswig (2010:157–59) discussed grater bowls from the Soconusco as a vessel type specific to the Middle Formative Conchas phase. Although Conchas phase grater bowls and Tlacuache grater bows share the feature of interior incisions, Conchas examples are larger, have more complex silhouette shapes, and sometimes lack incisions extending across the entire interior bottom of the vessel. This suggests that their uses may have differed. Also, unlike the Conchas phase examples, Tlacuache grater bowls lack applique supports. Geometric designs on Formative period vessels in various regions sometimes appear on the exterior of vessels, while Tlacuache grater bowls bear their incisions on the interior base and sometimes on the interior wall all the way to the rim. David Grove (1984:42, 80–81, 103) also discussed bowls with interior incising from Chalcatzingo. Notably, these bowls appear to have been decorative rather than utilitarian (at least as pertains to their incised designs) and have rounded bottoms. Like the Conchas phase examples, the Chalcatzingo vessels do not represent good analogs for La Consentida's grater bowls, which were in some cases extensively used to grate, and even to pour some substance or substances.

Formative period grater bowls with interior incising are found in highland Oaxaca, but are rare, executed in gray rather than brown paste, and occur in later phases such as San José and Guadalupe (see Flannery and Marcus 1994:figs. 12.74, 12.101). Perhaps the best-known examples of grater bowls among later Oaxacan ceramics can be found in the G-12 type of the Pe (500–100 BCE) and Nisa (100 BCE–200 CE) phases (see Caso et al. 1967:fig. 130b; A. Joyce 2010:150, 187, fig. 5.7c). Much later examples also occur in the Xoo phase (500–800 CE), and these again tend to be

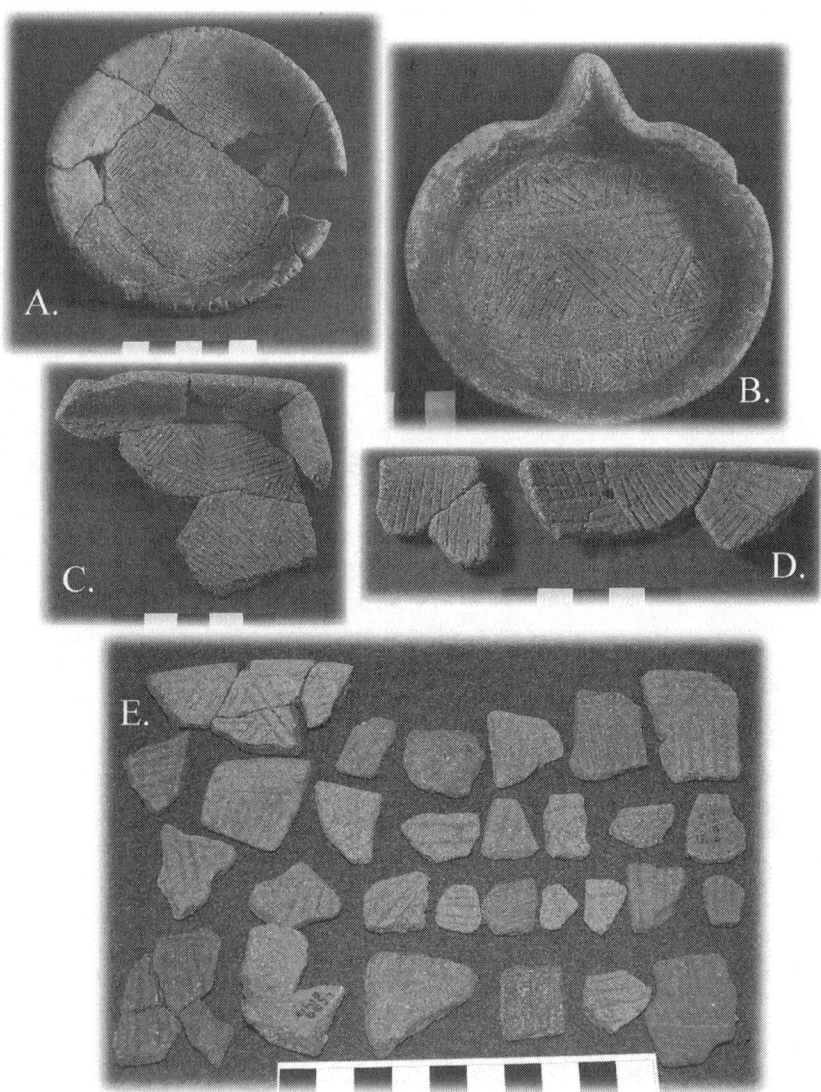

**FIGURE 8.3.** Remnants of grater bowls: (A) bowl with extensive use-wear recovered as offering with Burial B9-I111; (B) complete grater bowl with spout recovered as offering with Burial B11-I113; (C) partial square bowl from LC12 C-F2 domestic occupational debris at edge of Structure 1; (D) semispherical bowl rim from debris within Structure 2 domestic building; (E) grater bowl fragments from various contexts (mostly fill) in Ops. LC09 A and LC09 B

gray wares (Martínez López et al. 2000:165–66). The bowl shown in figure 8.3.A does have a similar pattern of rim notching to some nongrater hemispherical bowls of the Tierras Largas phase (Flannery and Marcus 1994:fig. 8.9). This vessel also seems to bear extensive use-wear. Some of the interior-incised bowls at Tlatilco (e.g., Piña Chan 1958:figs. 38.a and 38.b, see also geometric designs demonstrated in fig. 47) have incisions similar to Tlacuache grater bowls, though the examples from La Consentida appear smaller in diameter and lack tripod supports. As with the Tlatilco examples, Zohapilco bowls with interior incisions in geometric patterns (Niederberger 1976:lám. XXXVI, lám. XLV.22, lám. LI) are somewhat similar to Tlacuache grater bowls. Despite minor similarities with vessels from other regions, the La Consentida grater bowls seem to be relatively distinctive. Regarding relationships with West Mexican ceramics (see discussion below), I. Kelly (1980:31) pointed out that both the Capacha and El Opeño phase assemblages lack grater bowls.

Given their consistencies in form and in the placement of their interior incisions, it appears that the Tlacuache grater bowls served some food-processing or crafting need. Because the complex, incised designs in these bowls were carefully executed (likely drawing inspiration from the geometric patterns of woven basketry), their formal qualities or artistic value must also have been significant. Because two examples (figures 8.3.A and 8.3.B) were recovered with child burials, it seems possible that grater bowls were used to process food for weaning infants.[3] This interpretation is consistent with the age of onset for linear enamel hypoplasias in the La Consentida population, which appear to mark the advent of dietary stress at roughly three years of age (Hepp et al. 2017). Absorbed residue analysis may be the only way to definitively identify the uses of the grater bowls, however (see Morell-Hart et al. 2014; Seinfeld et al. 2009). Also, future investigations of nearby probable Early Formative period sites such as Cabeza de Vaca may indicate whether grater bowls are represented in the ceramic collections of other early coastal communities.

The Ojochi (1750–1550 cal BC) and Bajío phase ceramics of the Gulf Coast region are combined into a single phase by some researchers (e.g., P. Arnold 2003; Rodríguez Martínez and Ortiz C. 1997:82). These ceramics include long-necked bottles, worked sherd discs, zoned and impressed banded decoration, globular jars, decorated tecomates, and possible phytomorphic vessels that are generally similar to a few of the Tlacuache assemblage types (see YPM ANT 255088, 255093, 255099, 255101, 255105, 255109; 255207, 255221; appendix 2).[4] Terry Powis and colleagues (2011:8597, 8599) noted that an Ojochi phase bottle and a Bajío phase "necked jar" and "open bowl" tested positive for cacao use. Bottles emulating plants such as squashes, bearing "geometric designs painted in red," and which were used for cacao consumption have also been identified in the coastal Honduran Ocotillo phase (Joyce and Henderson 2007:645). Bottles from La Consentida generally resemble these cacao vessels, though residue analysis is necessary to identify their uses. As demonstrated by Powis and colleagues (2007, 2008), some early Soconusco ceramics were also used for cacao, emphasizing that chocolate was widely consumed in the Early Formative regardless of specific vessel types used to contain it.

Additional parallels between the La Consentida ceramics and those from other Early Formative sites are evident. The probable effigy vessel shown in figure 7.9.B, for instance, appears to be similar to one discussed by Román Piña Chan (1958:32) from a burial at Tlatilco. A few bottles from Tlatilco (e.g., Piña Chan 1958:fig. 34.i, j, k; fig. 35.v, w; fig. 37.ñ, o, p, r, s; fig. 39.y, z, a1, b1, fig. 43.r, fig. 44.k, fig. 46.f) also resemble Tlacuache bottles. Though some of the bold geometric designs of the Tlatilco decorated wares are reminiscent of those from the Tlacuache phase, the Olmec-inspired iconography found on some of the Tlatilco vessels is absent at La Consentida, as the site's entire occupation appears to predate the Olmecoid horizon. Similar bottles were also recovered at Zohapilco, and likely date to the Manantial phase (approx. 1250–1050 cal BC) (Niederberger 1976:lám. XXXVI.11, 12).

## CERAMIC EVIDENCE FOR A PACIFIC COASTAL INTERACTION AREA

Styles of ceramic decoration identified among La Consentida's Tlacuache phase ceramics imply interaction with regions as distant as West Mexico. Decades ago, Isabel Kelly (1980:37) suggested that archaeologists should focus more attention on what she believed was a Formative period Pacific coastal interaction sphere in western Mesoamerica, which perhaps brought ceramic technology and decorative inspiration northward out of lower Central and South America. Citing evidence for a broadly dispersed ceramic tradition with ties to the Capacha phase, Kelly believed that West Mexican decorative motifs likely had Formative period counterparts in coastal zones further to the southeast. She noted (1980:37) that Capacha may have been merely one of several "landfalls along the Pacific coast" for this tradition, and that a lack of information for early deposits in other coastal areas (such as Oaxaca and Guerrero) represented a challenge to understanding that potential interaction network. A few of the decorative elements visible in La Consentida's Tlacuache ceramics may be related to this poorly defined western macrotradition, which includes the slightly later Capacha and El Opeño assemblages.

Although decorated ceramics are fairly rare at La Consentida, midden, sheet midden, and eroded/redeposited midden contexts (e.g., LC09 B-F17, LC12 A-F7, LC12 D-F10 through F8, and LC12 E-F16 through F9) have provided a good sample of the various styles of decoration at the site (see chapter 4, appendix 2, and especially table A2.28). One of the most compelling pieces of evidence for including La Consentida in a broad Pacific coastal interaction area with distant western traditions can be found in the "sunburst" decorations on some vessels of Colima's Capacha phase bottles and jars (e.g., figure 8.4) and on several decorated fragments from La Consentida (figures 8.5 and 8.6). At La Consentida, sunburst designs appear on probable bottles, as is most clear in the example pictured in figure 8.5 (for an illustration of this vessel fragment, see figure A2.8.C). While different in form than the elaborate "stirrup" bottles of the Capacha phase, decorated Tlacuache bottles nonetheless bear a strikingly similar design to some of the Capacha wares (see I. Kelly 1974, 1980:figs. 15–19, 21, 24, 25; Mountjoy 1994, 1998:fig. 2). The Capacha phase stirrup bottles may come from later Middle Formative period deposits, and often lack good contextual information due to extensive looting of tombs and other burials (I. Kelly 1974, 1980; Mountjoy 1994). While the most elaborate forms are not recognized in the Tlacuache collection, a few fragments from composite silhouette or "belted" vessels, such as the sherd pictured in figure 8.7.A, indicate that more complex vessel forms existed at La Consentida but are not well understood due to fragmentation and small sample sizes. The impressed, teardrop-shaped dots or dashes visible in the figure 8.7.B conical bowl base fragment are also similar to some of the Capacha designs and to Middle Formative ceramics from the Jalisco's Mascota Valley (see I. Kelly 1974, 1980:figs. 18, 21, 26, 29; Mountjoy 2012:figs. 119, 280).

Recently, Mountjoy (1994, 2006, personal communication, 2015) has voiced skepticism regarding the early dates originally attributed to Capacha by I. Kelly and has suggested that the phase belongs to the Middle Formative period. Kelly (1974, 1980:4, 18–19) herself described the dismal conditions under which the carbon dating for the phase was secured. Mountjoy (personal communication, 2015) agrees that similarities between the Tlacuache and Capacha "sunburst designs" are suggestive of possible interaction between the two regions. La Consentida's early dates indicate that a direct association between Capacha and Tlacuache is unlikely, even if Kelly's initial dates are accepted without Mountjoy's modifications. Similarly, the West Mexican El Opeño phase seems to postdate occupations at La Consentida (Oliveros Morales and de los Ríos 1993). I am not suggesting that La Consentida ceramics represent direct contact with or importation of ceramics from West Mexico, or vice-versa. Rather, I agree with Kelly (1980:37; see also Anawalt 1998) that certain decorative styles among Pacific coastal traditions beg further investigation into a possible exchange and interaction network including these distant regions and possibly serving as examples of early Red-on-Buff ceramics (Clark 1991; Winter 1992). The earliest ceramics from much of Pacific coastal Mesoamerica (west of the Isthmus of Tehuantepec) are poorly understood, and it may be that a more systematic study of them would indicate that ceramic traditions in the intervening areas between Oaxaca and West Mexico share even more in common with the Tlacuache phase (see Brush 1965, 1969; I. Kelly 1980; Mountjoy 1994; E. Williams 2007).

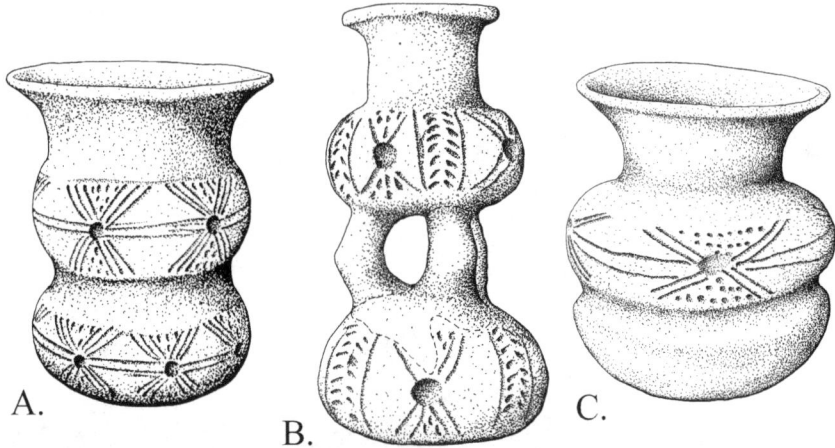

**FIGURE 8.4.** Capacha phase vessels: (A) belted jar—redrawn from Mountjoy (1994:40); (B) "stirrup" or "double" jar—redrawn from Mountjoy (1994:41); (C) belted jar—redrawn from Schmidt Schoenberg (2006). No scales available

**FIGURE 8.5.** Decorated probable bottle fragment from LC12 A-F7 eroded/redeposited midden (also see Figure A2.8.C)

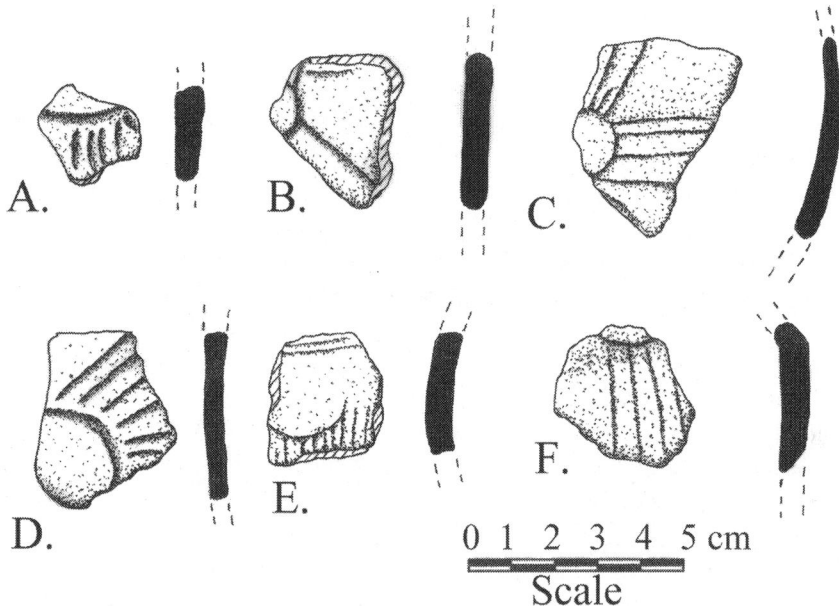

**FIGURE 8.6.** Other decorated fragments with sunburst-like designs: (A) from Structure 2 domestic area; (B) from LC09 B-F14 fill near LC09 B-F15 hearth; (C) from LC09 B-F14 fill near burials and LC09 B-F15 hearth; (D) from Structure 2 domestic area; (E) from Structure 2 domestic area; (F) from fill near burials in Op. LC09 B

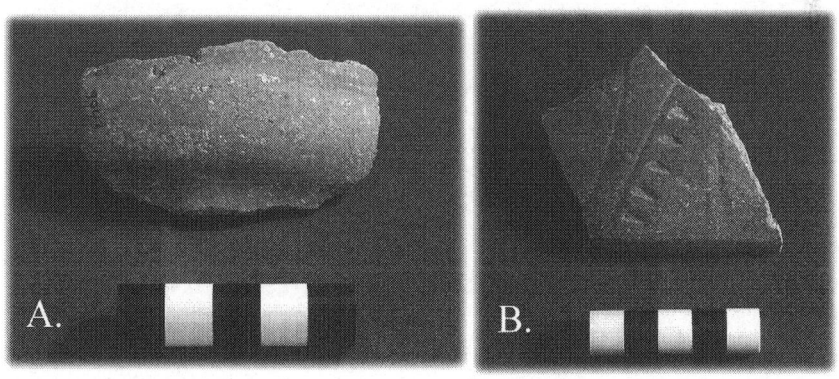

**FIGURE 8.7.** Potentially diagnostic ceramic fragments: (A) portion of composite silhouette or belted bottle or jar from burial B11-I13, bears bands of red slip or paint; (B) decorated conical bowl base fragment from LC09 B-F17 midden

Although the Tlacuache sunburst motif is similar to that found on some Capacha wares, a more general similarity can be seen among the simple, bold, geometric and impressed decorative style of the La Consentida vessels and those of both the Capacha and El Opeño phases (e.g., I. Kelly 1980:fig. 30; Mountjoy 1994; Oliveros Morales 1974; E. Williams 2007). Unfortunately, the friable nature of the sandy medium brown paste from which many of the finest decorated Tlacuache wares were constructed means that sherds tend to be small and are often eroded, leaving designs rarely visible in their entirety. Nevertheless, when they are at least somewhat well preserved, these vessels (e.g., figures 8.5–8.10) are notable for their finely slipped and burnished surfaces and geometric, impressed designs. The decorative motifs seem to have more in common with West Mexican ceramics than they do with Barra phase (Clark and Blake 1994) or Tierras Largas phase wares (Flannery and Marcus 1994).

Patterns of ceramic decoration and general vessel form also indicate some intriguing similarities with areas even more distant than West Mexico. James Ford's (1969; see also Anawalt 1998) extensive comparison of Formative cultures in the New World provides some useful points of comparison between the La Consentida artifacts and those of other early traditions in the Americas. Decorated sherds from Valdivia, for example, bear a resemblance to some of the La Consentida ceramics (Ford 1969:fig. 14; figs. 8.10.D and 8.10.E). Bottles from early Machalilla contexts in Ecuador and the Tehuacán Valley's Ajalpan phase deposits also appear similar to the La Consentida examples (Ford 1969:fig. 18.i, chart 16; MacNeish et al. 1970:figs. 6, 26; see figure 8.11 and appendix 2). Patricia Rieff Anawalt (1998) summarized the evidence for contact between West Mexico and Ecuador during Early Formative through Postclassic times, which may include patterns of attire and figurine iconography, in addition to ceramics. Given the available data, it is not possible to make strong claims about possible connections between La Consentida and distant areas such as Machalilla and Valdivia, though I. Kelly (1980) found such potential crossties intriguing.

## OBSIDIAN TRADE

Following the 2009 excavations at La Consentida, forty obsidian artifacts from Ops. LC09 A and LC09 B were selected for sourcing analysis (Hepp 2011b; D. Williams 2012:92–97). These artifacts came from Early Formative period fill, redeposited midden, and burial fill contexts, as well as from near the LC09 A-F4 and LC09 B-F15 hearths. These samples were submitted for X-ray Fluorescence (XRF) analysis at the University of Missouri Archaeometry Laboratory (MURR) (Glascock 2011; Hepp 2011b, 2014; D. Williams 2012:97; see figure 8.12.A).[5] The

**FIGURE 8.8.** Burnished or decorated conical bowl fragments from LC12 E-F9 and E-F11 midden

**FIGURE 8.9.** Decorated bottle fragments: (A) from near LC12 A-F4-s1 and A-F3 fill interface; (B) partial bottle fragment from top of LC12 E-F9-s1 midden

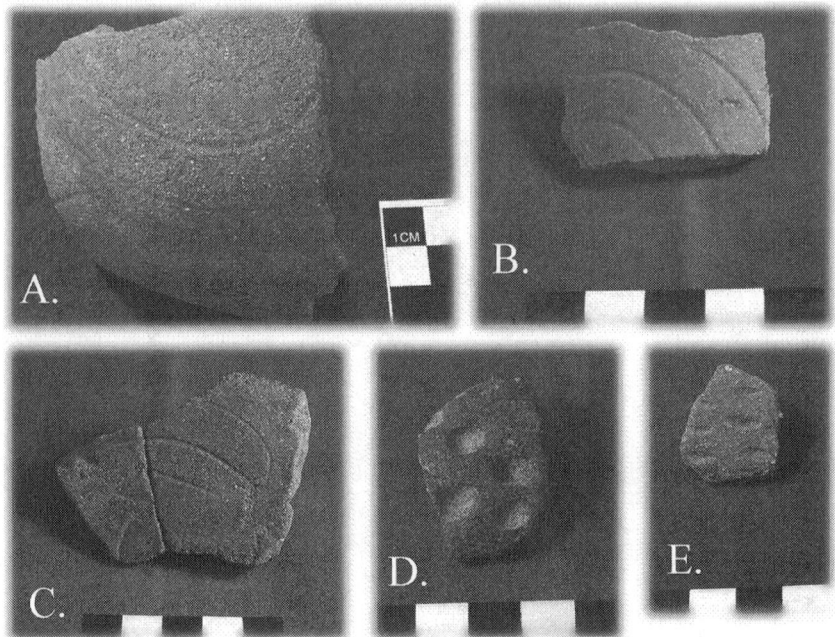

**FIGURE 8.10.** Miscellaneous decorated ceramics: (A) probable bottle base from burial B2-I3; (B) probable bottle neck fragment from LC12 A-F18-s2 deep fill context; (C) possible platter or dish fragment from domestic occupational layer at transition between LC12 C-F2 and C-F7 fill deposits; (D) sherd with impressed dots from LC12 C-F8 fill near domestic occupation layers; (E) sherd with impressed dashes from LC12 G-F16 fill just outside Structure 2.

results of this XRF study are consistent with an analysis of five pieces of obsidian collected during test excavations at La Consentida in 1988 (Joyce et al. 1995). Figure 8.12.B summarizes the sources of the total of forty-five samples analyzed by these two studies. These XRF data indicate La Consentida's involvement in an extensive trade network stretching to Central Mexico and even to areas near the Gulf Coast (figure 8.13). The results also provide an opportunity for comparison with studies of obsidian elsewhere in Oaxaca. In the Nochixtlán Valley, for example, Jeffrey Blomster and Michael Glascock (2010:189) determined that somewhat later Early Formative communities imported their obsidian from several sources, including Paredón, Otumba, Guadalupe Victoria, El Chayal, and Ixtepeque.

Notably, the lack of obsidian imported from Central America indicates that La Consentida had different interregional relationships than did communities in the

**FIGURE 8.11.** Partially reconstructed bottle recovered as offering with burial B2-I3

Mixteca Alta, the Valley of Oaxaca, the southern Isthmus of Tehuantepec, or the Soconusco region of Chiapas and Guatemala during the Early Formative period (Blomster and Glascock 2010:189; Clark and Salcedo Romero 1989; Pires-Ferreira 1978, 2009:293; Zeitlin 1982). Robert Zeitlin (1982:266–67) found that obsidian in use in the southern Isthmus of Tehuantepec during the Early Formative included material from Guadalupe Victoria and the Guatemalan source of El Chayal. Blomster and Glascock (2010:189) demonstrated that Cruz A and Cruz B phase communities in the Mixteca Alta imported up to 5 percent of their obsidian from El Chayal. In highland Oaxaca, Blomster and Glascock (2010:192) noted a transition away from the Early Formative period use of the "low quality" Guadalupe Victoria obsidian and toward an emphasis on central Mexican sources such as Paredón in later Formative times. The greater emphasis demonstrated at La Consentida for Guadalupe Victoria over Paredón material is therefore consistent with the site's early date. Blomster and Glascock (2010:192) also noted discrepancies among regions of highland Oaxaca, where sites in the Nochixtlán

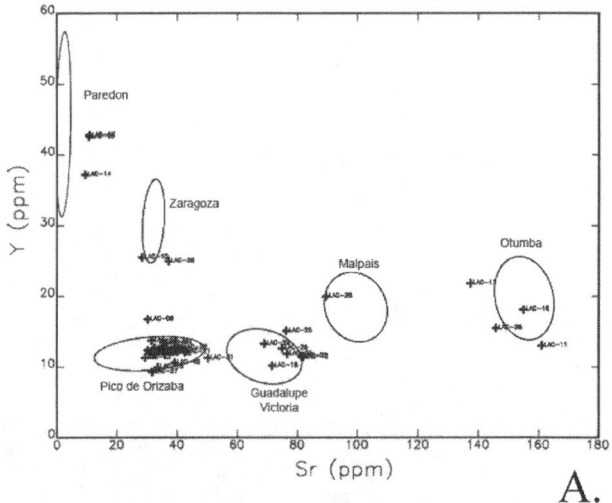

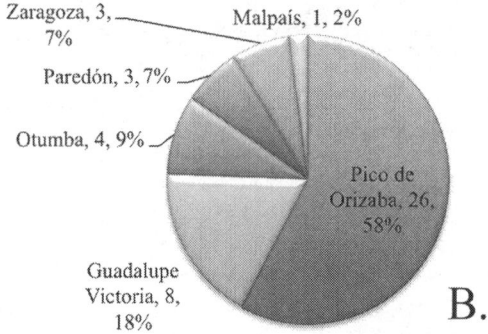

**FIGURE 8.12.** Results of X-ray fluorescence analysis indicating the use of six obsidian sources: (A) results from forty obsidian samples from 2009 excavations. Plot indicates Y (Yttrium) versus Sr (Strontium), and ellipses indicate discrete sources with 90 percent confidence; (B) pie chart summarizing results from forty-five obsidian samples, including five previously published by A. Joyce and colleagues (1995:6)

Valley used little West Mexican material (such as that from Ucaréo) while Valley of Oaxaca sites such as San José Mogote used more obsidian from western sources in addition to that from Zaragoza and Otumba. The lack of West Mexican obsidian at La Consentida is intriguing given styles of ceramic decoration discussed

**FIGURE 8.13.** Map with locations of La Consentida and the site's six known obsidian sources

above, which suggest that the regions were somehow in contact and shared elements of decorative style.

## OTHER IMPORTED MATERIALS AND FOREIGN-INSPIRED ICONOGRAPHY

Although the best data for La Consentida's networks of interregional interaction come from ceramic style comparisons and obsidian sourcing, it is worth making brief mention of some additional indications of connections with other areas. Greenstone beads recovered at the site (figures 7.11.D–7.11.F, table 7.1) represent probable prestige items traded among people in numerous regions of Mesoamerica during the Formative period (Carballo 2009:492; A. Joyce 1991a:141, 2013b:24; Tremain 2014). It is not yet clear whether La Consentida's greenstone comes in the form of jadeite, serpentine, or some combination of materials, but figures 7.11.D–7.11.F demonstrate its considerable variability in terms of color and texture. Some greenish stone items recovered at La Consentida may also be made of fine-grained basalt. Although greenstone distributions recorded thus far at La Consentida do not easily lend themselves to discussions of hierarchical social inequality, the presence of apparently diverse stone types suggests down-the-line interaction with distant regions such as Central Mexico, the Gulf Coast, and Guatemala (Gendron et al. 2002; Pool 2013; Reilly 1995). Other worked stone—such as small, one-handed manos from La Consentida—are similar to those at Zohapilco (Niederberger 1976:lám.

XXVIII.2, XXIX.1) and Tierras Largas phase sites in the Valley of Oaxaca (Winter and Sánchez Santiago 2014a:10–11; see chapters 5 and 6). La Consentida's ground stone is generally similar to pieces from Archaic and Early Formative period contexts in the Tehuacán Valley, particularly in the forms of manos (e.g., MacNeish et al. 1967:fig. 89) and mortars and metates (MacNeish et al. 1967:figs. 96, 98, 99). I do not suggest that manos were imported to La Consentida, but rather that they demonstrate stylistic and perhaps functional similarities with those from elsewhere.

Figurines and musical instruments may also indicate La Consentida's relationships with distant regions. I have already discussed one figurine (figure 7.7.A) that resembles Cruz A examples from the Mixteca Alta (Jeffrey Blomster, personal communication, 2015). At Zohapilco, Niederberger (1976:lám. II.16–18) found ceramic avian artifacts from various Formative period phases that are similar to La Consentida's bird imagery (e.g., figures 7.16–7.18). As discussed in chapter 7, one of the earliest anthropomorphic figurines at Zohapilco (Niederberger 1976:lám. XCV, Foto 16) perhaps shares stylistic similarities with La Consentida's simplest anthropomorphs (figures 7.4.A and 7.4.B). Another figurine, which appears to represent a monkey, is pictured in figure 8.14. The shape of this artifact's head is consistent with that of New World primates such as capuchins and spider monkeys (Marroig and Cheverud 2005:fig. 2). A recent study (Ortiz-Martínez and Rico-Gray 2007) has suggested that spider monkeys today sometimes live as far north as the southern Isthmus of Tehuantepec. People of the western Oaxaca coast may have seen monkeys in nearby regions, been aware of monkeys elsewhere, or imported monkeys or monkey skins from outside the area. Based on the paste of this figurine, there is no reason to suspect that it was imported.

## SUMMARY

As I have mentioned elsewhere in this book, one significant interpretation that arises from La Consentida's early dates relates to current explanations for how ceramics originated in Mesoamerica. Clark (e.g., Clark and Blake 1994) has argued that some of Mesoamerica's earliest ceramics arrived as a fully realized technological and stylistic tradition from Central America. On the basis of carbon dates recovered in context with Tlacuache sherds, ceramics from at La Consentida may represent the first well-dated examples of a ceramic tradition contemporary with the Barra phase but formally dissimilar to it. I suggest that early ceramics of western Mesoamerica—including Tlacuache, Tierras Largas, Purrón, and West Mexican phases such as Capacha and El Opeño—exemplify what other archaeologists have termed the Red-on-Buff horizon (see Clark 1991; Winter 1992; Winter and Sánchez Santiago 2014b). This interpretation explains why La Consentida's

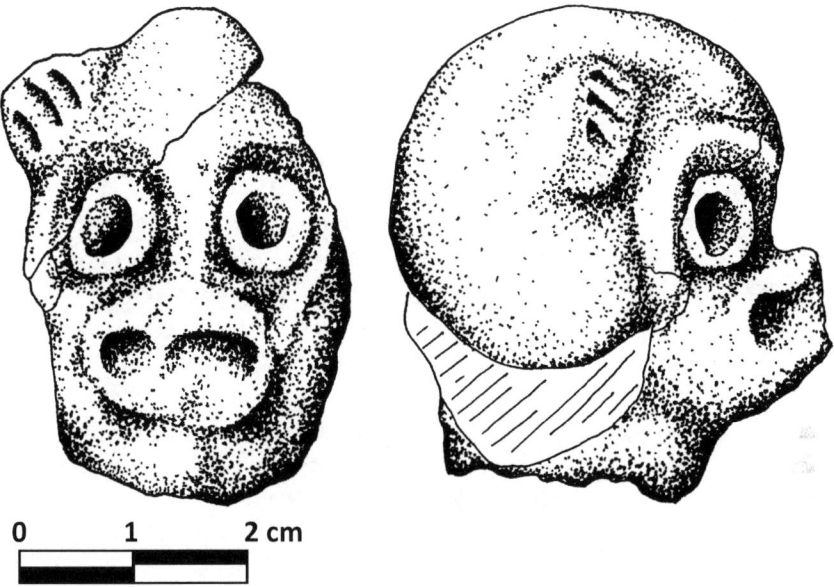

**FIGURE 8.14.** Probable monkey figurine fragment from occupational debris above Structure 2 domestic building

Tlacuache ceramics share little in common with the tecomate-emphasizing Barra phase (Clark and Blake 1994). In general, such marked differences between western Mesoamerican Red-on-Buff ceramics and the Locona horizon (i.e., Barra and Locona phase) ceramics of the Soconusco may represent ancient cultural and linguistic divides between speakers of Otomanguean and Mixe-Zoquean languages, as well as independent origins of ceramic traditions themselves (see Clark 1991; Josserand et al. 1984; Winter and Sánchez Santiago 2014b). Based on the available evidence, including the AMS radiocarbon dates from secure contexts as discussed in table 1.1, Tlacuache ceramics appear to be among the earliest known examples of the Red-on-Buff horizon in Mesoamerica.

The exchange of and interregional influence regarding ceramic styles, greenstone, iconographic imagery, and obsidian suggest a complex network of interregional relationships in which La Consentida was involved. At times, these probable exchange routes differ from one another. X-ray fluorescence sourcing has determined that La Consentida's obsidian was imported from Central Mexican sources, some of which are near the Gulf Coast. The lack of West Mexican and Central American obsidian sets the site apart from some of its Early Formative period contemporaries and sites occupied shortly thereafter (Blomster and Glascock 2010; Clark and Salcedo

Romero 1989; Zeitlin 1982). This pattern seems at odds with ceramic decoration styles that appear to have the most in common with West Mexico, as well as imported greenstone that may have come from Central America. What these various lines of evidence do clearly suggest is that La Consentida was a partner in broad interaction networks of the Early Formative period. It is not yet clear what goods La Consentida exported in exchange for its obsidian, though research in the areas surrounding the site is beginning to provide promising results. For example, Gracie Lock and colleagues (2014; see also Goman et al. 2005) noted that carbon dates in the salt flats adjacent to La Consentida suggest possible Early Formative salt procurement. Salt may have been a valuable trade good for exchange with networks providing imported obsidian and greenstone. As Richard Lesure and Thomas Wake (2011:84) argued, tecomates recovered at Early Formative period coastal sites may have been used for salt production. While tecomates are rare at La Consentida in comparison to early Soconusco sites, they are nonetheless present, and may have played a role in the site's resource exchange relationships. As the domestic and mortuary contexts excavated at La Consentida demonstrate, however, this was not a simple resource collection site for goods such as salt and shellfish. Instead, La Consentida represents the remains of an early village with an increasingly permanent settlement established during the Early Formative period.

Isabel Kelly (1980:37; see also Anawalt 1998) believed that the Capacha phase assemblage corroborated a hypothesis of James Ford (1969:166), who argued that the early ceramics of Pacific coastal Mexico should have more in common with early South American pottery from "Puerto Hormiga, Machalilla, or Valdivia," than with the early traditions of Central Mexico, such as that of the Tehuacán Valley. Kelly (1980:37) wrote that the sunburst motif appeared to be unique to Capacha, but as I have already discussed, she did predict the possible existence of other "landfalls" of this decorative tradition along the Pacific coast. I believe that the presence of the sunburst motif at La Consentida suggests that Kelly's predictions about a Pacific coastal interaction sphere need to be revisited. It may be that two contemporaneous ceramic traditions vied for influence in Early Formative Mesoamerica. This supports the model that the Locona ceramic horizon (exemplified by the Barra and Locona phases and coming out of Central America via the Soconusco) met with a contemporaneous Red-on-Buff horizon that included the Tlacuache, Tierras Largas, Purrón, and other western ceramic traditions and emphasized the use of jars, bowls, and bottles over that of tecomates (Clark 1991; MacNeish et al. 1970:21–25; Winter 1992; Winter and Sánchez Santiago 2014b; see appendix 2).

It is worthwhile, I think, to make a final point about identifying ancient networks of interaction and establishing chronologies on the basis of similarities in particular artifact classes. As discussed in this chapter, there are numerous similarities in ceramic styles between La Consentida's Tlacuache phase and those of other regions such as the Valley of Oaxaca, Central Mexico, and West Mexico. None of these other phases, however, contains all of the vessel forms and decorative styles identified in the Tlacuache assemblage. This finding serves as a warning against facile associations between the Tlacuache complex and other traditions such as Tierras Largas. Numerous Early Formative period pottery traditions exemplifying the Red-on-Buff horizon include similar styles of jars, bottles, hemispherical kidney-shaped bowls, and interior-incised bowls (Clark 1991; García Cook and Merino Carrión 2005; MacNeish et al. 1970:fig. 6; Winter 1992). Rather than indicating direct ties between the Tlacuache and Tierras Largas ceramic traditions (for example), these stylistic similarities indicate broad patterns of interaction and exchange across large geographic areas during the Early Formative period (see the proposed ceramic interaction map in Clark 1991:fig 8). Perhaps most significant, the ceramics from La Consentida appear consistent with the presence of two initial Early Formative period ceramic traditions (Locona and Red-on-Buff), the former coming north from Central America via the Soconusco and represented by the Barra phase, and the latter developing in or arriving at western Mesoamerica and exemplified by the Tlacuache complex as one of its earliest known variants (Clark 1991; Winter 1992).

# 9

## La Consentida

*A Community in Transformation*

In the preceding chapters and in the appendices to follow, I have sought to accomplish two main goals. First, I have summarized the results of several seasons of fieldwork and laboratory analyses focused on La Consentida. Second, this book has been designed to address the central research question of the LCAP: *what were the nature of and relationships between practices of mobility, subsistence, and social organization at La Consentida during the Early Formative period*? I dedicated separate chapters to each component of this question. These chapters have often referred to one another, emphasizing the connectedness of the different aspects of the research question. Because of the segmented organization employed here, however, I have thus far not dedicated sufficient attention to discussing how domestic mobility, subsistence, and social change were interrelated. In this chapter, I will briefly summarize the main results of this study and pay special attention to the relationships among different aspects of socioeconomic transformation at the site.

### KEY CHRONOLOGICAL IMPLICATIONS

La Consentida's AMS radiocarbon dates (table 1.1, figure 1.2) indicate its very early chronological position (relative to other Mesoamerican sites) for hallmarks of the Formative period such as ceramic vessels, mounded earthen architecture, and cemeteries (see also table1.2). Ceramics from La Consentida appear to be among the earliest in Mesoamerica. The site's ceramic assemblage is not redundant with any previously identified pottery tradition and therefore justifies the establishment of a new phase and ceramic complex, which I am calling Tlacuache (see chapter 8 and appendix 2). One of the most intriguing aspects of this very early pottery is that it complicates existing models for the arrival of ceramic technologies to Mesoamerica (e.g., Clark and Blake 1994). While I do not argue that the Tlacuache phase supplants the Barra phase (1900–1700 cal BC) as (perhaps) Mesoamerica's oldest ceramic tradition, I do contend that the secure carbon dates associated with pottery from La Consentida demonstrate that Tlacuache ceramics are *at least as early* as the Barra phase. Some of the strongest evidence for this argument includes sherds directly associated with the LC09 A-F4 hearth (1900–1690 cal BC) and a sherd with dated carbonized food (1880–1625 cal BC), presumably from a seasonal species, adhering to a jar fragment from the LC12 H midden (see table 1.1). The first ceramics of the Soconusco region and those of the western Pacific coast may have begun influencing Mesoamerican ceramic traditions at about the same time, despite the marked formal dissimilarities between them and their likely different points of origin.

Styles of ceramic decoration provide tantalizing clues that La Consentida may represent a very early example of the Red-on-Buff horizon (Clark 1991; Winter 1992:27–28). The Tlacuache phase may also represent one component of a little-known Pacific coastal interaction sphere also exemplified by slightly later Formative period ceramic traditions as distant as those of the Capacha and El Opeño phases and may even bear evidence of influence from South America (Anawalt 1998; Ford 1969:37; I. Kelly 1980:166; Mountjoy 1994; Oliveros Morales and de los Ríos 1993; E. Williams 2007). Styles of more "utilitarian" ceramics such as undecorated globular jars and hemispherical bowls suggest that La Consentida maintained affiliations with its highland neighbors, such as those in the Valleys of Oaxaca and Tehuacán (see chapter 8 and appendix 2). Other ceramic similarities (such as those with Tlatilco, Zohapilco, the Ojochi/Bajío phase, and the Ocotillo phase of Honduras) contextualize Tlacuache ceramics within a broader Mesoamerican interaction sphere likely related to the Red-on-Buff horizon and perhaps to the Otomanguean language area (P. Arnold 2003; Clark 1991; Clark and Blake 1994; Flannery and Marcus 1994; Ford 1969; R. Joyce and Henderson 2007; I. Kelly 1980; Niederberger 1976; Piña Chan 1958; Ramírez Urrea 1993; Rodríguez Martínez and Ortiz C. 1997:82; Winter 1992; YPM ANT 255207).[1] Imported greenstone and possibly foreign-inspired figurine imagery also evince the broad nature of La Consentida's interregional connections. Obsidian-sourcing data demonstrate that the site participated in an expansive trade network reaching as far as Central Mexico and regions near the Gulf Coast (Glascock 2011; Hepp 2011b; D. Williams 2012).

Mounded earthen architecture at La Consentida may predate any other currently known examples from Mesoamerica (see chapters 1, 4, and 5). Based on changes over time in the thickness and form of earthen construction layers, the community appears to have increased its emphasis on building mounded earthen architecture after an initial period of relatively modest labor efforts. Evidence for domestic structures in the later deposits demonstrates that the people living at the site shortly before its abandonment were building houses, at least some of which contained recycled stone metate fragments, perhaps as part of their foundations or walls. Population and labor estimates indicate that a group of about eighty people, of whom probably half or slightly fewer were capable of heavy labor at a given time, could have constructed all of La Consentida's architecture in 250 years or less. Labor estimates for a hypothesized early version of Platform 1 demonstrate that the site probably saw its first phases of earthen architectural construction and occupation over a relatively brief period of time (see chapter 5).

Ceramic and ground stone data indicate that La Consentida's occupants modified their food-processing technologies over the course of site occupation. They transitioned from more portable, multipurpose tools early in site's history to heavier tools with apparently more specialized purposes by closer to site abandonment (see chapters 5 and 6 and especially figure 5.7). This transition from portable to nonportable ground stone technology is consistent not only with the adoption of sedentism but also with transitions in food-processing practices (see P. Arnold 2009). Human skeletal remains show evidence of increasing dental attrition over time, which is consistent with the adoption of a grainy diet processed with stone manos and metates (Hepp et al. 2017; chapter 6). This pattern was concurrent with stable isotopic signatures in human teeth suggesting that the community had a transitional diet that incorporated maize and that may have included more maize consumption than that of contemporaneous Soconusco and Gulf Coast populations (e.g., Blake et al. 1992; Chisholm and Blake 2006; Killion 2013). The increasing use of nonportable ground stone at the site was concurrent with an apparent decline in the use of ceramic bottles (see appendix 2). These bottles were perhaps used to serve beverages at feasts, as evidenced by their decoration and recovery context in probable feasting middens. These patterns suggest that culinary preferences at the site changed over time. Specifically, the community may have shifted from consuming maize in a nongrainy (i.e., perhaps liquid) form to the consumption of maize flour processed with ground stone manos and metates (Hepp et al. 2017; see also R. Joyce and Henderson 2007).

Several lines of evidence demonstrate the importance of complex heterarchical social distinctions in La Consentida community. Changes over time in the form of earthen fill strata suggest increasing sedentism and possibly growing labor organization (see chapters 4 and 5). Anthropomorphic figurines, possible effigy vessels, and masks suggest people fulfilling diverse roles such as dancers, musicians, and possibly ballplayers and ritual specialists. Such artifacts display diverse types of dress, adornment, and body modification that include headgear and costumes likely associated with specialized social roles (see chapter 7). Comparison of figurines from the site with those found in other Early Formative contexts (e.g., Guernsey 2012:121–22; Lesure 1999a) suggests that some may portray community leaders and/or elders. Remnants of actual jewelry and related artifacts, including stone and ceramic beads (figure 7.11), a probable bone needle perhaps used to fashion clothing (figure 7.12), and a possible hematite mirror fragment (figure 7.13) suggest that figurines sometimes reflected actual practices of dress and comportment, rather than just the prescription of social ideals or mere artistic fancy. The dancing and music implied by ceramic aerophones (e.g., figure 7.16), a possible dancer figurine (figure 7.19), and ceramic mask fragments (figure 7.20) further emphasize the roles of community members specializing in ritual and public performance (see also figures 7.1.C, 7.17, and 7.18) The recovery of a ceremonial cache (LC12 A-F15) containing shell, ceramic sherds, a bird ocarina, a fossil bull shark tooth, and the complete skeleton of a *Heloderma horridum* reptile suggests the presence of sophisticated ritual practice at the site, and perhaps thus the presence of a ritual specialist (figures 7.14 and 7.15). A comparison of the relative quantities of decorated ceramics recovered in several middens (table A2.28), in conjunction with faunal remains recovered in those middens, suggests public feasting (see appendix 2). As discussed by numerous scholars (e.g., Clark 2004a; Clark and Blake 1994; Hayden 1990, 1995, 2009; Hill and Clark 2001; R. Joyce and Henderson 2007), feasting was likely one of a suite of significant communal events integral in sowing the seeds of social inequality in transegalitarian societies, and decorated ceramics such as those of the Barra and Tlacuache phases likely played a role in the public displays such feasts entailed (see chapter 2). These various lines of evidence indicate that the La Consentida community was complex in a heterarchical sense but provide little support for an interpretation of ascribed hierarchical inequality. As discussed in chapter 2, previous research in Mesoamerica has tended to focus on formalized hierarchy rather than on heterarchy in the origins of complex social systems. In future research, I hope to more fully explore the ways in which heterarchically diverse communities such as La Consentida were vital in the origins of later social hierarchies in Mesoamerica.

## RELATIONSHIPS BETWEEN TRANSITIONS IN SETTLEMENT, DIET, AND SOCIAL ORGANIZATION

If there is one consistent theme suggested by the data presented in this book, I propose that it is *change*. Alterations over time in practices of domestic mobility are logically tied to concurrent shifts in subsistence. In fact, evidence for a change in one is often some of the best evidence for a change in the other. Some authors (e.g., P. Arnold 2009:404) have therefore chosen to discuss an Early Formative "settlement-subsistence strategy," rather than to separate these socioeconomic factors. As La Consentida's community members shifted toward the permanent occupation of the site, they also began to transform their food-processing technologies and the culinary choices behind the foods they ate, particularly regarding the processing of maize (see chapter 6). It may have been the removal of the limiting mechanism of domestic mobility as a size and weight constraint on technologies such as ground stone metates that prompted the population to turn its focus to more efficiently processing maize flour. Although perhaps an important part of the diet since the site was first occupied, the transition from consuming maize in probable liquid form to the processing and consumption of maize flour ground on stone manos and metates had significant impacts on material culture and dental health at La Consentida (see chapter 6). It is difficult, if not impossible, to know exactly which of these changes came first. The apparently late advent of metates at the site may indicate that the adoption of a diet based on maize flour came after the transition to sedentism. I would further argue, in part because other early traditions such as Tierras Largas also included metates (see Ramírez Urrea 1993:fig. 86) and because not all metates and grinding platforms were necessarily used strictly for maize processing (see P. Arnold 2009:404–5), that the mere presence of these artifacts does not constitute absolute, *a priori* evidence of a fully agricultural diet. In the La Consentida case, isotopic analysis of human teeth demonstrates that maize was consumed at the site from its earliest occupation and apparently that community members consumed more maize than did some other Early Formative coastal Mesoamericans. Increases in the size of ground stone tools and the presence of dental carries, mandibular abscesses, and dental attrition among later burials suggest that the La Consentida subsistence strategy became increasingly agricultural over time (see Hepp et al. 2017; chapter 6).

While perhaps easiest to distinguish from discussions of mobility and subsistence, evidence for social organization is also intimately related to these other topics. As the community became more sedentary and began to eat a more traditionally agricultural, flour-based diet, they may also have grown in population size. The argument that shifting to a more agricultural diet may lead to higher rates of reproduction (perhaps despite increasing morbidity per capita) is hardly new (see Binford 1968; Hodges 1987; Larsen 1987). To the already fundamental social reorganizations of living in permanent proximity to one's neighbors was added the need to find novel ways to facilitate peaceful coexistence and to organize community events such as labor for earthen architectural construction, dances, music, feasts, and ritual activities. The numerous anthropomorphic figurines at La Consentida bear testament to the desire to explore these new social dynamics. Through their use, the site's occupants sought to discover ways to distinguish certain identities while also remembering important ancestors and probably forming a more cohesive idea of community (sometimes referred to as *communitas*) than had been necessary or, in fact, even *possible* during the preceding Archaic period (see Hepp and Joyce 2013; Hill and Clark 2001; Turner 1969:131–65). The occurrence of musical instruments, ceramic masks, a figurine perhaps depicting shamanic transformation (figure 7.10.B), and a complex ritual offering (LC12 A-F15) suggests the presence of ritual practitioners who may have possessed the sorts of specialized knowledge attributed to Mesoamerican nobility in later times (Barber and Hepp 2012; Clark 1997; Hepp et al. 2014; Hepp and Joyce 2013; A. Joyce 2010:61–63; Schele and Freidel 1990; see chapter 7).

Transitions to living more permanently in a village community, to eating a more agricultural diet, and to fulfilling increasingly diverse social roles (heterarchically distinct and with perhaps some of the first glimmers of the hierarchical social inequality that would become more formalized in later Mesoamerican history) occurred more or less simultaneously. One could not have existed without the others. Determining what actually *caused* these changes could in some ways be the topic of an entirely different study, but I will attempt in the conclusion that follows to pull together some of the diverse strands of evidence presented in this book as a way to discuss these changes at La Consentida in a synthesized fashion.

## CONCLUDING THOUGHTS

As Lesure and Blake (2002) argued, the sweeping socioeconomic changes of the Early Formative period probably took on different forms in different areas, and we should thus expect the archaeology of this time to be a patchwork of diverse forms of evidence for those changes that varied for a wide range of social, economic, and ecological reasons. My impression of the evidence I have gathered for this book is that the Archaic-Formative period transition at La Consentida must have been cumulative and, to a certain degree, self-perpetuating. As a group changed its dietary practices, for example, they might also gradually change the tools they used to process their food, the cultural values behind how that food was consumed, and the amount of time they spent tending to it and nurturing it (particularly concerning domesticates). In increasingly permanent settlements where horticultural goods could be better tended, new social dynamics involving ever more permanent neighbors necessitated innovative modes of social interaction and community organization, with the production of anthropomorphic imagery as one avenue to negotiating novel relationships. In a different (hypothetical) Early Formative community, dissimilar ecological conditions may have meant that maize and other domesticates were less important resources, at least as subsistence staples. Such a community could thus remain semimobile until a later date (e.g., P. Arnold 2009). In some places, decorated pottery employed in public events such as feasts may have helped tie villagers to their communities and provided an avenue for the display and negotiation of social status (e.g., Clark and Blake 1994; Hayden 1990, 1995, 1998, 2009), while in others, semimobile populations may have retrofitted new ceramic technologies to the ancient practices of domestic mobility they maintained (P. Arnold 1999; Rosenswig 2015). While it is possible that communities remaining semimobile (after their neighbors had shifted to more dedicated sedentism) might also establish some brand of social complexity differently or at a later date than their neighbors, such an inference must be made cautiously. One thing we can learn from the archaeology of complex hunter-gatherers and builders of monumental constructions elsewhere in the world is that a group need not have a permanent address in order to construct sophisticated ideologies, ways of transmitting knowledge, edifices, subsistence regimes, and social distinctions (J. Arnold 1993, 1996; Banning 2011; M. Cohen 1985; Gibson 2000; L. Kelly 2015; Peters and Schmidt 2004). If the reader detects that I am hedging my bets when it comes to addressing issues of causation in Early Formative period social change, it is no accident. My speculative examples differ from some models proposed previously (e.g., Coe and Flannery 1967; Flannery 1968b, 1972a, 1986) because they do not rely on "prerequisites" or specific ecological conditions for social change, nor on a monolithic assumption that ceramics must always indicate sedentism (Clark and Blake 1994) or that Early Formative period individuals and social collectives lacked the foresight to see what some of their social machinations might imply for future generations (R. Joyce 2004a).

In other words, I am deliberately not proposing a new model for Early Formative period social change, precisely because a monocausal explanation seems inappropriate given diverse material records, social conditions, and ecological settings across Mesoamerica. Although a community in one area may have begun its transition toward the settled farming of the later Formative period through a shift to more maize consumption, as is suggested by diachronic study of isotopic indicators in coastal Oaxacan populations (A. Joyce et al. 2017), another group may have grown more sedentary and socially complex through the influence of a different set of ecological and social variables and the use of different cultigens such as manioc and malanga (see Sheets et al. 2012). The apparent lack of evidence for significant maize consumption in the Early Formative period Soconusco, when compared to La Consentida, confirms the need for flexible and multivariate explanations of social change. Finally, I do agree with Clark (Clark and Cheetham 2002; Clark et al.

2007) that the impetus for some of the major socioeconomic changes identified for the Early Formative should actually be sought in the late Archaic period. Nobody woke up in an Archaic period campsite on the morning of 2000 cal BC and decided it was time to begin the Early Formative period. Something about the social relationships, economies, and ecological contexts of the late Archaic precipitated massive (though perhaps gradual) socioeconomic change. The geographical breadth of these conditions and of their eventual implications for other aspects of village life is attested to by evidence for shared ceramic vessel forms and far-reaching exchange routes across Mesoamerica (see chapter 8). It is also likely that some transitions took the form of punctuated equilibria as socioeconomic thresholds were surpassed. Unfortunately, Archaic period sites in many parts of Mesoamerica, including coastal Oaxaca, are notoriously difficult to locate (Borejsza et al. 2014). The investigation of that portion of the region's history will have to wait until Archaic period sites are rediscovered and studied in detail. On the basis of available paleoecological data (e.g., Goman et al. 2005, 2013), however, I believe it will only be a matter of time.

La Consentida represents the traces of a community in transformation. Radiocarbon dates demonstrate that the site has some of the earliest ceramics, mounded earthen architecture, and cemeteries currently known in Mesoamerica. Evidence for the community's settlement practices—including that from ground stone tools, platform construction phases, and domestic architecture—suggests that its people may have been seasonally migratory when they founded the site and that they transitioned to dedicated sedentism thereafter. Dietary evidence from studies of ground stone, faunal remains, stable isotopes, and skeletal pathologies suggests that the community's subsistence practices transitioned from a broad-based diet including maize consumption (perhaps in beverage form) to a more agricultural one (emphasizing maize flour consumption) by the time of site abandonment. Mortuary and figurine data suggest that social distinctions among community members were real, though likely heterarchical and/or achieved in their form. Evidence for ascribed status differences in Mesoamerica by later in the Formative period implies that the first sparks of social differentiation at sites such as La Consentida ultimately led to more formalized social hierarchies (Clark and Blake 1994; R. Joyce 2004a). Indeed, evidence from La Consentida's earthen architectural sequences (chapters 4, 5, and 7), as well as from diverse forms of anthropomorphic iconography and other evidence of ritual specialization (chapter 7), suggests that La Consentida may have experienced nascent stages of hierarchical social inequality, though the evidence remains scant.

As a community transitioning from egalitarian to hierarchical social organization, the residents of La Consentida maintained interregional contacts to obtain utilitarian and prestige goods, participated in crafting activities (likely for their own subsistence as well as for export in exchange for goods such as obsidian and greenstone), practiced communal feasting utilizing decorated serving wares, and perhaps participated in ball games. Radiocarbon dates demonstrate that these practices were occurring very early in Mesoamerica's Formative period history. As I have asserted several times, La Consentida presents evidence for some of Mesoamerica's earliest well-dated pottery, mounded earthen architecture, and cemeteries. In future research, I will continue to explore the ramifications of the Tlacuache phase ceramics, which appear to represent an example of the Red-on-Buff horizon that was contemporaneous with the first pottery adopted in the Soconusco region (Clark 1991; Winter 1992:27–28). These two early ceramic horizons may serve as material evidence for macroregional patterns of southeastern and northwestern Mesoamerican cultural diversity (such as divisions between the Otomanguean and Mixe-Zoquean language families) that have long been the subject of research and speculation by archaeologists, linguists, and sociocultural anthropologists, but which remain poorly understood in terms of their ancient histories (e.g., Josserand et al. 1984; G. Lowe 1977; Winter and Sánchez Santiago 2014b).

In this book, I have sought to demonstrate that data from La Consentida have the potential to contribute to ongoing debates concerning the origins of and relationships between sedentism, agriculture, and social complexity in Mesoamerica. These issues are generally recognized as being among the most important research topics confronted by archaeologists worldwide. In fact, La Consentida may serve as an example of a "transegalitarian" society, which I described in chapter 2 (see Blake and Clark 1989, 1999; Clark 2004a; Hayden 1990, 1995, 2009).We may never find conclusive evidence to support a single model explaining these socioeconomic transitions, and perhaps that is in part because no one model can hope to encompass all of Mesoamerica's cultural and ecological diversity. Although I have resisted the temptation to propose my own argument for Early Formative social changes writ large, it seems appropriate to at least propose explanations for those changes at La Consentida. I believe that the available data regarding transformations in settlement practices, subsistence, and social organization at this early Mesoamerican village demonstrate the strengths of both ecological and agency models for social change. These data suggest to me that at La Consentida, an Early Formative shift toward food production was indeed an important component of social development, as was proposed decades ago by researchers such as Flannery (Coe and Flannery 1967; Flannery 1972b, 2002; Marcus and Flannery 1996). Perhaps the surplus resources afforded by agriculture and a relatively rich selection of natural coastal resources relieved some individuals from the need to participate in all subsistence concerns, and they thus were able to set about negotiating new social dynamics and establishing specialized skills and knowledge sets, which are demonstrated by evidence for ritual specialization at the site. This is not a new argument but is instead a very old idea in anthropology that can be traced back at least to the work of Gordon Childe (1936; see also Çilingiroğlu 2005; Weisdorf 2005), who saw the Neolithic as a "revolution" later described by archaeologists as implying a "Neolithic package" of material and social traits. If the reader will pardon the pun, I agree with the general consensus in modern archaeology that this "package" must be unpacked (e.g., Çilingiroğlu 2005; Thomas 1991). If we eliminate the concept of agriculture itself as somehow prerequisite to burgeoning social complexity (contra Coe and Flannery 1967), and instead consider all communities (including those of hunter-gatherers) in which surplus resources permit social competition, accumulation, and growing complexity, then perhaps this idea was put best by Brian Hayden (1990:33) thirty years ago:

> Among hunter/gatherers the eclipse of rigid egalitarianism and sharing that was brought about by the emergence of economic competition (made possible in turn by the effective exploitation of highly productive r-selected resources) is possibly the single most important development in cultural evolution in the last 2 million years. It can be linked to the emergence of food production, hierarchical societies, craft specialization, slavery, intensive warfare, and many other important cultural traits.

In other words, Coe and Flannery (1967) had it *partly* right. At La Consentida, a shift toward agriculture was key to social and settlement change, though these transformations could have been equally supported by the exploitation of natural resources in some other (hypothetical) ecological setting. The heterarchical social distinctions at La Consentida are probably equally significant for understanding how such early communities articulated with hierarchical social inequalities in later Mesoamerica. Evidence for interregional exchange lends support to other models (e.g., Drennan 1983; Flannery 1968a; Hirth 1978; Lesure 2004; Zeitlin 1979) for the importance of interaction and increasingly shared ideologies in the rise of Mesoamerica. To these concerns of subsistence and long-distance interaction must be added the significance of human agency at individual and collective scales, as articulated with the material word composed of the resources we use and the tools we make to exploit them, and as operating within the structure of diverse social contexts that helps to produce the infinitely complex nature of human action (e.g., Blake and Clark 1999; Bourdieu 1977; Clark 2004a, 2004b; Giddens 1979; Hayden 1990, 1995, 2009; Hodder 2012; R. Joyce 2004b; B. Olsen 2010; B. Olsen et al. 2012; Sewell 1992). In a sense, what the data from La Consentida tell me is not that Mesoamericanist studies need a new "meta-theory" for how social complexity in the region came about. Instead, among the many previous models for social change proposed in archaeology can be found the key elements for explaining the patchwork tapestry of material evidence (predicted by Lesure and Blake [2002]) for complex social processes among diverse populations in varied ecological contexts.

Continuing to refine our explanations for the Archaic–Early Formative transition is a worthwhile goal and one that we must strive to attain if we are to understand the fascinating social developments that took place throughout pre-Hispanic history. On an interregional and even continental scale, La Consentida represents the remains of an early community as it transitioned from an Archaic to a Formative period lifestyle, an event that represents a watershed moment in human history shared by ancient peoples across much of the New World and beyond (P. Arnold 2009; Blake and Clark 1999; Clark 1991, 2004a; Ford 1969; R. Joyce 2004a; I. Kelly 1980; Lesure and Wake 2011; Pool 2013).

# Appendix 1

## Description of Excavated Deposits

Excavations at La Consentida in 1988, 2009, and 2012 uncovered natural and cultural strata resulting from the preoccupational and occupational history of the site. In this appendix, I describe these excavated deposits as supporting evidence for a chronological summary of occupation presented in chapter 4. For each operation, I provide a table listing the cultural and natural strata identified in excavations, as well as profiles for some excavations. For a schematic map indicating the locations of each operation, refer to figure 3.4. Not every stratum mentioned in this text is visible in the profile drawings. For additional excavation profile drawings, refer to my dissertation (Hepp 2015:95–181).

Op. LC09 A was a 5 ×5 m grid located just west of the road bisecting Platform 1 (figure 3.4, table A1.1). The operation was located on the southern margin of the platform and was positioned to explore GPR anomalies identified in 2008 (Barber 2009; figure 3.1.A). Artifacts exposed in the roadcut adjacent to LC09 A were more frequent than in many other areas, which appeared to suggest midden deposits. Four 1 × 1 m units were opened here, and their primary focus was vertical penetration in order to reveal Platform 1 construction stratigraphy and search for midden or other domestic features. Approximately 7.4 m³ of sediment was excavated in this operation. Although no midden was identified, an Early Formative hearth (A-F4-s1), intrusive into initial Platform 1 fill (A-F5), was uncovered. This hearth provided an AMS date of 3482 ± 40 (AA92453; carbon-rich sediment; δ13C, −24.0), or 1900–1690 cal BC, as discussed below.

TABLE A1.1. Description of natural strata and features from Op. LC09 A (figures A1.1 and A1.2)

| Stratum / Feature no. | Munsell color and sediment description | Probable date | Formation | Notes |
|---|---|---|---|---|
| F1 | 5YR 2.5/2 Topsoil | Formative–Modern | Occupational debris | Extensive root activity and modern soil formation. |
| F2 | 7.5YR 3/3 Clayey loam | Early Formative | Fill | Less root activity than Stratum A-F1. |
| F3-s1 | 10YR 5/6 Compact silt with rock and shell | Early Formative | Fill | Sediment containing rock and shell inclusions. Overlies A-F4-s1-3. |
| F3-s2 | N/A | Early Formative | Fill | Burned sediment containing granodiorite and gneiss inclusions. Redeposited as fill. |
| F3-s3 | 2.5YR 4/3 Compact silt | Early Formative | Fill | Contains decomposing roots. Overlies A-F4-s1, s2, s3, and F5. |
| F4-s1 | N/A | Early Formative (1900–1690 cal BC) | Hearth | Large hearth ringed with burned earth and stone. Intrusive into A-F5. Contained compact shell deposit. |
| F4-s2 | N/A | Early Formative | Secondary hearth | Small hearth-like feature attached to A-F4-s1. |
| F4-s3 | N/A | Early Formative | Ash from hearth | Deposit of ash eroding from A-F4-s1 hearth and intruding into stratum A-F5 and A-N1. |
| F5 | | Early Formative | Fill | Compact tan silt with decomposing roots. A-F4-s1 hearth intrudes into this stratum. Contains small inclusions of granodiorite and gneiss. |
| N1 | 2.5YR 6/4 Sand | Early Formative or before | Fluvial deposit | Fine-grained river sands. Probable point bar deposit. |
| N2 | 2.5Y 6/3 Silt | Early Formative or before | Alluvial deposit | Laminated, overbank silt with no rock or sand. |
| N3 | 2.5Y 5/4 Sand | Early Formative or before | High-energy fluvial deposit | Coarse river sands. |

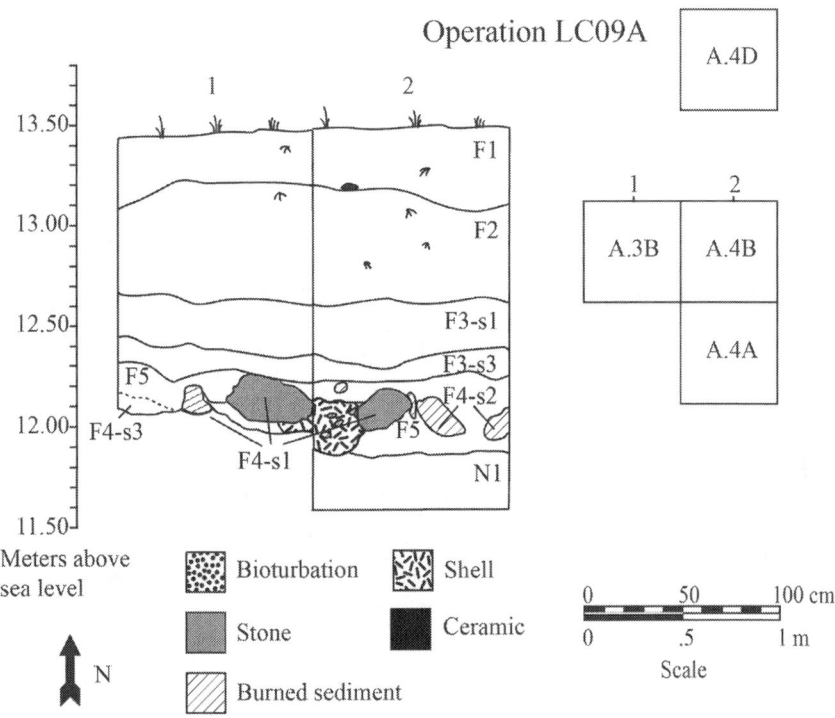

**FIGURE A1.1.** Op. LC09 A excavation profile (Units A.3B and A.4B)

Op. LC09 B was first excavated during the 2009 pilot project and was revisited by the research team in 2012. The operation began as a 2 × 2 m grid on the northern side of Platform 1 (figure 3.4). This location was chosen because of the possibility of uncovering middens at the edge of Platform 1, as well as to investigate a GPR anomaly (figure 3.1B). Due to the identification of numerous human burials, more units were opened in this operation than originally planned. The excavation team opened six 1 × 1 m penetrating units during the 2009 field season. The 2012 excavations reopened these initial units and expanded the area to a total of thirteen 1 ×1 m units. Excavations during the two field seasons analyzed 21.0 m³ of sediment (table A1.2). In the main area of LC09 B (Units 0Y, 1Y, –1Z, 0Z, 1Z, 0A, 2A, 1B, 2B, and 3B), the high density of lithic and ceramic artifacts, the artifact diversity, and lenses containing refitting ceramics suggest that some of the sediment was likely composed of redeposited midden. In addition to uncovering human remains (B1-I1, B1-I2, B2-I3, B3-I4, and B4-I5), Op. LC09 B excavations identified a hearth (B-F15 [figure 4.4]) intrusive into an early anthropogenic fill layer (B-F16). A carbon sample from this

208 DESCRIPTION OF EXCAVATED DEPOSITS

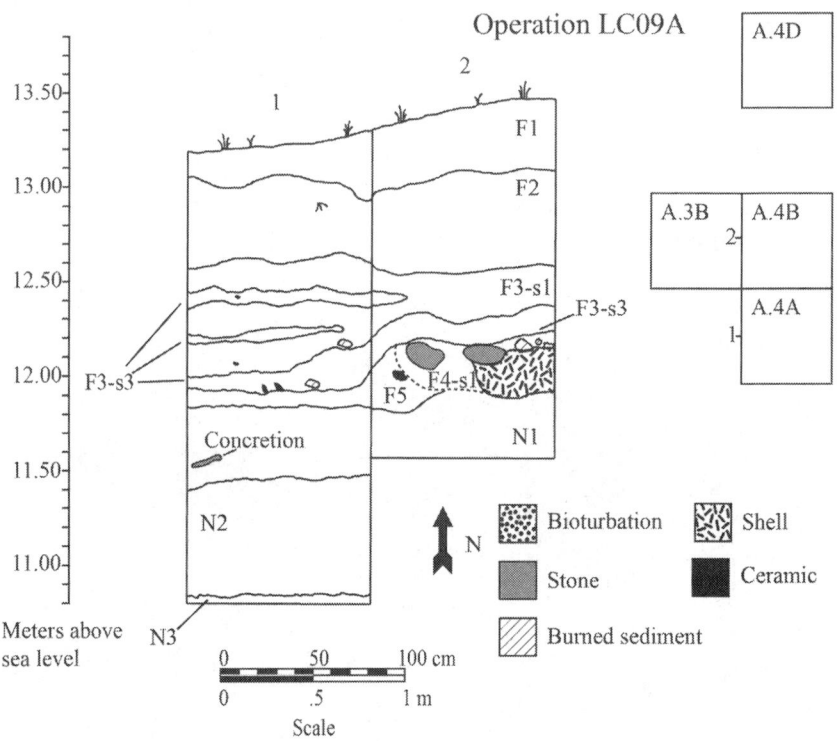

**FIGURE A1.2.** Op. LC09 A excavation profile (Units A.4A and A.4B)

hearth returned an AMS radiocarbon date of 3358 ± 43 (AA92454; carbon-rich sediment; $\delta 13C$, −25.2), or 1755–1525 cal BC, as discussed below. In the north extension of LC09 B (Units 1E, 1H, and 1J), excavations uncovered a diffuse deposit of midden with wood ash (B-F17-s1 and B-F17-s2).

Op. LC12 A was excavated as a transect bisecting the northern slope of Platform 1 and Substructure 1 (figure 3.7, table A1.3). The purpose of this transect was to expose an uninterrupted section of a large earthen architectural feature from the modern surface down to sterile deposits, in order to better understand construction sequences. The specific location of LC12 A was selected because Substructure 1 is the tallest at the site and excavating a deep section of its northern margin permitted analysis of the building sequence in the area of the site's largest construction. On the basis of preliminary work at La Consentida in 1988, it was proposed (Winter 1989) that parts of the site might have been human modifications to a natural hill or bedrock outcrop. Excavations at Ops. LC12 A and LC12 B have disproven this

TABLE A1.2. Description of natural strata and features from Op. LC09 B (figures A1.3–A1.5)

| Stratum / Feature no. | Munsell color and sediment description | Probable date | Formation | Notes |
|---|---|---|---|---|
| F1 | 7.5YR 4/3 Silty clay loam | Formative–Modern | Modern soil formed in occupational debris | Extensive root activity and modern soil formation. Appears gray in color in parts of Units 1E, 1H, and 1J. |
| F2 | 7.5YR 4/3 or 10YR 5/4 Clay or Silty clay | Early Formative | Fill | Less root activity and more sand and clay than F1, but otherwise similar. Appears gray, feels loamy, and increases in compactness with depth in Units 1E, 1H, and 1J. |
| F3 | 7.5YR 4/3 Silty sandy clay | Early Formative | Fill within probable burial pit | Intrusion into F10. Contains highly prismatic paleosol material. Looser consistency than F10. Soil formed after intrusion. Associated with human burial (B-1). |
| F4 | 10YR 4/3 Sandy clay | Early Formative | Fill within intrusive pit | Intrusion into F10. Contains highly prismatic paleosol material. Soil formed after intrusion. |
| F5 | 10YR 4/3 Sandy clay | Early Formative | Fill within intrusive pit | Intrusion into F10. Contains highly prismatic paleosol material. Looser consistency than F10. Soil formed after intrusion. |
| F6 | 10YR 4/3 Sandy clay | Early Formative | Fill within intrusive pit | Intrusion into F10. Contains highly prismatic paleosol material. Soil formed after intrusion. |
| F7 | 10YR 4/3 Silty sandy clay | Early Formative | Fill within intrusive pit | Intrusion into F10. Contains highly prismatic paleosol material. Looser consistency than F10. Soil formed after intrusion. |
| F8 | 10YR 4/3 Silty clay | Early Formative | Fill within probable burial pit | Probable burial pit intrusive into F10. Contains highly prismatic paleosol material. Large pieces of bone identified in Unit 0Y wall, suggesting additional burial. Soil formed after intrusion. |
| F9 | 7.5YR 2.5/2 | Early Formative | Fill within intrusive pit | Intrusion into F10. Contains highly prismatic paleosol material. Soil formed after intrusion. |

*continued on next page*

**Table A1.2**—*continued*

| Stratum / Feature no. | Munsell color and sediment description | Probable date | Formation | Notes |
|---|---|---|---|---|
| F10 | 10YR 4/3 Clay | Early Formative | Surface formed in ancient fill | Redeposited midden within which a prismatic paleosol has formed, indicating a period of abandonment of this part of the site. Less prismatic in structure (less well formed) than F12. Some sand and silt. Underlies F1 and F2 and has greater artifact density than those deposits. Identified in Units 0Y, 1Y, −1Z, 0Z, 1Z, 0A, 2A, 1B, 2B, and 3B. |
| F11 | 10YR 4/2 or 10YR 5/4 Sandy clay | Early Formative | Occupational debris | Dark, grayish layer of fill or eroded fill identified in Units 1H and 1J. Likely slope wash at edge of ancient platform. Perhaps produced during a period of brief abandonment of this area of the site. High degree of bioturbation. |
| F12 | 10YR 4/3 Sandy clay | Early Formative | Surface formed in ancient fill | Redeposited midden within which prismatic paleosol has formed, indicating a period of abandonment of this part of the site. Represents a stable ancient surface. More prismatic in structure (better developed) than F10. Contains high quantity of daub, ceramics, and obsidian. Identified in Units 0Y, 1Y, −1Z, 0Z, 1Z, 0A, 2A, 1B, 2B, and 3B. |
| F13-s1 | 7.5YR 4/3 or 10YR 5/4 (in unit 1J) Silty clay | Early Formative | Surface formed in ancient fill | Thick fill layer. Appears gray in color. Has formed a prismatic paleosol indicating a period of abandonment of this part of the site. Identified in Units 1E, 1H, and 1J. May correspond to F10 or F12. In Unit 1J, transitions to F13-s2. |
| F13-s2 | 7.5YR 4/3 or 10YR 5/4 (in unit 1J) Silty clay | Early Formative | Surface formed in ancient fill | In Unit 1J, this stratum transitions to a darker, more clayey sediment, due perhaps to postdepositional flooding or slope wash. |
| F14 | 10YR 5/4 or 7.5YR 5/4 Silty clay | Early Formative | Fill | Lightly colored Platform 1 fill. Identified in all but Units 1H and 1J. Contains some sand. Similar to F16, but more compact. |

*continued on next page*

TABLE A1.2—*continued*

| Stratum / Feature no. | Munsell color and sediment description | Probable date | Formation | Notes |
|---|---|---|---|---|
| F15 | 7.5YR 2.5/1 Carbon and sandy loam | Early Formative (1755–1525 cal BC) | Hearth | Hearth containing dark, carbonized sediment. Similar in form to LC09 A-F4-s1, though lacks shell and contains more carbonized sediment. Intrusive into F16. |
| F16 | 10YR 6/4 Sandy clay loam Or 10YR 4/3 Silty clay | Early Formative | Fill | Almost a sandy loam. Less compact than F14. Platform 1 fill and occupational surface below F15 hearth. Some burials intrude into this stratum. Contains carbon flecking. |
| F17-s1 | 10YR 6/4 Sandy loam | Early Formative | Ashy fill and midden layer | Thick, ashy layer containing diffuse but apparently primary midden with ceramics, obsidian, bone, and shell. Likely deposited after F18. Identified in Units 1H and 1J, at the base of the platform. Underlies F13. The midden also appears to be deposited in a pit that disturbs underlying natural stratum N4. |
| F17-s2 | 10YR 6/4 Sandy loam | Early Formative | Ashy fill and midden layer | Sandy loam substratum within ashy midden deposit. Possibly indicates use of fill material to cover midden for hygienic reasons. Likely deposited immediately after F18. |
| F17-s3 | 10YR 6/4 Sand | Early Formative | Fluvial sand inclusions within ashy midden layer | Deposits of channel sand within F17-s1 ashy midden layer. Appears to be same material as N4-s1, suggesting intrusions into or mixing with underlying natural sediments. Likely deposited immediately after F18. |
| F18 | 7.5YR 5/4 Silty sandy clay | Early Formative | Fill | At times blends into N1 but is generally harder and siltier. N1 is possible source material. Contains daub and sherds. Initial Platform 1 fill atop natural deposits, or possibly debris from occupation atop N1. Ceramics of earliest style. Sediment contains some river pebbles. May have a small sheet midden deposit at interface with F16, which may also be associated with F17 midden. |

*continued on next page*

**TABLE A1.2**—*continued*

| Stratum / Feature no. | Munsell color and sediment description | Probable date | Formation | Notes |
|---|---|---|---|---|
| N1 | 10YR 6/4 Sand or sandy loam | Early Formative or earlier | Probably natural, lower-energy alluvial deposit | Sandier than F18. At times blends into F18. Similar to N4-s1 but finer and containing some clay. May be uppermost part of natural river deposit. Essentially sterile. Identified in Units 1E and 1B. |
| N2 | 2.5Y 6/3 Silt | Early Formative or earlier | Natural, moderate energy overbank deposit | Laminated, uniform silt with no rock or sand. Identified in Unit 1B. |
| N3 | 10YR 7/4 Sandy silty clay | Early Formative | Natural alluvium | Apparently produced by low energy overbank deposition. Culturally sterile. Identified in Unit 1E. |
| N4-s1 | 10YR 6/4 Sand | Early Formative or earlier | Natural, high energy fluvial deposit | Natural fluvial sand. Fine in some spots. Culturally sterile. Identified in Units 1E, 1H, and 1J. Contains some silty clay loam inclusions with shell. |
| N4-s2 | 10YR 6/4 Sand concretion | Early Formative or earlier | Natural, high energy overbank deposit | Concretion formed within sand deposit. Probably postdepositional. Identified in Unit 1H. |
| N4-s3 | 10YR 6/4 River sand | Early Formative or earlier | Natural, high energy overbank deposit | Loose substratum of natural fluvial sand. Identified in Unit 1J. |
| N4-s4 | Silty clay loam | Early Formative or earlier | Inclusion within overbank deposit | Silty clay loam inclusions with shell. Possibly intrusive. |
| N5 | 10YR 6/4 Gravely river sand | Early Formative or earlier | Natural, very high energy fluvial deposit | Gravely texture indicates very high fluvial energy. Identified in Unit 1H. |

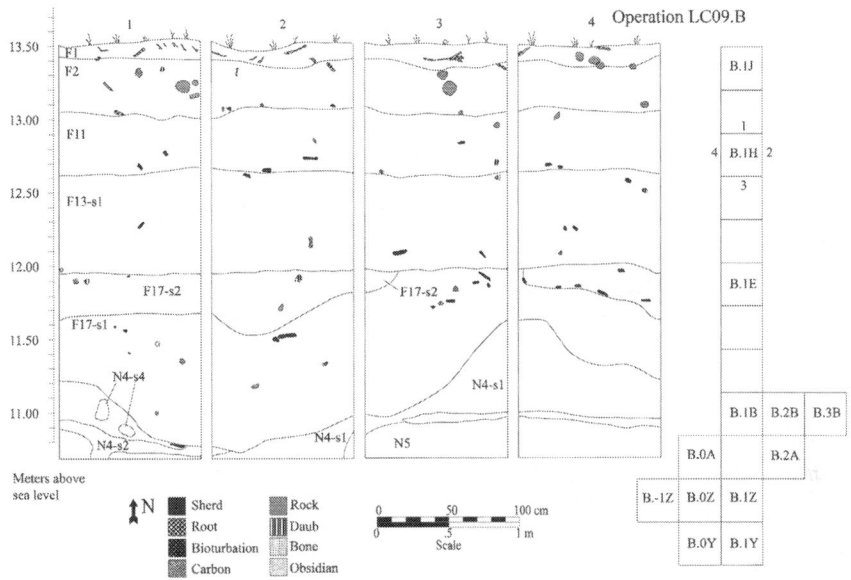

**FIGURE A1.3.** Op. LC09 B excavation profile (Unit B.1H)

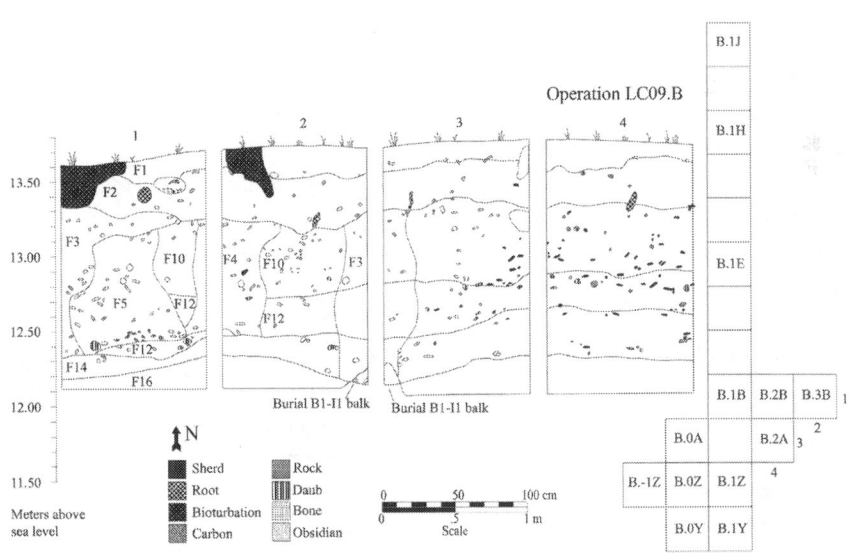

**FIGURE A1.4.** Op. LC09 B excavation profile (Units B.3B and B.2A)

214   DESCRIPTION OF EXCAVATED DEPOSITS

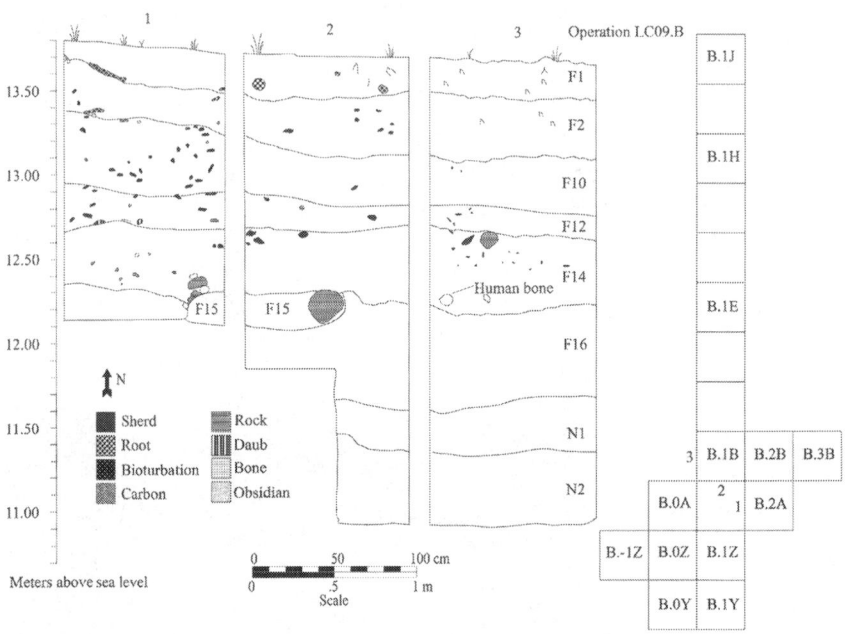

**FIGURE A1.5.** Op. LC09 B excavation profile (Units B.1B and B.2A)

initial hypothesis. When human burials were identified at the edge of Platform 1 and Substructure 1 in Op. LC12 A, the transect was expanded to investigate mortuary practices at the site. In all, thirty penetrating 1 × 1 m units and one .5 × 1 m unit were opened in this operation. These excavations analyzed approximately 59.2 m³ of sediment.

Op. LC12 B was planned as a 5 × 7 m horizontal excavation atop Substructure 1, the tallest point on Platform 1 (figure 3.4, table A1.4). Ten 1 × 1 m units were opened in LC12 B, though all but two of these remained very shallow. The discovery of architectural remnants, including large chunks of burned daub with post impressions, indicated the presence of a building atop the mound. Ceramic evidence from the upper 30 cm of sediment indicates that this structure was part of a small Early Classic period reoccupation of La Consentida after the site was abandoned during the Formative period. Below the dense artifact layer associated with this late structure, the deep layers of nearly artifact-free construction fill contained only early artifacts, suggesting that the majority of Substructure 1 was constructed in the Early Formative period and that reoccupation in the Early Classic took place atop a preexisting mound. No other evidence of reoccupation after the Formative has been discovered, though a few Classic and Postclassic artifacts scattered at or near the

TABLE A1.3. Description of natural strata and features from Op. LC12 A (figures A1.6–A1.8)

| Stratum / Feature no. | Munsell color and sediment description | Probable date | Formation | Notes |
|---|---|---|---|---|
| F1 | 10YR 3/3 Silty clay loam Or 7.5YR 4/4 Clay Or 5YR 4/3 Silty clay | Formative–Modern | Modern soil formed in occupational debris | Extensive root activity and modern soil formation. Dark humus staining. |
| F2 | 10YR 3/3 Silty clay loam Or 7.5YR 4/4 Silty Clay | Early Formative | Occupational debris/fill | Similar to F1, but with fewer roots and a less blocky texture. Dark humus staining. |
| F3 | 10YR 4/4 Silty clay | Early Formative | Fill | Lacks humus staining of F1 and F2. Distinguished from F4 by color and from F7 by blockier texture. |
| F4-s1 | 10YR 5/4 Silty clay Or 7.5YR 5/6 Clay Or 2.5YR 4/1 Clay | Early Formative | Fill/ resurfacing layer | Blockier in texture than F7. Becomes darker and chunkier clay towards north end of trench. Possibly affected by flooding in low-lying areas. May correspond to LC12 B-F4. |
| F4-s2 | 7.5YR 5/6 Silt | Early Formative | Small inclusion in fill | Small deposit of dissimilar material within F4-s1 fill. |
| F20 | 10YR 4/4 Silty clay | Early Formative | Fill within intrusive pit | Pit intrusive into F4-s1 fill. Likely corresponds to human burial. |
| F5 | 10YR 5/3 Silty clay | Early Formative | Fill within intrusive pit | Pit intrusive into F4-s1 fill. Contains human bone and likely corresponds to an unexcavated burial. |

*continued on next page*

TABLE A1.3—*continued*

| Stratum / Feature no. | Munsell color and sediment description | Probable date | Formation | Notes |
|---|---|---|---|---|
| F6 | 7.5YR 5/3 Silty clay | Early Formative | Fill within probable burial pit | Probable pit for human burial. May correspond to nearby B6-I8, though human bone in unit wall suggests unexcavated burials nearby. Disturbs stratum F4-s1 and penetrates into stratum F7. |
| F7 | 7.5YR 5/4 Silty clay Or 2.5YR 4/3 Clay | Early Formative | Fill consisting of eroded/redeposited midden | Fill layer specific to the edge of Platform 1, in area of human burials. Artifact density and sherd size suggest redeposited midden. More prismatic structure than surrounding sediments such as F4 and F10 suggests an ancient surface. Becomes gradually more clayey at north end of trench. This deposit decreases the angle of Platform 1 and Substructure 1 by adding material at the base of these features. |
| F8 | 7.5YR 5/4 Sandy clay | Early Formative | Fill within intrusive pit | Pit intrusive into F7 and possibly also intrusive into F10-s1. Contains human bone and likely corresponds to burial. |
| F9 | 7.5YR 5/4 Silty clay | Early Formative | Fill within intrusive pit | Pit intrusive into F7 fill. Contains human bone and likely corresponds to burial. |
| F10-s1 | 10YR 5/4 Silty clay Or 5YR 4/6 Silty clay Or 7.5YR 4/4 Silty clay | Early Formative | Fill/possible resurfacing layer | Well-sorted and less crumbly than overlying strata. Slightly darker than F11-s1. Varies in texture and color as it tapers toward the edge of Platform 1. Contains possible eroded midden materials, including refitting decorated bottle fragments, at the northern edge of platform. Becomes dark in color and low in artifact density at north end of trench. May correspond to LC12 B-F5, and may represent a resurfacing layer. |

*continued on next page*

TABLE A1.3—*continued*

| Stratum / Feature no. | Munsell color and sediment description | Probable date | Formation | Notes |
|---|---|---|---|---|
| F10-s2 | 10YR 5/4 Silty loam | Early Formative | Intrusion, bioturbation, or fill from different source material | A series of small intrusions within F10-s1. These are composed of loose, crumbly sediment, and may be a result of bioturbation or basket loads of fill from a different source than F10-s1. |
| F11-s1 | 10YR 5/4 Silty clay loam | Early Formative | Fill | May be the first deposit that clearly marks the construction of the Substructure 1 mound. Identified in Units 0A through 0I. Well-sorted, less blocky, and containing fewer roots than overlying strata. Lighter in color than F10-s1. May correspond to LC12 B-F6-s1. |
| F11-s2 | 10YR 6/4 Sandy loam | Early Formative | Probable inclusion of different source material | Small area of possible intrusion or basket load of different sediment within F11-s1 fill. |
| F11-s3 | 10YR 5/4 Sandy loam | Early Formative | Probable inclusion of different source material | A series of small intrusions or basket loads of different sediment within F11-s1. Composed of a calcified sandy loam. |
| F11-s4 | 10YR 5/4 Sandy clay loam | Early Formative | Probable inclusion of different source material | Small deposit of dissimilar material within F11-s1 fill. Possible basket load from a slightly different source material. |
| F12 | 10YR 5/4 Silty clay loam | Early Formative | Fill | Contains a small amount of fluvial gravel. Source material from higher energy deposits than overlying strata. May have occupational surface on top contemporary with that of F13. |
| F13 | 10YR 5/4 Silty clay loam | Early Formative | Possible occupational surface | Contains more rock and burned daub than overlying sediments. Slightly gray color may indicate ash. Possible occupational surface. May also be fill from a different source than surrounding strata. |

*continued on next page*

**TABLE A1.3**—*continued*

| Stratum / Feature no. | Munsell color and sediment description | Probable date | Formation | Notes |
|---|---|---|---|---|
| F14 | 7.5YR 5/3 Silty clay | Early Formative | Fill within pit containing ritual cache | Intrusive pit dug into F17-s2 for the placement of F15 ritual cache. Sediment is prismatic. Not visible in profile. |
| F15 | Offering | Early Formative | Ritual cache | Cache placed within F14 intrusive pit. Contains animal remains (complete *Heloderma horridum* (Mexican beaded lizard) skeleton, fish bones, shell [including oyster], turtle bones, and fossil bull shark tooth), along with sherds and a likely associated bird ocarina. Not visible in profile, though depth is noted. |
| F16 | 10YR 5/4 Silty clay | Early Formative | Fill within intrusive pit | Pit intrusive into F17-s2 fill. Likely corresponds to human burial. |
| F17-s1 | 10YR 6/4 Silty clay Or 5YR 4/6 Sandy/gritty clay | Early Formative | Fill | Compact and contains possible calcium carbonate staining. Early Platform 1 fill into which F14 intrudes. Slopes sharply downward toward the north in central portion of Op. LC12A excavations. Similar to F17-s3, but softer and with fewer inclusions. |
| F17-s2 | 10YR 6/4 Silty clay Or 5YR 4/6 Sandy/gritty clay | Early Formative | Fill | Compact and contains possible calcium carbonate staining. Early Platform 1 fill into which F14 intrudes. Slopes sharply downward toward the north in central portion of Op. LC12A excavations. |
| F17-s3 | 10YR 5/4 Calcium and sand | Early Formative | Deposits within fill | Series of sand deposits within F17-s2. Contain calcium concretions. These may represent basket loads of dissimilar material deposited as part of the F17-s2 fill event. |
| F17-s4 | 10YR 6/3 Sandy clay loam | Early Formative | Small inclusion in fill | Small deposit of dissimilar material within F17-s2 fill. |

*continued on next page*

TABLE A1.3—*continued*

| Stratum / Feature no. | Munsell color and sediment description | Probable date | Formation | Notes |
|---|---|---|---|---|
| F17-s5 | 10YR 6/4 Silty clay | Early Formative | Small inclusion in fill | Small deposit of dissimilar material within F17-s2 fill. |
| F18-s1 | 10YR 5/3 Silty clay | Early Formative | Fill | Substratum of F18 fill that may be disturbed by downslope erosion affecting northern edge of Platform 1. |
| F18-s2 | 10YR 5/4 Silty clay loam Or 7.5YR 5/6 Silty clay loam Or 10YR 6/4 Gritty clay | Early Formative | Fill | Silty clay loam fill containing deposits of F18-s3 and F18-s4. Likely represents basket loads of dissimilar material used as part of the F18 fill event. Becomes thinner toward the north edge of Platform 1 and varies somewhat in consistency. Stratum may represent an extension of Platform 1 to the north. Anthropomorphic figurine fragment (figure 7.6.A) found at interface between this deposit and A-F17-s2. |
| F18-s3 | 10YR 4/4 Or 10YR 3/4 Silty clay | Early Formative | Fill | Silty clay deposits within F18-s2. Likely represents basket loads of dissimilar fill material. |
| F18-s4 | 10YR 4/4 Sand | Early Formative | Fill | Sand deposits within F18-s2 fill layer. |
| F19 | 10YR 3/4 Clay | Early Formative (1885–1665 cal BC) | Fill with probable occupation layer | Homogenous clay layer containing large chunks of fired daub. Possible occupational layer at interface with F18. Becomes gritty clay in spots, such as in Unit 0V. |
| N1 | 10YR 3/4 Silty clay | Early Formative or earlier | Natural with probable occupation layer | Natural clay layer containing rhizoliths. Probable brief occupation layer at interface between N1 and F19, which predates fill episodes. |
| N2 | 10YR 4/4 Loamy sand | Early Formative or earlier | Natural alluvial deposit | Precultural layer containing small clay pockets. |

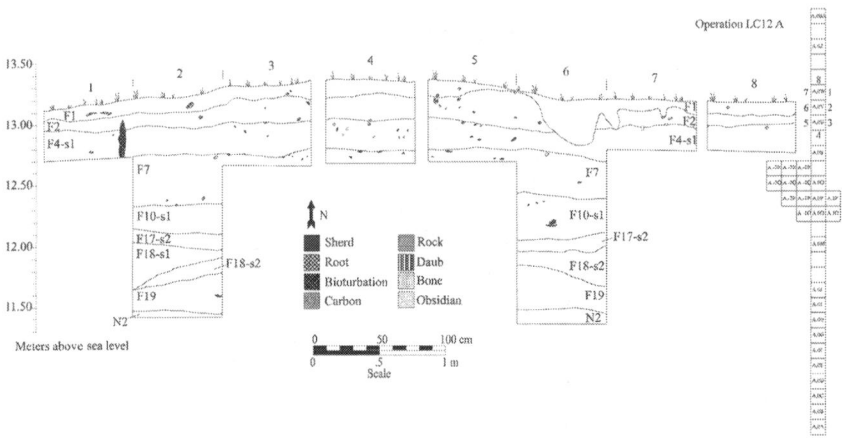

**FIGURE A1.6.** Op. LC12 A excavation profile (Units A.0U, A.0V, and A.0W)

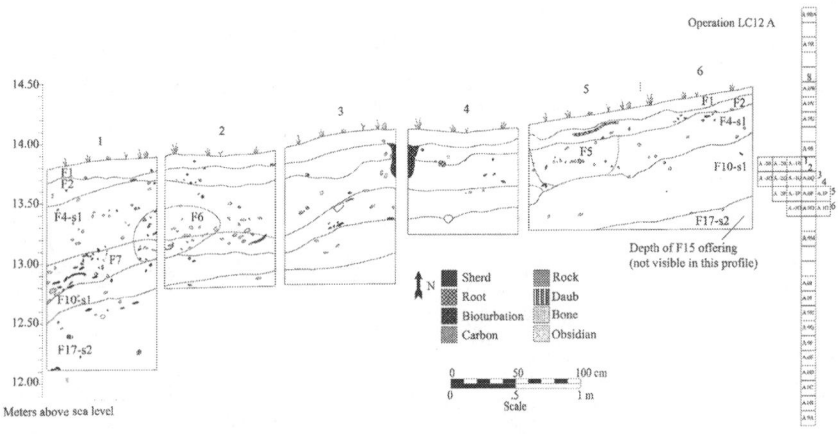

**FIGURE A1.7.** Op. LC12 A excavation profile (Units A. −1R, A.0Q, A.1P, A.0O)

modern surface indicate that the site was known to and periodically visited by later occupants of the region. In all, 6.4 m³ of sediment was excavated in Op. LC12 B.

Op. LC12 C began as a 5 × 7 m excavation area atop the northern portion of Substructure 2 (figure 3.4, table A1.5). Most units in this area were shallow, horizontal excavations intended to uncover domestic remains. Despite heavy bioturbation associated with the modern soil, including extensive *camote de agua* root and acacia tree activity (see figure 3.6), the research team identified several superimposed

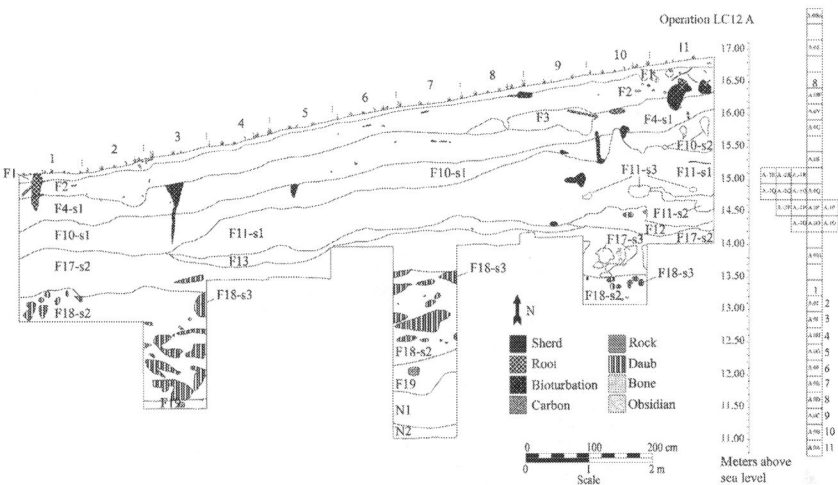

**FIGURE A1.8.** Op. LC12 A excavation profile (Units A.0A–A.0J)

occupational surfaces with in situ artifacts and the foundation of a small structure (Structure 1; see figure 5.1). Ceramic and ground stone evidence suggests that this was probably a domestic area occupied during the latter part of the Early Formative period occupation. The exposure of architectural features resulted in the excavation of some units outside the original 5 × 7 m area. The team excavated fifteen 1 × 1 m units, totaling 9.0 m³ of sediment, in Op. LC12 C.

Op. LC12 D was a 2 × 2 m excavation area at the northeastern edge of Substructure 2, in an area where Platform 1 is low and nearly even with the modern alluvial plain (figure 3.4, table A1.6). The purpose of these excavations was to try to recover stratified midden deposits associated with domestic areas atop Substructure 2. In all, two penetrating 1 × 1 m units, totaling 6.1 m³ in volume, were excavated at this operation. Excavation uncovered thin sheets of midden containing ceramic fragments and faunal remains such as bone and shell.

Op. LC12 E, like Op. LC12 D, began as a simple 2 × 2 m excavation area at the northern edge of Substructure 2, somewhat upslope on Platform 1 and between Substructures 2 and 3 (figure 3.4, table A1.7). This operation was repeatedly expanded because the research team located a stratified midden with dense shell lenses and an ashy matrix. Although this midden is not located at the edge of Platform 1, its position between two substructural mounds indicates that it may relate to events that took place on one or both of those features. Its overall size and the presence of decorated ceramic vessel fragments (see appendix 2) tend to suggest communal feasting as the source of the deposit, rather than domestic activities (for

TABLE A1.4. Description of natural strata and features from Op. LC12 B (figure A1.9)

| Stratum / Feature no. | Munsell color and sediment description | Probable date | Formation | Notes |
|---|---|---|---|---|
| F1 | 5YR 3/3 Silty clay loam | Classic Period and modern | Modern soil formed in occupational debris | High degree of bioturbation and surface disturbance. Contains large burned daub chunks indicative of architectural context. |
| F2 | 7.5YR 3/4 Silty clay loam | Classic Period and modern | Modern surface and occupational debris | Contains rocks and large burned daub chunks mixed in with sediment. |
| F3 | 5YR 3/3 Sandy clay | Classic Period | Classic period occupational layer | Very high ceramic content from Early Classic period occupation. Contains daub indicative of architecture. Also contains intrusions, some of which may be rodent burrows or tree root bioturbation. |
| F4 | 7.5YR 5/4 Silty clay loam | Formative period | Fill | Artifact content low, but all ceramics are Formative period. May correspond to LC12 A-F4-s1. |
| F5 | 7.5YR 5/6 Silty clay | Formative period | Fill | Much less compact than shallower deposits. Artifact content low, but all ceramics are Formative in date. May correspond to LC12 A-F10-s1. |
| F6-s1 | 7.5YR 5/6 Silty clay loam | Formative period | Fill | Thick fill layer with sandy clay loam inclusions (F6-s2). May correspond to LC12 A-F11-s1. May represent one of two sources of F6 fill, perhaps brought as basket loads. |
| F6-s2 | 10YR 6/4 Sandy clay loam | Formative period | Fill | Small pockets of different sediment within F6-s1 matrix. Artifact content low, but all ceramics are Formative in date. May represent one of two sources of F6 fill, perhaps brought as basket loads. |

discussions of feasting, see Clark and Blake 1994; chapters 6 and 7; appendix 2). Ceramic vessel and figurine fragments indicate that the majority of these deposits date to very early in the occupational history of the site (see appendix 2). A figurine (figure 7.7.A) found in Feature E-F10 for example, may be similar, especially in the form of its eyes, to slightly later Cruz A phase figurines from highland Oaxaca (Jeffrey Blomster, personal communication, 2015). The ceramics from around and

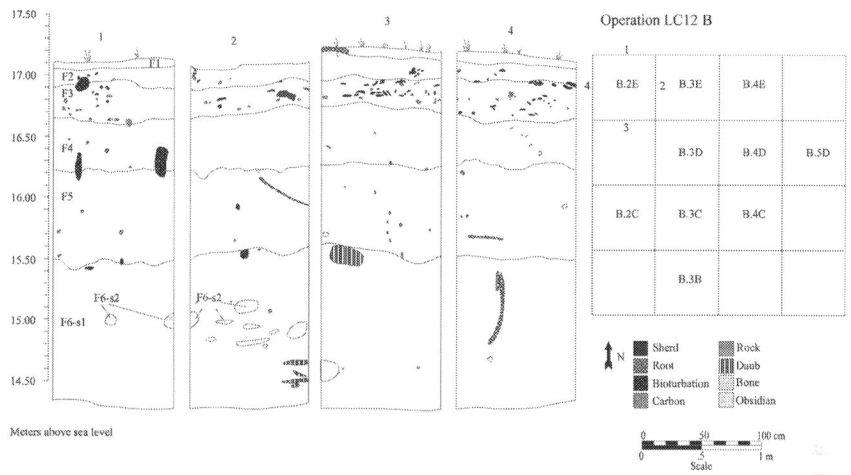

**FIGURE A1.9.** Op. LC12 B excavation profile (Unit B.2E)

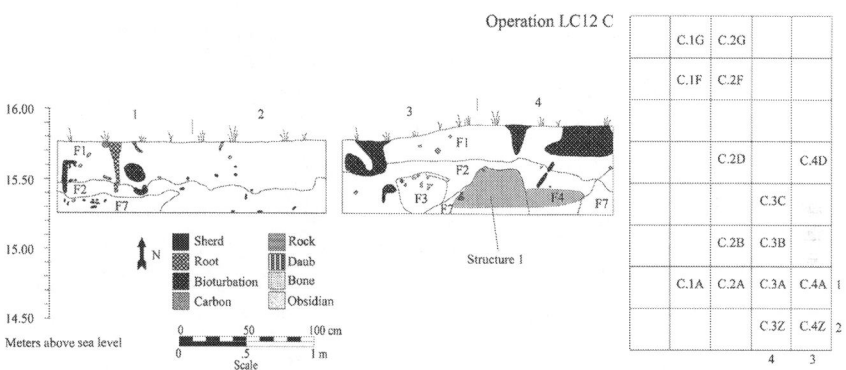

**FIGURE A1.10.** Op. LC12 C excavation profile (Units C.3Z, C.4A, and C.4Z)

above this figurine are all burnished examples of the site's earliest style. After exploring this midden, excavations at LC12 E expanded to the west to ascertain the relationship of the midden with a thin stratum filled with bright orange daub that may have come from a structure located slightly upslope and to the west (E-F4). In this western extension, the research team uncovered architectural evidence in the form of daub wall-fall with wattle impressions. Ceramic evidence suggests that this shallower deposit dates to slightly later in the occupation of the site. In all, eleven 1 × 1 m units were excavated in this operation. These units totaled 18.4 m³ in volume.

**TABLE A1.5.** Description of natural strata and features from Op. LC12 C (figures A1.10 and A1.11)

| Stratum / Feature no. | Munsell color and sediment description | Probable date | Formation | Notes |
|---|---|---|---|---|
| F1 | 10YR 3/2 Loose silty clay loam | Formative–Modern | Modern soil formed in occupational debris | Modern surface with high artifact content. Bioturbation and surface disturbance significant. |
| F2 | 10YR 3/2 Loose silty clay loam | Formative period | Occupational debris | Sediment impacted by modern bioturbation. Possible domestic occupation layer. |
| F3 | 10YR Sandy clay loam | Unknown | Fill within possible intrusive pit | Small possible intrusive pit containing ceramic and stone debris. May be a tree root intrusion. |
| F4 | 10YR 3/2 Silty clay containing stone architecture | Formative period | Architectural structure floor and foundation fragments | Silty clay fill sediment with spots of darker staining. Possibly a preserved floor. Surrounds manos and metate fragments used as part of foundation or wall construction material. Associated with Structure 1. See figure 5.1. |
| F5 | 10YR 3/2 Silty clay | Formative period | Fill within intrusive pit | Fill within pit feature in F4 structure floor. See figure 5.1 for plan and profiles. |
| F6 | 10YR 3/2 Silty clay | Formative period | Fill within intrusive pit | Fill within pit feature in F4 structure floor (not visible in profile). |
| F7 | 2.5Y 4/2 Hard sandy clay loam | Formative period | Fill with occupation layer | Fill with occupation layer and lens of in situ artifacts on top. F4 structure floor overlies this deposit. |
| F8 | 10YR 4/3 Clay loam | Formative period | Fill with occupation layer | Fill layer with probable occupation layer and lens of in situ artifacts on upper surface. |
| F9 | 10YR 6/6 Sandy clay loam | Formative period | Fill with occupation layer | Thick fill layer with stone and bioturbation. Has probable occupation layer and lens of in situ artifacts on upper surface. |

Op. LC12 F was similar in its initial intended purpose to LC12 D and LC12 E. At the western edge of Substructure 2, the excavation team searched for stratified midden deposits with one 1 × 1 m unit totaling 2.7 m$^3$ in volume (table A1.8). Although a dense deposit of ceramic artifacts was recovered here, the sherds were

DESCRIPTION OF EXCAVATED DEPOSITS   225

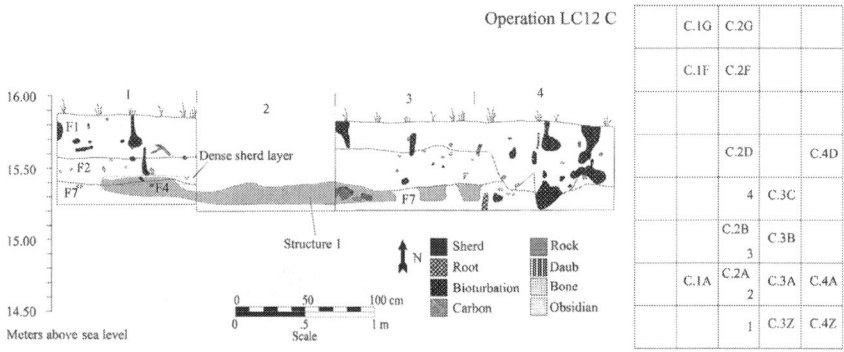

**FIGURE A1.11.** Op. LC12 C excavation profile (Units C.3A, C.3B, C.3C, and C.3Z)

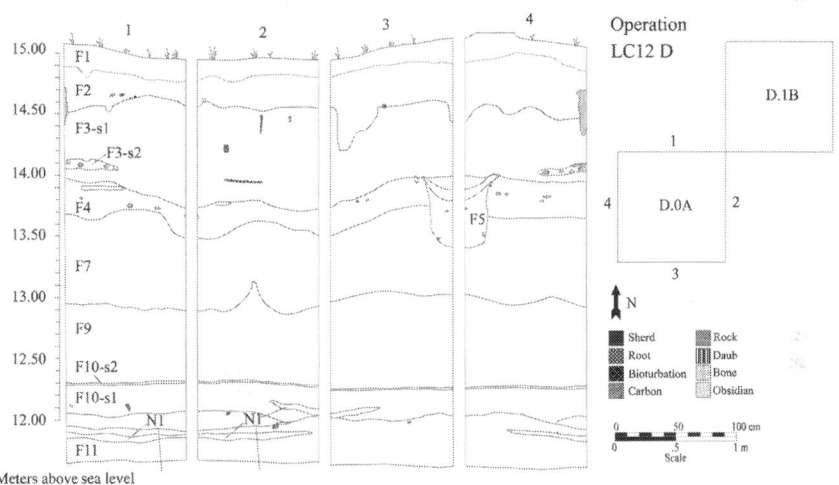

**FIGURE A1.12.** Op. LC12 D excavation profile (Unit D.0A)

mostly eroded and were thus deemed likely redeposited. The excavated column of LC12 F demonstrates fill layers used to build a central portion of Platform 1 at the base of Substructure 2.

Op. LC12 G was planned as a 5 × 7 m shallow, horizontal excavation at the southern end of Substructure 2 (figures 3.4 and 5.3, table A1.9). Because the floor and several floor features from a probable domestic structure (Structure 2) were identified in this operation (see figures 5.2 and 5.3), the excavation area was extended to the west and north. Twenty shallow 1 × 1 m units were excavated in this area,

TABLE A1.6. Description of natural strata and features from Op. LC12 D (figure A1.12)

| Stratum / Feature no. | Munsell color and sediment description | Probable date | Formation | Notes |
|---|---|---|---|---|
| F1 | 10YR 3/3 Silty clay loam | Formative–Modern | Modern soil formed in occupational debris | Extensive root activity and modern soil formation. Dark humus staining. |
| F2 | 7.5YR 4/4 Clay loam | Early Formative | Occupational debris/fill | Similar to F1, but with fewer roots and a less blocky texture. May be the B-horizon of the modern soil. Color appears gray. |
| F3-s1 | 10YR 5/6 Clay | Early Formative | Fill | Well-sorted fill with intrusions and some root activity. |
| F3-s2 | 5YR 5/6 Clay loam | Early Formative | Inclusion of ceramics and daub | Small yellowish red deposit of dense ceramic and daub fragments within F3-s1. |
| F4 | 10YR 4/3 Gritty, silty clay | Early Formative | Shell-rich fill deposit | Thin band of probable resurfacing fill with shell mixed into sediment. F5 is intrusive into this deposit. |
| F5 | 10YR 5/4 Shell dump | Early Formative | Fill consisting of crushed shell in a pit | Pit intrusive into F4 with extremely dense shell content. Very little sediment aside from crushed shell. |
| F6 | 7.5YR 5/6 Sandy clay loam | Early Formative | Fill | Small deposit containing mineral concretions and granodiorite. May represent a basket of fill from a different source. |
| F7 | 5YR 5/6 Gritty, silty clay loam | Early Formative | Fill | Layer of well-mixed, loose fill. Only subtly different from F9. |
| F8 | 10YR 4/4 Sandy clay loam | Early Formative | Possible occupational surface | Thin probable occupation surface with grayish color, likely from ash staining. Compact and contains stones. May also represent minor hearth-cleaning episode. Probable sheet midden atop deposit. |
| F9 | 7.5YR 5/6 Gritty, silty clay | Early Formative | Fill | Only subtly different from F7, with a gradual transition between the two. More compact in texture than F7. Probable sheet midden atop deposit. |

*continued on next page*

DESCRIPTION OF EXCAVATED DEPOSITS   227

TABLE A1.6—*continued*

| Stratum / Feature no. | Munsell color and sediment description | Probable date | Formation | Notes |
|---|---|---|---|---|
| F10-s1 | 10YR 5/4 Sandy loam | Early Formative | Fill | Deposit of well-sorted fill with few or no roots. |
| F10-s2 | 10YR 4/4 Sandy loam | Early Formative | Probable occupation surface | Probable occupational surface atop F10-s1. Contains same sediment as F10-s1, but is grayer in color due to ash and carbon. Contains gravel. Probable sheet midden atop deposit. |
| F11 | 10YR 5/4 Silty loam | Early Formative | Fill | Very loose fill layer mixed with natural river sand. Probable sheet midden atop deposit. |
| N1 | 10YR 5/4 River sand | Early Formative | Natural river sand | Natural river sand deposited between F11 and F10-s1. Gray and black flecks mottle color. Medium grain size. Perhaps resulting from a strong flood or hurricane. |

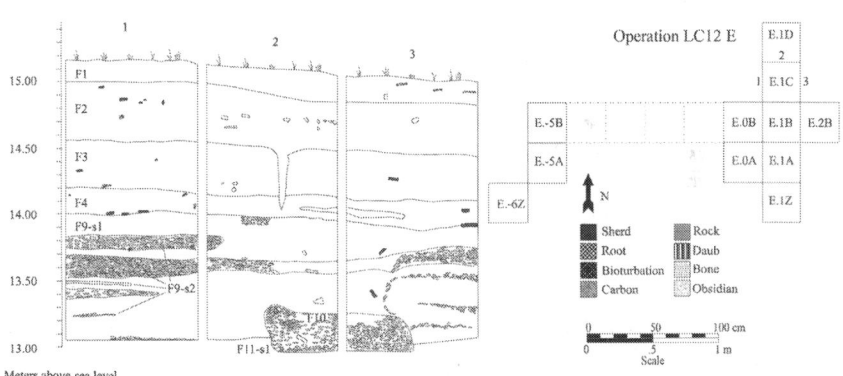

FIGURE A1.13. Op. LC12 E excavation profile (Unit E.1C)

totaling 10.7 m³ in volume. Ceramic evidence, including vessel fragments and part of a probable effigy vessel (figure 7.9.B), suggests that the structure was occupied shortly before Formative period site abandonment (refer to appendix 2 for a discussion of subtle change over time in Tlacuache phase ceramics).

Op. LC12 H was a 2 × 2 m excavation area at the base of Substructure 2, where both Platform 1 and the substructure drop off to the level of the natural floodplain.

TABLE A1.7. Description of natural strata and features from Op. LC12 E (figures A1.13 and A1.14)

| Stratum / Feature no. | Munsell color and sediment description | Probable date | Formation | Notes |
|---|---|---|---|---|
| F1 | 10YR 3/2 Silty clay | Formative–Modern | Modern soil formed in occupational debris | Extensive root activity and modern soil formation. Dark humus staining. |
| F2 | 10YR 3/2 Silty clay | Formative–Modern | Modern soil formed in occupational debris | Decreased root activity and modern surface formation. |
| F3 | 10YR 5/6 or 10YR 5/4 Sandy clay | Early Formative | Fill | Fill with less root activity than F1 and F2. Contains dense ceramic deposits and some stones. Also contains probably natural intrusions. |
| F4 | 10YR 5/4 or 10YR 5/8 Sandy clay loam | Early Formative | Fill containing architectural debris | Fill layer containing a large quantity of daub. Also contains dense ceramic deposits near interface with F3, especially in Units −5.A and −6.Z. |
| F5 | 10YR 6/6 Silty clay | Late Early Formative | Fill within a possible intrusive pit | Fill within a possible pit intrusive into F4. May have been an animal burrow. |
| F6 | 10YR 5/6 Sandy clay | Late Early Formative | Possible floor with wall fall | Thin layer of sandy clay with large, possibly in situ sections of daub from wall fall. May be a floor of a collapsed structure. Likely associated with F4. |
| F7 | 10YR 5/8 Clay | Late Early Formative | Probable floor | Floor or narrow band of fill below occupational surface. May represent floor below wall fall. Contains shell. |
| F8 | 10YR 6/4 Clay | Early Formative | Possible floor or fill below F7 | Fill below occupational surface. May represent earlier floor. Contains shell. Lenses of daub may suggest previous construction phases. |
| F9-s1 | 10YR 6/3 Sandy clay with shell | Early Formative | Midden deposit | Thick, ashy layer containing decorated vessel fragments, animal bone, and dense shell lenses (F9-s2). |
| F9-s2 | 2.5Y 6/4 Silty clay loam with shell | Early Formative | Midden deposit | Dense lenses of ash and shell within midden deposit. |

*continued on next page*

TABLE A1.7—*continued*

| Stratum /Feature no. | Munsell color and sediment description | Probable date | Formation | Notes |
|---|---|---|---|---|
| F9-s3 | 7.5YR 6/4 Sandy loam | Early Formative | Midden deposit | Small midden substratum with daub. |
| F9-s4 | 10YR 6/4 Silty loam | Early Formative | Midden deposit | Small, ashy midden substratum. |
| F10 | 10YR 6/3 Silty clay loam with shell | Early Formative (1885–1635 cal BC) | Possible hearth or shell dump | Dense deposit of shell and ash in possible hearth intrusive into F11. Anthropomorphic figurine (figure 7.7.A) found in this deposit. |
| F11-s1 | 10YR 6/3 Sandy clay formed into concretion | Early Formative | Calcium concretion formed within midden | Stratum of midden that has formed into a hard calcium concretion, perhaps due to water percolating through ash and shell. Seems to cap much of the deeper midden in the area. May result from same formation processes as F11-s3. |
| F11-s2 | 10YR 6/6 Sandy clay | Early Formative | Midden deposit | Small deposit within F11. May represent small filling episode or even sandy fill used to cover midden for hygienic reasons. |
| F11-s3 | 10YR 6/3 Sandy clay formed into concretion | Early Formative | Calcium concretion within midden | Small pockets of calcium-rich concretion within midden deposit. May result from same formation processes as F11-s1. Not visible in profiles. |
| F11-s4 | 10YR 6/4 Sandy clay loam | Early Formative | Midden deposit | Stratum of midden with shell and ash. Below F11-s1 cap of calcium concretion. |
| F12 | 10YR 6/3 Sandy loam | Early Formative | Midden deposit | Stratum of midden with shell and ash. Contains fired daub. |
| F13 | 10YR 6/4 Silty clay loam | Early Formative | Midden deposit | Stratum of midden with shell, ash, and stones. |
| F14 | 10YR 6/4 Clay or silty clay loam | Early Formative | Midden deposit | Stratum of midden with shell, ash, and carbon flecking. Varies in consistency from clay to silty clay loam. |

*continued on next page*

TABLE A1.7—*continued*

| Stratum / Feature no. | Munsell color and sediment description | Probable date | Formation | Notes |
|---|---|---|---|---|
| F15 | 10YR 6/3 Clay | Early Formative | Midden deposit | Stratum of midden with shell, ash, and carbon flecking. Not visible in profiles. |
| F16 | 10YR 5/6 Silty clay with sand | Early Formative | Midden deposit | Stratum of midden with shell, ash, and carbon flecking. Slightly sandier than F14. |

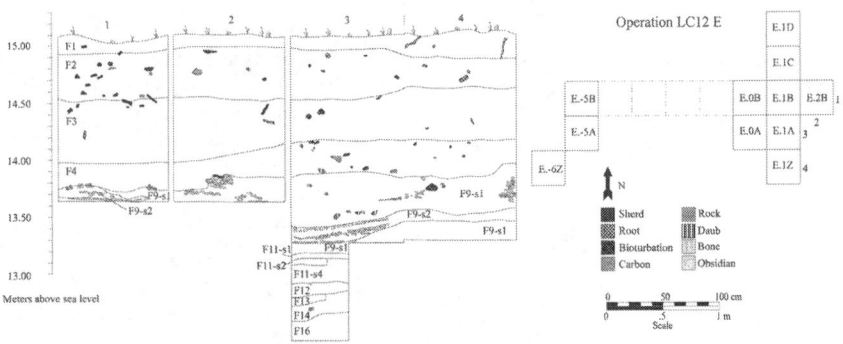

**FIGURE A1.14.** Op. LC12 E excavation profile (Units E.1A, E.1Z, and E.2B)

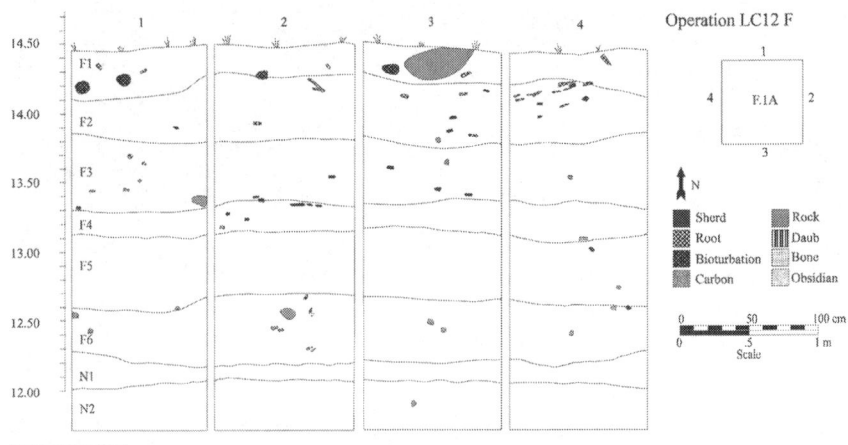

**FIGURE A1.15.** Op. LC12 F excavation profile (Unit F.1A)

DESCRIPTION OF EXCAVATED DEPOSITS 231

TABLE A1.8. Description of natural strata and features from Op. LC12 F (figure A1.15)

| Stratum / Feature no. | Munsell color and sediment description | Probable date | Formation | Notes |
|---|---|---|---|---|
| F1 | 10YR 5/3 Silty clay | Formative— Modern | Modern soil formed in occupational debris | Modern surface soil forming in F2 fill. Large deposits of ash likely represent modern burning event at the surface. Root activity and bioturbation significant. High artifact density and diversity, even at surface. |
| F2 | 10YR 4/3 Silty clay | Formative period | Fill | Likely uppermost layer of fill within which F1 has formed. Semicompact sediment. Surface root activity continues. Increased artifact content in comparison to F1. |
| F3 | 10YR 4/4 Silty clay | Formative period | Fill | Compact sediment with stone and ceramics. Thicker fill deposit. High artifact density. |
| F4 | 10YR 5/4 Silty clay | Formative period | Fill with possible occupational surface | Thin fill layer. Artifacts at F4 interface with F3 likely represent occupation layer before subsequent fill episode. High artifact density. |
| F5 | 10YR 6/8 Clay | Formative period | Fill | Thick fill deposit with low artifact density. |
| F6 | 10YR 5/8 Clay | Formative period | Fill | Clay layer with some stone and calcium concretions. Likely initial fill deposit for Platform 1 in this area. |
| N1 | 10YR 7/2 Sand | Early Formative or earlier | Natural river deposit | Natural fluvial sand below fill deposits. Grayer in color than N2. Low artifact density. |
| N2 | 10YR 4/6 Sand | Early Formative or earlier | Natural river deposit | Natural fluvial sand. More yellow in color than N1. Culturally sterile by the bottom of excavation. |

Two deep 1 × 1 m units were excavated here, demonstrating a significant depth of cultural materials below the modern alluvial plain (table A1.10). This circumstance likely results from the aggradation of the floodplain since site abandonment. These excavations totaled 5.1 m$^3$ in volume. The high percentage of jars in the Op. LC12 H midden deposits (see appendix 2) suggests food preparation for communal feasting events possibly associated domestic contexts located at the southern end of Substructure 2, where Op. LC12 G was excavated.

TABLE A1.9. Description of natural strata and features from Op. LC12 G

| Stratum /Feature no. | Munsell color and sediment description | Probable date | Formation | Notes |
|---|---|---|---|---|
| F1 | 10YR 3/2 Silty loam | Formative–Modern | Modern soil formed in occupational debris | Modern surface soil overlying F16 fill stratum. Root activity and bioturbation significant. Relatively high artifact density. |
| F2 | 10YR 4/3 Clay with darker, organic staining | Formative period | Earthen floor | Stained and compacted earthen floor of small (approximately 3 × 3 m) probable residential structure. Several pit features with charcoal staining, ceramics, obsidian, and a few stones penetrate F2. Pit features and probable postholes extend below floor. Only a limited section visible in stratigraphic profile. See figure 5.3 for complete plan and profile drawings. Associated with Structure 2. |
| F3 | 10YR 4/3 Clay with charcoal staining | Formative period | Pit feature in F2 floor | Pit feature dug into F2 floor. Fill within pit more charcoal stained than floor but composed of similar material. Contained a few stones, ceramics, and one piece of bone. |
| F4 | 10YR 4/3 Clay with darker, organic staining and stones | Formative period | Pit feature in F2 floor | Pit feature dug into F2 floor. Fill within pit contained stones at top of feature. Contained charcoal staining, a few ceramics, and a ground stone fragment. |
| F5 | 10YR 4/3 Clay with charcoal staining | Formative period | Pit feature or entryway at edge of F2 floor | Feature dug into F2 floor. Fill within pit more darkly charcoal stained than surrounding sediments. Contained several alluvial pebbles and a few ceramic fragments. Had ridge of yellowish-brown sediment at center/bottom of feature. Possible entryway, based on location. |
| F6 | 10YR 4/3 Clay with charcoal staining | Formative period | Pit feature in F2 floor | Pit feature dug into F2 floor. Slightly more charcoal stained than floor. Contained a few ceramics and no stones. |

*continued on next page*

TABLE A1.9—continued

| Stratum /Feature no. | Munsell color and sediment description | Probable date | Formation | Notes |
|---|---|---|---|---|
| F7 | 10YR 4/3 Clay with darker, organic staining | Formative period | Pit feature in F2 floor | Pit feature dug into F2 floor. Fill within feature same as overlying floor fill, but slightly darker in color. Contained a few ceramics and bones, but no stone. |
| F8 | 10YR 4/3 Clay with charcoal staining | Formative period | Probable pit feature in F2 floor | Probable pit feature in F2 floor. Identifiable as a circular charcoal stain. Contained a few ceramics. |
| F9 | 10YR 4/3 Clay with darker, organic staining | Formative period | Pit feature in F2 floor | Pit feature dug into F2 floor. Fill within feature slightly darker in color than adjacent floor surface. Contained a few ceramics but no stones. |
| F10 | 10YR 4/3 Clay with darker, organic staining | Formative period | Possible posthole/ pit feature just outside structure | Pit feature just outside remains of domestic structure. Fill slightly more charcoal stained than surrounding sediments. Large piece of daub suggests association with architecture. Disturbed by roots. Contained a few ceramics. Possible posthole. |
| F11 | 10YR 4/3 Clay with charcoal staining | Formative period | Probable post mold | Probable post mold associated with structure. Extends below level of F2 structure floor, ranging from about 15.00 to 15.13 masl. Despite possible rodent and plant bioturbation, carbonized wood, vertical form, and positioning around edge of structure suggest status as post mold. |
| F12 | 10YR 4/3 Clay with charcoal staining | Formative period | Probable post mold | Probable post mold associated with structure. Extends below level of F2 structure floor, ranging from about 15.00 to 15.13 masl. Despite possible rodent and plant bioturbation, carbonized wood, generally vertical form, and positioning around edge of structure suggest status as post mold. |

*continued on next page*

**TABLE A1.9**—*continued*

| Stratum / Feature no. | Munsell color and sediment description | Probable date | Formation | Notes |
|---|---|---|---|---|
| F13 | 10YR 4/3 Clay with charcoal staining | Formative period | Probable post mold | Probable post mold associated with structure. Extends below level of F2 structure floor, ranging from about 15.00 to 15.13 masl. Despite possible rodent and plant bioturbation, carbonized wood, generally vertical form, and positioning around edge of structure suggest status as post mold. |
| F14 | 10YR 4/3 Clay with charcoal staining | Formative period | Probable post mold | Probable post mold associated with structure. Extends below level of F2 structure floor, ranging from about 15.00 to 15.13 masl. Despite possible rodent and plant bioturbation, carbonized wood, generally vertical form, and positioning around edge of structure suggest status as post mold. |
| F15 | 10YR 4/3 Clay with charcoal staining | Formative period | Probable post mold | Probable post mold associated with structure. Extends below level of F2 structure floor, ranging from about 15.00 to 15.13 masl. Despite possible rodent and plant bioturbation, carbonized wood, generally vertical form, and positioning around edge of structure suggest status as post mold. |
| F16 | 10YR 4/3 Clay | Formative period | Fill | Uppermost Substructure 2 fill below F2 domestic structure floor. Artifact content relatively high. |

*Note*: Op. LC12 G excavations were shallow, and profile illustrations from this area are not very informative. For more information, refer to the discussion of the Structure 2 context uncovered in this area (see chapter 5 and especially figures 5.2 and 5.3), and description of this context elsewhere (Hepp 2015:147–53).

**TABLE A1.10.** Description of natural strata and features from Op. LC12 H (figures A1.16 and A1.17)

| Stratum / Feature no. | Munsell color and sediment description | Probable date | Formation | Notes |
|---|---|---|---|---|
| F1 | 10YR 3/2 Silty clay loam | Formative—Modern | Modern soil formed in occupational debris | Extensive root activity and modern soil formation. Dark humus staining. |
| F2 | 10YR 4/3 Silty clay | Formative period | Fill and occupational debris | Mixing with F1. Fewer roots. |
| F3 | 10YR 4/4 Sandy clay loam with ashy inclusions | Formative period | Fill and occupational debris mixed with midden deposit | Ashy inclusions in this stratum indicate mixing with underlying F4 ashy midden deposits. Also contains some intrusions from overlying F2. |
| F4-s1 | 10YR 4/3 Ashy clay loam | Formative period | Midden deposit | Uppermost stratum to contain significant quantities of primary midden deposit. Gradually transitions to F4-s2, with a more visible divide in Unit 0A. Contains some intrusions from overlying F3. |
| F4-s2 | 2.5Y 5/2 Very ashy clay | Formative period (1880–1625 cal BC) | Midden deposit | Primary midden deposit with ceramics. Denser with ash and ceramics than F4-s1. Dated with burned food adhering to jar fragment (table 1.1) |
| F4-s3 | 2.5Y 5/3 Ashy sandy clay | Formative period | Midden deposit | Primary midden deposit with ceramics. Denser with ash and ceramics than F4-s1, though with perhaps less ash and more sand than F4-s2. |
| N1 | 2.5Y 4/2 Sand | Early Formative or earlier | Natural river deposit | Natural fluvial sand below fill and midden. Culturally sterile in Unit −1Z, few artifacts in Unit 0A. Water table encountered during excavation. |

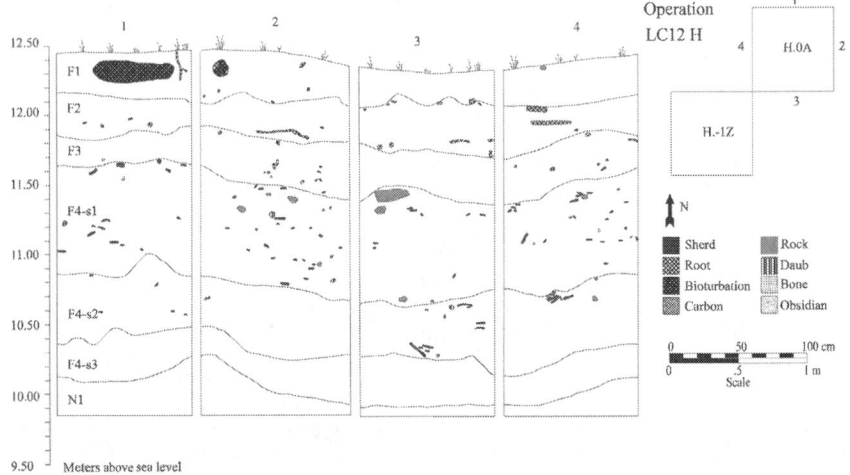

**FIGURE A1.16.** Op. LC12 H excavation profile (Unit H.0A)

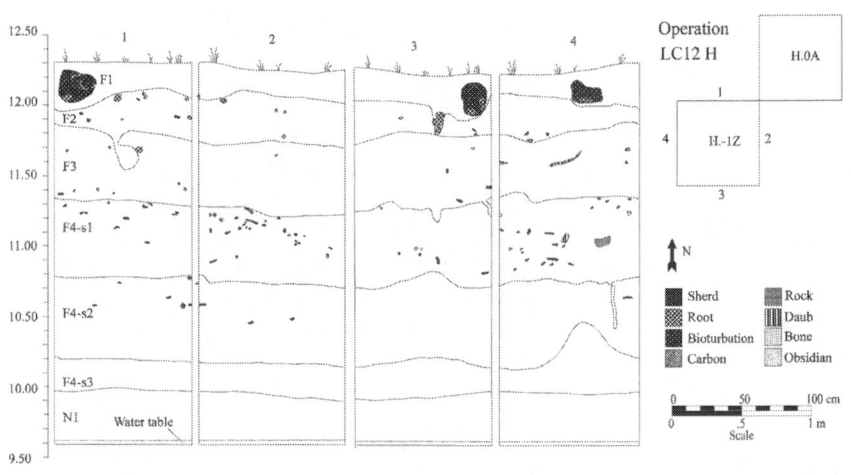

**FIGURE A1.17.** Op. LC12 H excavation profile (Unit H. −1Z)

# Appendix 2

## The Tlacuache Ceramic Assemblage

In this appendix, I present a summary of the ceramic vessel assemblage of the Tlacuache phase (approx. 1600–1350 BCE or 1950–1500 cal BC), which includes narrative descriptions and illustrations of the various vessel types. This appendix refers to the Tlacuache phase as a whole and also presents results of context-specific analysis of ceramics from the site. The Tlacuache phase is so far well represented only at La Consentida, though the nearby site of Cabeza de Vaca also contains Early Formative contexts and medium brown ware ceramics similar to those discussed here (Joyce et al. 2009b:542–45; see figure 1.3). Additionally, redeposited artifacts, survey data, and anecdotal information suggest that other nearby sites with deeply stratified occupation levels (such as Charco Redondo and Río Grande) may contain Early Formative materials that have yet to be excavated (Gillespie 1987; Grove 1988; Zárate Morán 1995). It is, therefore, possible that future investigations at these nearby sites will promote refinement of the Tlacuache phase typology.

As discussed in chapter 3, ceramic typological analysis presented here follows the methodological framework set by previous studies in highland Oaxaca (e.g., Caso et al. 1967; Martínez López et al. 2000) and also applied in the lower Río Verde Valley (Baillie 2012:62–130; Hedgepeth 2009:77–150; A. Joyce 1991a:121–73). To summarize, ceramics are categorized first according to paste. The Tlacuache phase assemblage consists of medium and coarse brown ware ceramics, which I have differentiated according to a visual assessment of their inclusion sizes. The most frequent inclusions are angular pieces of white stone such as quartzite, though organic temper, sand, gravel, shell, and grog (recycled and crushed ceramic material) are also present. Although a good deal of variation exists within coarse wares (which have large and ubiquitous inclusions) and medium wares (which tend to have a sandy paste), ceramics at La Consentida do generally fall into one of these two categories. Although there is color variation (with some sherds appearing gray or reddish, and with reds being especially common among the coarser pastes), all Tlacuache phase vessels are considered to be brown wares fired in oxidizing, open-air settings rather than in formal kilns. Open-air ceramic firing is still practiced today by native potters on the coast of Oaxaca and Guerrero (Ahern 2010:31). All Tlacuache phase wares have a micaceous paste, with mica fragments tending to be very small and likely occurring as a natural component of clays deposited by the Río Verde, rather than representing an intentional temper additive (A. Joyce 1991a:834–36). A few fine brown and gray ware sherds have been recovered near the modern surface at La Consentida, but these are considered later refuse and are not discussed in detail here.

Due to a shortage of storage space for archaeological materials and regulations of the Centro INAH Oaxaca, many nondiagnostic ceramic fragments (especially undifferentiated body sherds) were discarded following initial cleaning and analysis. The categorization of all vessel fragments recovered during the LCAP according to paste characteristics prior to this vetting process permits discussion of the relative emphases of La Consentida's potters on the paste types represented in the Tlacuache phase assemblage. In summary, medium brown ware pastes are the most common type in all excavated contexts at the site. Due to the highly friable nature of Tlacuache ceramics, weight totals are a more meaningful measure for comparison than are sherd counts, and I will thus primarily use weights by gram. Table A2.1 demonstrates the relative frequency of pastes for all sherds recovered at the site during the 2009 and 2012 excavations.

Following their categorization according to paste, I organized the diagnostic Tlacuache phase ceramics into basic vessel types, including jars, bowls, bottles, and tecomates. Each of these categories included various subtypes, such as conical versus hemispherical bowls, globular jars with in-leaning versus out-curving necks, and bottles with short versus long necks. As described below, minor variations in the form of rims, lips, or bases, as well as aspects of surface treatment, allow further refinement of this typology. As I will also discuss, there seem to have been minor changes in the Tlacuache phase assemblage over time. Because most vessel categories are represented throughout the excavated contexts at La Consentida, it is impossible at present to divide the ceramics into rigidly defined subphases. Instead, I will describe subtle chronological variations notable in the assemblage with the hope that future research may promote the further refinement of this categorization.

THE TLACUACHE CERAMIC ASSEMBLAGE   239

**TABLE A2.1.** Weights and relative percentages of ceramic paste types recovered at La Consentida in 2009 and 2012

| Ceramic paste type | Total weight (g) | Percentage of overall assemblage |
|---|---|---|
| Fine brown | 895.0 | 0.18 |
| Fine gray | 251.8 | 0.05 |
| Medium brown | 382923.2 | 75.06 |
| Coarse brown | 126052.6 | 24.71 |

A basic formal analysis demonstrates the relative frequencies of Tlacuache phase vessel forms within paste categories. Due to time constraints, only those diagnostics ceramics from primary contexts—including middens, burials, and a ritual offering—were analyzed for this study. Note that while table A2.1 refers to all ceramics excavated at La Consentida in 2009 and 2012, tables A2.2–A2.4 refer to a smaller collection of artifacts, as they discuss only *diagnostic* fragments from the Op. LC12 D sheet middens (associated with interfaces between LC12 D-F11 and F8); the LC12 E-F16–F9 midden, the LC12 H-F4 midden; the domestic areas in and around Structures 1 and 2; the occupational surfaces atop LC12 A-N1, LC12 A-F19, and LC12 A-F13; a possible small midden deposit at the LC09 B-F16 and LC09 B-F18 interface; and the LC09 B-F17 midden. When all diagnostic vessel fragments from primary contexts are compiled into a single summary for each paste category (tables A2.2 and A2.3), it becomes clear that there was a good deal of overlap in the types of vessels constructed of coarse and medium pastes, but also that there are significant and likely meaningful differences between paste categories. As demonstrated in table A2.2, coarse ware vessels of the Tlacuache phase are primarily globular jars. In some cases, ceramic remains are complete enough to be identified as jars or bowls, but not as a particular subcategory of vessel within that type. When globular jars, collared jars, and generic (i.e., unidentified) jars are combined, they account for 86 percent of all diagnostic coarse wares analyzed for this book. Combined subcategories of bowls make up 9 percent, bottles 5 percent, and other miscellaneous vessel types make up the remainder. In summary, coarse ware vessels are usually jars. These are often undecorated and were likely used for utilitarian purposes such as cooking, storage, and water transportation. Medium brown ware vessels (table A2.3) also consist of a high proportion of jars (69 percent when subtypes are combined) but are more diverse in form, consist of a higher frequency of bowls (23 percent) and bottles (6 percent) than coarse wares, and were more likely to be decorated and perhaps used in public events such as feasting (see chapter 7 and table A2.28).

Together, diagnostic remains from combined medium and coarse ware ceramics indicate an emphasis on jars, bowls, and bottles in order of decreasing frequency

**TABLE A2.2.** Frequencies of diagnostic coarse brown ware vessel fragments from several contexts

| Coarse brown ware vessel type | Weight (g) | Percentage of overall assemblage |
|---|---|---|
| Bottle | 648.1 | 4.8 |
| Generic jar | 1296.5 | 9.5 |
| Grater bowl | 52.7 | 0.4 |
| Tecomate | 79.6 | 0.6 |
| Generic bowl | 90.2 | 0.7 |
| Collared jar | 30.4 | 0.2 |
| Globular jar | 10422.6 | 76.4 |
| Conical bowl | 401.4 | 2.9 |
| Semi/hemi bowl or dish | 620.4 | 4.5 |
| Total | 13641.9 | |

**TABLE A2.3.** Frequencies of diagnostic medium brown ware vessel fragments from several contexts

| Medium brown ware vessel type | Weight (g) | Percentage of overall assemblage |
|---|---|---|
| Bottle | 1242.8 | 6.3 |
| Bule | 20 | 0.1 |
| Collared jar | 213 | 1.1 |
| Cylindrical vessel | 20.6 | 0.1 |
| Generic bowl | 532.4 | 2.7 |
| Generic jar | 2616 | 13.2 |
| Grater bowl | 274.8 | 1.4 |
| Tecomate | 243.6 | 1.2 |
| Globular jar | 10872.9 | 54.9 |
| Conical bowl | 2261 | 11.4 |
| Semi/hemi bowl or dish | 1490.8 | 7.5 |
| Total | 19787.9 | |

(table A2.4). Special types such as tecomates and grater bowls are much less frequent. In the following sections, I will describe each of the main vessel categories within the coarse and medium brown ware types and provide example illustrations where appropriate.

**TABLE A2.4.** Frequencies of diagnostic vessel fragments of both medium and coarse brown paste from several contexts

| Combined paste vessel type | Weight (g) | Percentage of overall assemblage |
|---|---|---|
| Bottle | 1890.9 | 5.7 |
| Bule | 20 | 0.1 |
| Collared jar | 243.4 | 0.7 |
| Cylindrical vessel | 20.6 | 0.1 |
| Generic bowl | 622.6 | 1.9 |
| Generic jar | 3912.5 | 11.7 |
| Grater bowl | 327.5 | 1.0 |
| Tecomate | 323.2 | 1.0 |
| Globular jar | 21295.5 | 63.7 |
| Conical bowl | 2662.4 | 8.0 |
| Semi/hemi bowl or dish | 2111.2 | 6.3 |
| Total | 33429.8 | |

## COARSE BROWN WARE VESSELS

### Bowls

Coarse brown ware bowls (see figure A2.1) occur in generally the same vessel types (conical and hemispherical) as medium ware bowls, which are described below. Conical coarse ware bowls include mostly out-leaning and out-curving wall forms. Coarse ware conical bowls are less common than medium ware examples but are generally similar in form. Varying wall angles and heights mean that some conical coarse ware bowls are relatively deep vessels while others are more akin to low dishes. Rims are often direct but may be out-leaning, out-curving, or in-leaning. Lip variants among the conical coarse ware bowls include flat, rounded, beveled, and interior or exterior thickened shapes. These vessels tend to have thicker walls than the medium ware examples, with a wall thickness of 10 mm or more being common. Conical coarse ware bowls exhibit decoration and fine surface treatment less frequently than do their medium ware counterparts, suggesting that they were more rarely employed as serving vessels. Some surfaces of these coarse ware bowls are nicely smoothed, burnished, and slipped, though they tend not to exhibit the high degree of surface finishing that the medium ware examples often do.

Among coarse ware hemispherical bowls, subcategories are very similar to those for medium brown ware hemispherical bowls. These coarse ware hemispherical bowls are less common than medium ware examples and tend to bear nicely slipped and burnished surfaces less frequently than do their medium brown ware

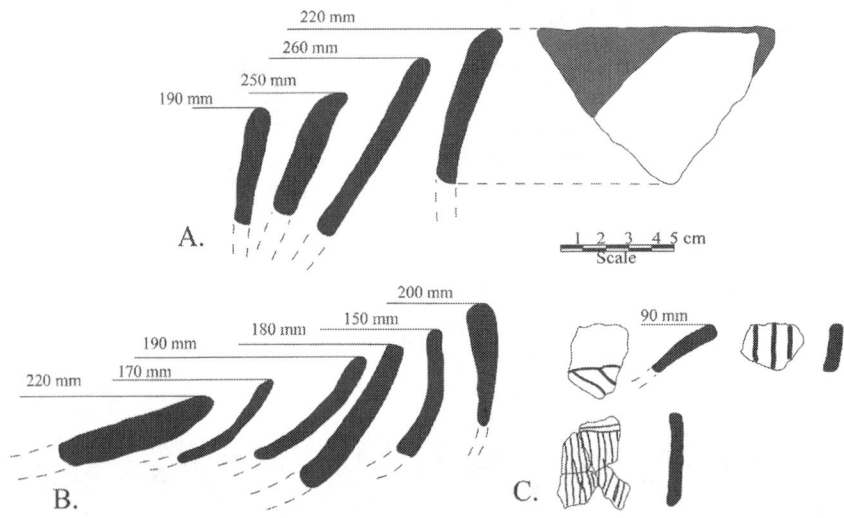

**FIGURE A2.1.** Coarse brown ware bowl fragments: (A) conical bowls, one with red painted design; (B) semi- and hemispherical bowls; (C) grater bowls

counterparts. As with other bowls, the rims of these vessels are often direct but may be out-leaning, out-curving, or in-leaning. Lips are usually flat or rounded but may in some cases be beveled. Thickening is common, particularly on the exterior of the rim and lip. Some of the coarse ware hemispherical bowls are very shallow, almost like a plate or low-walled dish. The pastes and slips of coarse ware hemispherical bowls are more likely to be reddish or orangish in color than those of medium ware examples. Hemi- and semispherical bowls are considered variants of a general vessel type, with the only difference being that hemispherical examples represent half of a sphere in their form, while semispherical forms are less than half of a sphere.

Grater bowls are typically ashtray-sized vessels. Their most common form is that of a conical bowl with interior incisions, often in intricate geometric patterns. In rarer cases, they may be semispherical. Rims are typically direct but may be slightly out-curving. Lips vary from flat to rounded. In rare cases, lips may bear ticking or scalloping, though it is not clear whether this is restricted by paste type. In general, coarse and medium brown ware grater bowls are identical in form, with the only exception being that coarse ware examples have larger inclusions in their paste. In a few instances, spouts molded into the rim of grater bowls indicate that they were used to process something that could then be poured as a liquid. For this analysis, grater bowls have been considered as an independent vessel category, though they might also have been analyzed as variants of other bowl types such as that of conical bowls.

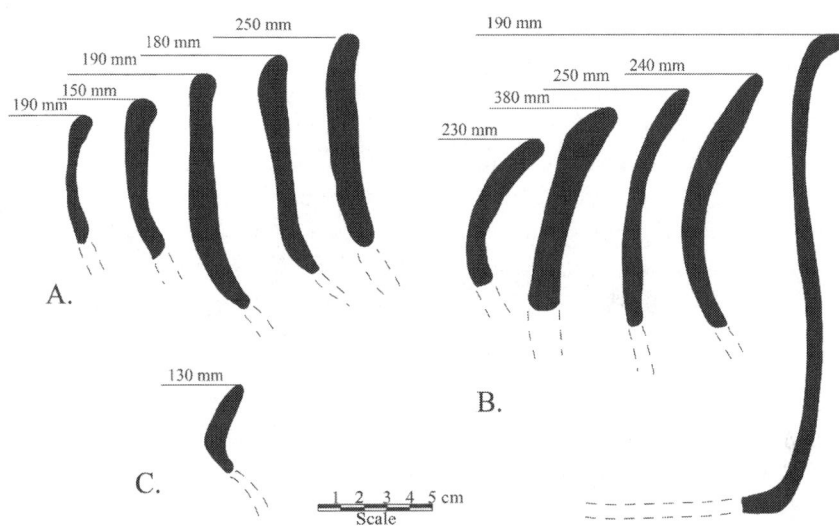

**FIGURE A2.2.** Coarse brown ware jar fragments. (A) in-leaning neck globular jars; (B) out-leaning/out-curving neck globular jars; (C) collared jar

## Jars

Coarse brown ware jars (see figure A2.2) are one of the most abundant types among Tlacuache phase vessels, probably due to their extensive use for daily cooking activities. Basic vessel divisions include globular and nonglobular body shapes, among which globular examples are by far the most common. Among globular coarse ware jars, neck variants include out-leaning, out-curving, and in-leaning examples. In-leaning neck jars are generally distinguished from in-leaning wall bowls (where basal fragments are absent) by the lack of interior finishing. This method of identification is supported by the presence or absence of interior surface treatment on vessel fragments that do have a preserved basal angle. Coarse ware jars are often reddish in color and may have an extremely coarse paste with heavy inclusions of angular white stone or gravel. Sometimes these inclusions may be exposed on even well-preserved interior and exterior surfaces. In-leaning neck jars seem to be almost exclusively coarse wares. These vessels often have interior thickening of the rim, round or flat lips, and the interior of the neck appears to be scraped in some instances. Slip and paste color variants within this vessel type include red, orange, brown, and black. Paints are most often red. Some jars with in-leaning necks may have out-curving rims, which gives them a "collared" appearance, though collared jars are more common among medium ware vessels, as discussed below.

Out-curving neck coarse ware jars may have a smoothed or unfinished neck interior. Rims tend to be direct, and often have subtle exterior or interior thickening. Lips vary but tend to be round or flat. Rims may also be out-curving. Some of the vessel necks are subtle in their out-curving shape. These vessels appear to mostly have had flat bases, though they may be rounded or angled at the edges, giving them an almost composite silhouette form near the base. Out-leaning neck coarse ware globular jars are similar to the out-curving neck varieties, and the distinction between the two forms is often subtle. Some of these vessels have only a slightly out-leaning neck, while other neck angles are more severe. These vessels are generally similar to in-leaning neck examples in that they are often formed from a very coarse reddish paste, which may have taken on some of its coloration during firing. The necks are often scraped, roughened, or unfinished on the interior. Lips tend to be round but can also be flattened or pointed (beveled on both interior and exterior), and often bear exterior or interior thickening.

### Bottles

Coarse brown ware bottles (see figure A2.3.A) most frequently have long, slightly out-flaring necks and globular bodies. They appear to be less common than medium brown ware bottles, which are described below. This discrepancy may explain why greater formal variation is identified among the medium brown ware bottles so far recovered at La Consentida than is the case for coarse brown ware examples. Coarse ware bottles sometimes bear slips in brown or red and may have impressed decoration. These bottles appear to mostly have had flat bases. In general, the coarse ware bottles fall within the range of vessel form variation described below for medium brown ware bottles.

### Tecomates

Coarse ware tecomates (see figure A2.3.B) are generally similar in their formal and surface treatment variation to the medium paste examples described below. This is one of the rarer vessel types identified among the Tlacuache phase assemblage. The interior of these vessels is often unfinished due to the difficulty of reaching past the restricted vessel mouths. Rim variants include those with interior thickening and/or exterior flattening or beveling. Vessel lips tend to be either round or flat. These vessels are usually less well-finished than medium brown ware examples, though they may bear burnishing and slipping in brown, gray, or orange. Due to the similarity of tecomate body sherds and those from globular jars, tecomates are generally only recognized when rim fragments are identified. For that reason, their relative

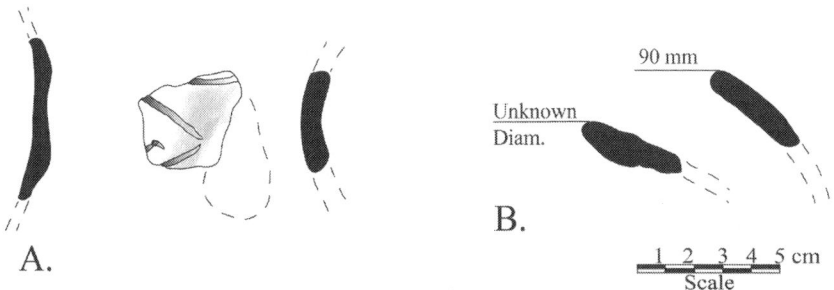

**FIGURE A2.3.** Rare coarse brown ware vessel fragments: (A) probable bottles with composite silhouette or decoration (a possible face)—for a better example of a probable effigy vessel, see figure 7.9.B; (B) tecomate rims

frequency in the Tlacuache assemblage may be understated, and their original vessel dimensions are difficult to ascertain.

## MEDIUM BROWN WARE VESSELS
### BOWLS

Medium brown ware bowls (see figures A2.4 and A2.5) are relatively common in the Tlacuache phase assemblage. In broad terms, they can be divided into conical and semispherical forms. Within those categories, conical bowls consist of out-leaning wall, out-curving wall, and in-leaning wall bowls. Wall angles and heights vary considerably, with some vessels being deeper bowls while others being more like low dishes. Similarly, significant variation in vessel rim diameter (of at least 70–330 mm) indicates that some bowls were small while others were quite wide. Some of these vessels have interior incisions that classify them as grater bowls. Grater bowls tend to have low walls, direct rims, and rounded lips. Occasionally these vessels have unusual attributes such as spouts or a square shape. Out-curving wall and out-leaning wall conical bowls can be further divided according to rim and lip styles. Rims are often direct and frequently have interior thickening. Lips may be flat, round, or beveled either uniformly or with greater beveling on the interior or exterior. Conical bowls tend to have flat bases, though some of the better finished and elaborate bowls in early deposits (particularly in the sheet middens associated with LC12 D-F10, F9, and F8 and in the LC12 E-F16–F9 midden) have a circular foot attached to the base. Cylindrical vessels are rare but also present in the Tlacuache collection and may be considered a type of bowl.

Surface treatment on conical bowls frequently includes smoothing, burnishing, and slipping in brown, gray, red, and orange. Interior smoothing and/or burnishing

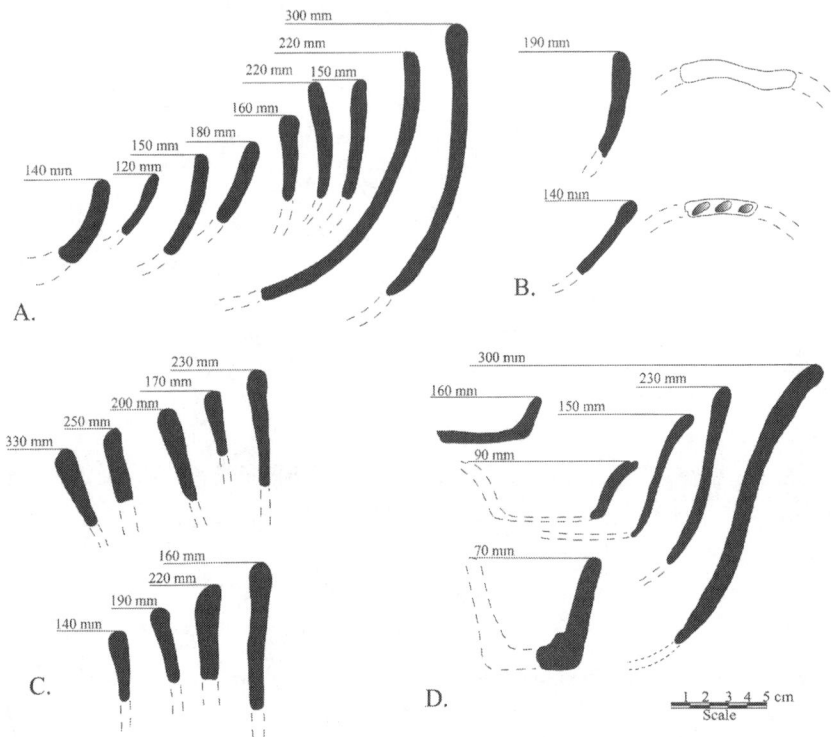

**FIGURE A2.4.** Medium brown ware bowl fragments: (A) semispherical bowls; (B) semispherical bowls with unusual rim or "kidney" shape; (C) cylindrical and in-leaning wall conical bowls; (D) out-curving and out-leaning wall conical bowls

is very common among these bowls and is a useful way to identify conical bowl fragments from jar fragments, even where basal angles are not preserved. As discussed above, this inference is supported by instances where vessels do retain basal angles. Decorations on conical bowls include impressed and occasionally incised geometric patterns and zones of paint or slip, most often in red. The fine surface finish on many conical bowls, along with occasional decoration, suggests that these vessels were used as fancy serving wares more frequently than were other vessel types such as jars. As discussed below, relative frequencies of these finely finished bowls and other decorated vessels aid in the identification of feasting middens. Among conical bowls, out-curving wall examples frequently have the most attention paid to their surface treatment and decoration. Out-curving wall conical bowls often have well-finished, smoothed, and burnished surfaces with interior thickened rims. They tend to occur in brown, orangish-red, or black colors.

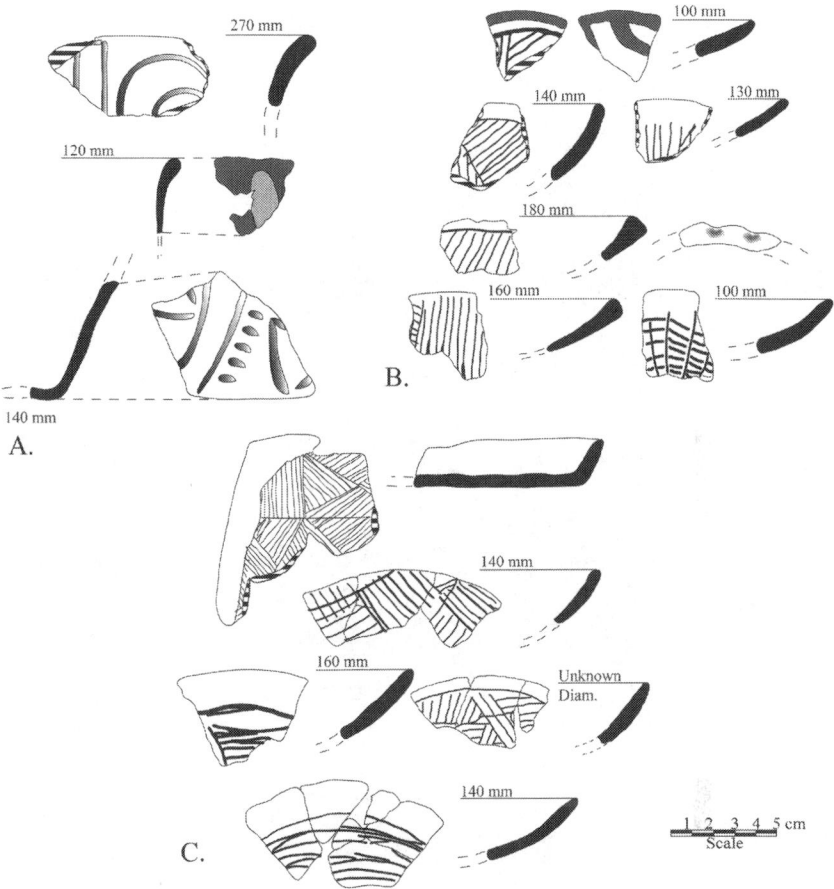

**FIGURE A2.5.** Medium brown ware decorated bowl fragments: (A) decorated conical bowls, one of which is painted with red and cream-colored pigment; (B) incised grater bowls, one of which bears red paint; (C) incised grater bowls

In rare instances, semispherical bowls have an eccentric rim form that leaves them "kidney-shaped," similar to some vessels of the Tierras Largas phase (Flannery and Marcus 1994:fig. 7.2), as well as some from Tlatilco (Piña Chan 1958:fig. 40.j, lám. 21), the Tehuacán Valley (MacNeish et al. 1970:fig. 6), and Zohapilco (Niederberger 1976:lám. LII.16, 25, lám. LIV.16, foto 37; see chapter 8). A few semispherical bowls of the Tlacuache phase have interior incisions that categorize them as grater bowls, though these are less common than conical grater bowls. Most semispherical bowls are undecorated and have direct rims. Some rare

**FIGURE A2.6.** Small medium brown ware collared jar with suspension holes. Recovered as offering with child burial B9-I11. Photograph shows vessel still containing sediment. Once removed, this sediment was found to contain a large sherd that may have served as a lid

examples have notched rims. These vessels tend to lack extensive surface treatment and almost always lack decoration.

## Jars

Medium brown ware jars (see figures A2.6 and A2.7) are a common vessel type in the Tlacuache assemblage. While their coarse ware counterparts seem often (by virtue of context and a lack of surface finish) to have been used in daily cooking and other utilitarian activities, medium ware jars appear to have enjoyed a more diverse set of uses in the La Consentida community, including as burial offerings and likely as serving vessels. Almost all of these vessels are globular, with a few examples lacking the rounded lower vessel wall that would define them as such. Among

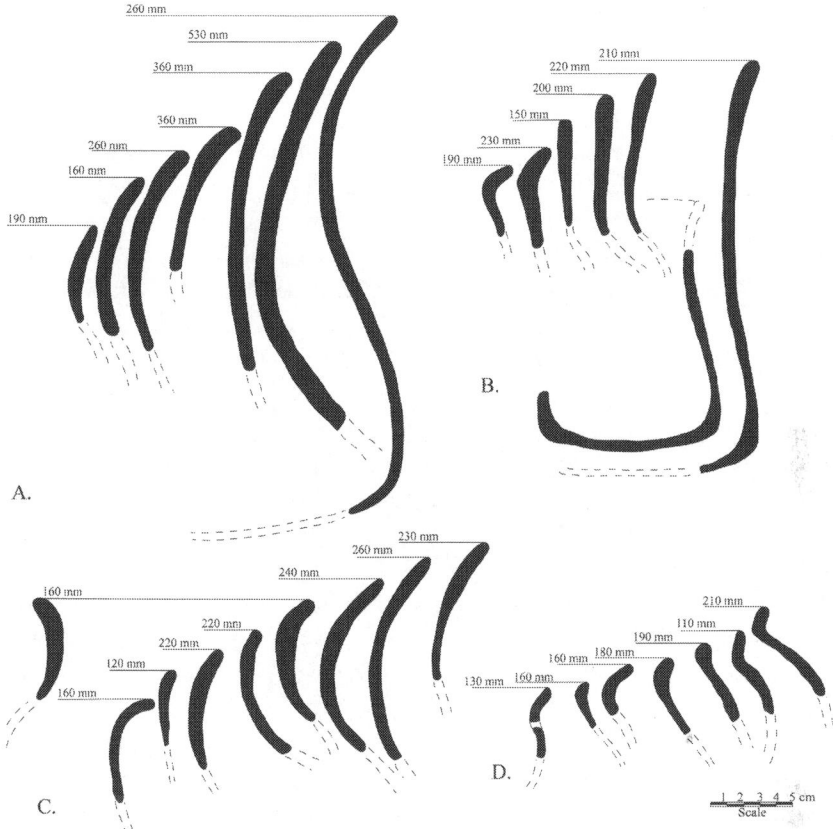

**FIGURE A2.7.** Medium brown ware jar fragments: (A) globular jars with out-curving necks; (B) globular and nonglobular jars; (C) globular jars with out-curving necks; (D) collared jars

globular medium paste jars, the basic formal distinctions are similar to those for the coarse ware vessels described above and include out-curving neck, out-leaning neck, and collared examples. In-leaning neck jars seem to have been almost exclusively a coarse ware vessel form.

Variations within the out-curving and out-leaning neck categories occur according to distinctions in interior rim thickening, lip form (which tends to be rounded or flat), and neck angle. Exterior rim thickening is rare but sometimes does occur. Some rims also have a deliberate interior flattening. Many of these vessels have flat bases, though the edges of the bases may be somewhat rounded, leaving the impression of a composite silhouette near the base of the vessel. Wall thickness varies considerably, with some examples reaching 28 mm in thickness at the vessel shoulder. Medium ware jars are sometimes very large, and the most well-preserved examples (from the LC12 H midden) bear exterior and sometimes interior smoothing, wiping, and even burnishing. These surfaces tend not to be as nicely finished as those of serving bowls, however.

Paste and slip color variants among globular medium ware jars include brown, orange, and dark gray. Interiors of the vessel necks are sometimes scraped or roughened below a narrow band of finishing and even burnishing at the vessel rim. Some of these vessels have impressed decoration that coincides with zones of pigment decoration, typically using a red slip or paint. Examples from the LC12 E midden, where remnants of fancy serving bowls and bottles were recovered, tend to exhibit more surface finishing and have a smaller overall vessel size than other jars. These vessels were likely for serving rather than cooking. Some small collared jars have suspension holes impressed into the clay before the vessel was fired. These vessels at least sometimes have rounded bottoms, a form permitted by the suspension of the jar rather than resting it upon the ground or a cooking surface. It is also possible that the "suspension holes" could have been used to securely affix a lid to the vessel. Importantly, these holes are not "repair holes" that were drilled through the wall of a vessel cracked in antiquity but were instead an intentional aspect of the vessel's form and intended use and appear to have been most common among small collared jars. The best-preserved example of a vessel with such suspension holes was recovered as an offering with burial B9-I11 (figure A2.6).

### Bottles

Medium brown ware bottles (see figure A2.8) are globular in body form and include straight long-necked, out-flaring long-necked, and out-flaring short-necked varieties. Out-flaring and straight long-necked bottles tend to differ from one another only subtly and are more common than short-necked examples. Among the straight neck bottles, further division is possible between those with flat versus rounded lips. All the bottles appear to have had flattish bases. Where preservation permits its identification, a brown or red slip is common among medium brown ware bottles, and orange slip is also present. They tend to have unfinished interiors below the top few centimeters of the rim, due to the restricted nature of the bottles' necks. Decoration on bottles includes impressed lines, bands, and geometric patterns, and these sometimes correspond to zones of paint or slip decoration, typically utilizing red pigment. Decorations sometimes go all the way to the vessel base and occasionally also occur on bottle necks. Decorations include "sunbursts," bands, and other simple geometric patterns (see chapter 8). The decorations on some bottles tend to suggest their use as fancy serving vessels.

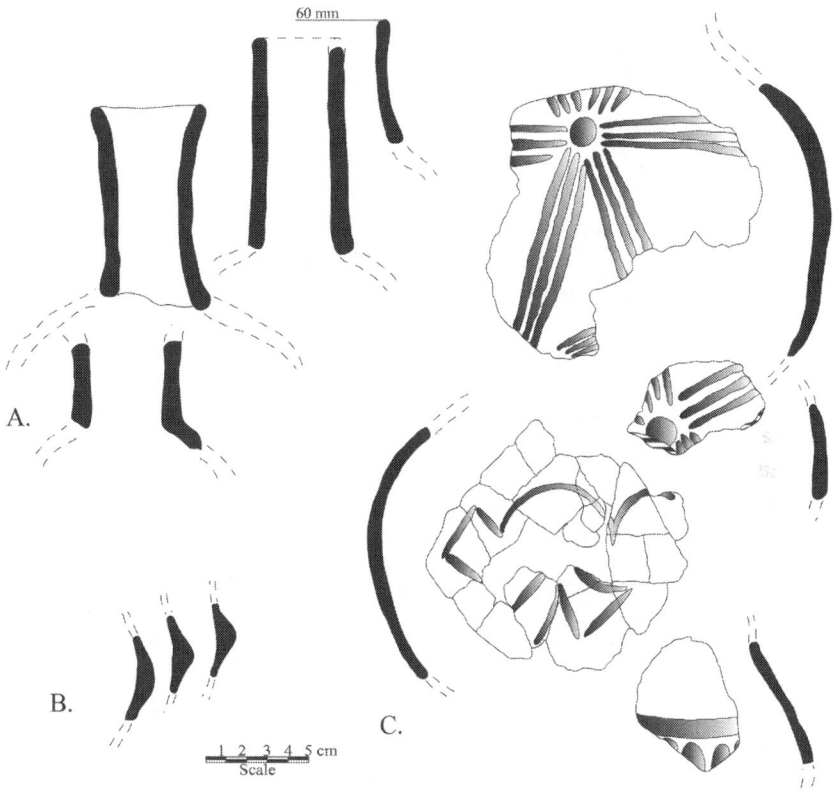

**FIGURE A2.8.** Medium brown ware bottle fragments: (A) bottle necks; (B) possible composite silhouette bottles; (C) decorated bottles

## TECOMATES

Medium brown ware tecomates (see figure A2.9) are a rare vessel category in the Tlacuache assemblage, though they do occur in diverse contexts at La Consentida and notably in each of the LC12 D-F4, D-F9, and D-F10, LC12 E-F16 through E-F9, and LC12 H-F4 midden deposits. Some tecomates had large rim diameters of up to at least 260 mm. Rim variations among tecomates include those with interior thickening and others with a very slightly out-curving lip. Variants within the tecomate form also include miniature tecomates and *bule*-shaped vessels. Tecomates sometimes had nicely slipped and burnished surfaces in brown, gray, orange, or black and occasionally bore impressed decoration in the form of bands around the rim and radiating grooves, somewhat akin to the "gourd-like" phytomorphic form of some Barra phase vessels (see Clark and Blake 1994). Some tecomate rim fragments also

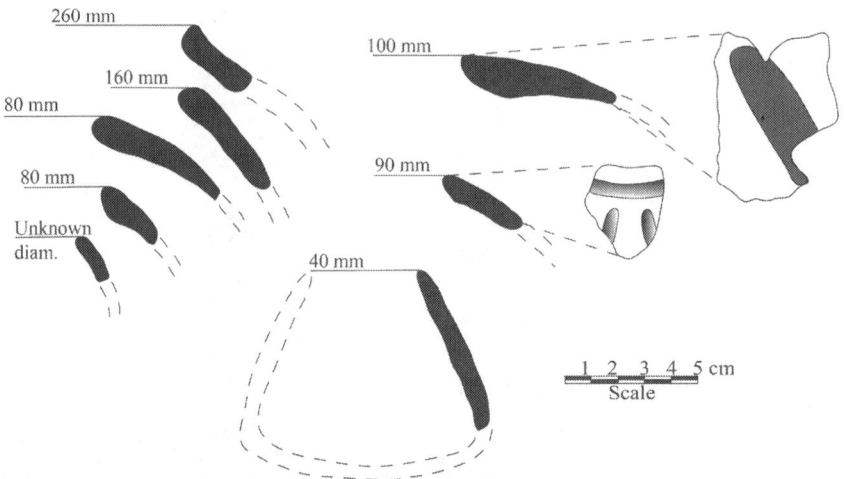

**FIGURE A2.9.** Medium brown ware tecomates and bule. Some examples bear red paint or phytomorphic decoration.

have traces of red pigment from a painted decoration that included bands radiating out at roughly a 45-degree angle from the vessel mouth. As discussed above for coarse ware tecomates, the similarity between body sherds from globular jars and those from tecomates means that the latter are only identified in cases where rims are recovered. For this reason, their relative frequency is likely underrepresented in descriptions of the Tlacuache assemblage.

### CHANGE OVER TIME WITHIN THE TLACUACHE PHASE ASSEMBLAGE

The vessel forms described above vary in their relative frequencies by excavated context at La Consentida. In general, however, the diverse forms of the Tlacuache phase seem to be present throughout the history of site occupation. Nevertheless, the relative frequencies of vessel types do appear to have changed over time, which may indicate that the ceramic styles at the site shifted subtly during the occupation. While evidence of these vessel form variations does not yet permit the division of the Tlacuache phase into formal subcategories, the observation of these changes may promote refinement of the ceramic chronology following future research at La Consentida and/or at contemporaneous sites nearby.

It was initially reported (Hepp 2011a) that La Consentida likely saw two phases of occupation, the first resulting in the production of an early version of Platform 1 composed largely of a yellowish silty clay and associated with ceramics bearing burnished and slipped surfaces. The second proposed occupational phase resulted, according to initial interpretations, in a more significant effort to construct Platform 1 and the earthen substructures atop it, as well as in the interment of several human burials in the area of Op. LC09 B and the use of ceramics with less-finished surfaces that were more susceptible to erosion. The diachronic assessment of ceramics from a deep fill column in Op. LC12 A (analyzing strata F19 through F1) in unit A.0E (see table A1.3 and figure A1.8) has cast some doubt on the presence of two distinct phases or subphases. Instead of a clear change over the course of site occupation in the vessel types present at La Consentida, the primary vessel categories—including globular jars, conical bowls, and bottles—are present throughout the sequence. Even the interpretation that the construction of Substructure 1 happened later in the occupational history of the site is now in doubt, due to the relatively early deposition of Stratum LC12 A-F11-s1 (see table A1.3, figure A1.8, and chapters 5 and 7). Instead of clearly discernable occupation phases, then, it seems more appropriate to discuss minor chronological variations in the relative frequency of certain specific vessel categories at the site.

One type of vessel that seems to have become more common over time during the occupation of La Consentida is the grater bowl. As demonstrated in table A2.25, grater bowls are up to about ten times higher in relative frequency in the domestic contexts associated with the Structure 1 (in Op. LC12 C) and Structure 2 (in Op. LC12 G) than they are in earlier contexts (e.g., LC12 H-F4 and LC12 E-F16 through E-F9). This observation is complicated by two other pieces of information. First, as discussed below, part of the variation in the presence of grater bowls is likely due to the context of their use within the site. As artifacts probably associated with some sort of crafting or food preparation in the domestic sphere, grater bowls are likely to be more common in domestic contexts than elsewhere. Second, two relatively early burials of children (B9-I11 and B11-I13) were accompanied by offerings of the best-preserved grater bowls yet recovered from La Consentida. Nonetheless, the earliest fill deposits throughout the site (e.g., LC09 A-F5, LC12 A-F19, LC12 D-F11) do seem to produce fewer grater bowl fragments than later domestic occupational, fill, and midden deposits (e.g., LC09 B-F10, LC12 G-F1, F2, F16, and LC12 C-F2). One probable change over time in the Tlacuache phase assemblage, therefore, is an increase in the production and use of grater bowls.

Even as grater bowls likely became more common over time at La Consentida, bottles seem to have decreased in frequency. As demonstrated in table A2.26, bottles were ten times more common among diagnostic ceramics in the early LC12 E-F16 through E-F9 midden than they were in and around the Structure 2 domestic context excavated in Op. LC12 G. As with the grater bowls discussed above, this observation is complicated by the probable public feasting nature of the Op. LC12

E midden as contrasted with the domestic nature of Structure 2. If bottles were used to serve and consume beverages in public settings, one would expect their recovery in higher numbers in feasting deposits than in households. Several other lines of evidence tend to support the decrease over time in the use of bottles, however. As discussed in chapter 6, dietary evidence in the form of ground stone food-processing tools and dental pathologies indicate that La Consentida's diet shifted from consumption of maize in a nongritty form (i.e., likely as a beverage or porridge) to the consumption of maize flour processed on stone manos and metates that increased tooth wear even as stable isotopic indicators support relative consistency in absolute $C_4$ plant consumption over time (e.g., tables 6.8–6.10). At the risk of somewhat circular reasoning, the apparent decrease over time in the use of ceramic bottles is consistent with a shift from the consumption of maize beverages (perhaps often in feasting contexts) served in bottles to the use of maize flour produced on larger ground stone tools that became more common over time (see figure 5.7.A). Depending on the tasks for which grater bowls were used, their apparent relative increase over time at the site may also be linked to the dietary shift that affected bottle use and ground stone tool variation. Ongoing studies include the analysis of sediment trapped in the interior incisions of the grater bowls in order to determine whether the vessels contain plant microfossils, though these results are not yet ready for reporting (see Morell-Hart et al. 2014).

Despite the complicating factors of contextual variation and preservation issues that cause me to question the initial attribution of two occupational phases to La Consentida, a few observations remain accurate. The very earliest ceramics at the site (associated with initial fill such as LC09 A-F5 and LC12 A-F19, natural strata such as LC12 A-N1 and LC12 H-N1, and early dated features such as the LC09 A-F4 hearth) are unequivocally dated to initial occupation and have well-executed surface treatment. Although of a soft, friable paste likely fired at a low temperature, these mostly medium brown ware vessel fragments tend to have nicely slipped and burnished surfaces. They occasionally bear decoration in geometric patterns of impressed lines on their exterior (perhaps produced with rounded bone tools such as the crocodilian jawbone tool pictured in figure 6.5.C), and they appear closer in terms of decorative style to later ceramics from West Mexico than they do to contemporaneous pottery from other early traditions such as the Barra phase (Clark and Blake 1994; I. Kelly 1980; see chapter 8). Later ceramics at La Consentida, while similar enough in overall vessel type variation to preclude separation into a second phase or subphase (on the basis of current data, at least) are often less well preserved than the early examples, lack the highly slipped and burnished surfaces of those earlier sherds, and are more frequently coarse ware vessels comprising higher relative frequencies of grater bowls and lower relative frequencies of bottles than their

**FIGURE A2.10.** A reconstruction of some of the Tlacuache phase vessels

precursors. Even when later ceramics of the Tlacuache phase are well preserved, their surfaces rarely attain the burnished luster of the earliest examples. They more frequently have pastes that are "sandwiched" in cross section, indicating perhaps a change in vessel firing techniques. Somewhat like the "embarrassingly well decorated" tecomates of the Barra phase, the first vessels of the Tlacuache phase were more finely made than seems necessary for mere utilitarian use and were likely at least partly intended for use in public events such as feasting (Clark and Blake 1994; Lesure and Wake 2011:85).

In its present form, the Tlacuache phase assemblage is diverse. It includes various kinds of jars, especially those with a globular shape. Bowls are mostly conical, but semispherical examples are also present. The interior incisions of small grater bowls suggest some kind of food preparation or crafting activity. Bottles may have been used to serve beverages, and decorated examples were likely used in public events such as feasts. Most vessels had flat bottoms, though some collared jars had round bottoms and likely were suspended by cords, perhaps from house rafters (see Clark 1991). Ceramic pastes varied from a sandy medium brown ware to a very coarse brown ware that often took on a reddish, oxidized color during open-air firing. These diverse vessels served the needs of a dynamic community. Refer to figure A2.10 for a reconstruction of many of the vessel forms based on all ceramic analyses to date.

## WITHIN-SITE CERAMIC VESSEL CONTEXT

A key step in understanding the uses of Tlacuache phase ceramics at La Consentida is identifying the contexts within which different vessel types occurred at the site.

Depending on how one categorizes these vessels, such intrasite patterns may vary. For example, the ratio of bowls to jars shifts if one chooses to lump both medium and coarse brown ware vessels into single categories of "bowl" and "jar" rather than keeping them divided according to paste. Through experimenting with these varying patterns, one can begin to identify what vessel categories might have been meaningful to the ancient occupants of La Consentida. Here, I discuss variations in the vessel types and relative quantities of decorated ceramics recovered in several key contexts. I pay particular attention to variation among middens, as the vessel types identified in those deposits serve as supporting evidence for arguments made in the body of this book, notably regarding public feasting and food preparation (see chapters 6 and 7). I consider various levels of vessel distinction (e.g., classification according to paste types as well as according to simple macrocategories such as "bowl" and "jar") in the hopes of identifying practices of vessel use meaningful to the ancient community. I report vessel type variation according to grams of diagnostic fragments recovered. This method causes some distortion of reported vessel values. Large jars, weighing more than most decorated bowls, will be overrepresented, for example. The method is more effective than using sherd counts, however, particularly because the sandy, medium brown ware pastes common in the Tlacuache phase assemblage are extremely friable. In very few circumstances are vessels complete enough to get an accurate idea of minimum vessel counts, so I have avoided this method for purposes of consistency. Note that any vessel type referred to as "generic" (e.g., "generic jars") is not a unique vessel category, but instead represents vessel fragments diagnostic enough to be identified as jars (for example) but not diagnostic enough for more specific identification (i.e., globular versus non-globular). The discussion of "generic" fragments permits a more thorough analysis of the contextual variations in basic vessel categories. Finally, I compare analyzed contexts using Chi-square and Fisher's exact test nonparametric statistics.[1]

## VESSEL CONTEXT
### Op. LC12 D Sheet Middens

Thin sheet midden deposits in Op. LC12 D were indicated by the presence of large, well-preserved ceramic sherds at the interfaces between deep fill strata (LC12 D-F11 through F8). These contexts produced 2153.9 g of diagnostic sherds. Other material types in these deposits included shell and burned bone. As demonstrated in tables A2.5–A2.7, sherds identified in these strata come from a diverse set of vessel categories including jars, bowls, and bottles. This context produced more tecomate fragments than many others, with tecomates representing 13 percent of medium brown wares (table A2.5). When vessel categories of different pastes in the Op. LC12 D middens are compared (tables A2.5 and A2.6), it is notable that most of the coarse wares are jars (92 percent), while the medium wares are more evenly divided among several vessel types. This is consistent with sitewide patterns of vessel variation by paste. Although jars alone make up 51 percent of diagnostic sherds recovered in the Op. LC12 D middens, the deposits appear to result from a mixture of uses rather than from a specific food preparation or serving practice (table A2.7). Relatively high percentages of serving wares such as bottles (15 percent) and conical bowls (17 percent) suggest the deposition of some feasting refuse. Although decorated ceramics could obviously occur in nonfeasting deposits such as domestic areas or cooking refuse middens, I refer to their relative frequencies here to support arguments for probable feasting deposits (see Clark and Blake 1994).

## Op. LC12 E Midden

The Op. LC12 E midden (LC12 E-F16–F9) was an ashy deposit containing shell, animal bone, and ceramic vessel fragments. Among the sherds recovered here were some of La Consentida's best examples from decorated vessels (see chapter 8). The Op. LC12 E midden produced ceramic remains from a variety of vessel types (tables A2.8–A2.10). Rapidly deposited decorated serving wares (with cross-fits across multiple excavated lots) and well-preserved faunal remains (see chapter 6 and discussion of decoration below) suggest that the LC12 E midden was likely at least partly the result of public feasting (see Clark 2004a; Clark and Blake 1994; Clark, Pye, and Gosser 2007:25; Hayden 1990, 1998, 2009).

## Op. LC12 H Midden

The Op. LC12 H midden (LC12 H-F4) consisted of a dense deposit of ash and ceramics at the base of the southern end of the Substructure 2 mound (see figure 3.4). The deposit was immediately notable for its large, well-preserved vessel fragments, some of which refit across about 60 cm of depth. As discussed elsewhere in this book, this pattern indicates the rapid deposition of the midden in only one or very few dumping episodes (see chapters 4 and 7). The context produced 11519.2 g of diagnostic vessel fragments. What the data displayed below (tables A2.11–A2.13) most clearly indicate is the heavy emphasis on jars in this deposit. While "generic jars" occur alongside globular jars, it is probable that most of the generic jar fragments are also from globular vessels that were not complete enough to be identified in more detail. Among medium brown wares in the deposit, jars account for 92 percent of diagnostic vessel fragments. Jars account for 98 percent of coarse wares identified in the midden. Among all pastes, jars comprise 94 percent of diagnostic

**TABLE A2.5.** Medium brown ware vessel types in the LC12 D-F11 through F8 middens

| Vessel type | Weight (g) | Percentage of context assemblage |
|---|---|---|
| Conical bowl | 373.3 | 30.9 |
| Bottle | 288.8 | 23.9 |
| Globular jar | 186.7 | 15.4 |
| Tecomate | 155.7 | 12.9 |
| Generic bowl | 99.7 | 8.2 |
| Generic jar | 45.3 | 3.7 |
| Semi/hemispherical bowl or dish | 42.1 | 3.5 |
| Grater bowl | 17.2 | 1.4 |
| Total | 1208.8 | |

**TABLE A2.6.** Coarse brown ware vessel types in the LC12 D-F11 through F8 middens

| Vessel type | Weight (g) | Percentage of context assemblage |
|---|---|---|
| Globular jar | 726.9 | 77.9 |
| Generic jar | 97.7 | 10.5 |
| Bottle | 39.1 | 4.2 |
| Semi/hemispherical bowl or dish | 39.1 | 4.2 |
| Collared jar | 30.4 | 3.3 |
| Total | 933.2 | |

**TABLE A2.7.** Combined paste vessel types from the LC12 D-F11 through F8 middens

| Vessel type | Weight (g) | Percentage of context assemblage |
|---|---|---|
| Globular jar | 913.6 | 42.7 |
| Conical bowl | 373.3 | 17.4 |
| Bottle | 327.9 | 15.3 |
| Tecomate | 155.7 | 7.3 |
| Generic jar | 143 | 6.7 |
| Generic bowl | 99.7 | 4.7 |
| Semi/hemispherical bowl or dish | 81.2 | 3.8 |
| Collared jar | 30.4 | 1.4 |
| Grater bowl | 17.2 | 0.8 |
| Total | 2142 | |

**TABLE A2.8.** Medium brown ware vessel types from the LC12 E-F16–F9 midden

| Vessel type | Weight (g) | Percentage of context assemblage |
|---|---|---|
| Globular jar | 1609.6 | 40.7 |
| Conical bowl | 905.1 | 22.9 |
| Semi/hemispherical bowl or dish | 564 | 14.2 |
| Bottle | 435.7 | 11.0 |
| Generic bowl | 138.8 | 3.5 |
| Collared jar | 125.2 | 3.2 |
| Generic jar | 69.4 | 1.8 |
| Tecomate | 40.2 | 1.0 |
| Grater bowl | 30.6 | 0.8 |
| Cylindrical vessel | 20.6 | 0.5 |
| Bule | 20 | 0.5 |
| Total | 3959.2 | |

**TABLE A2.9.** Coarse brown ware vessel types from the LC12 E-F16–F9 midden

| Vessel type | Weight (g) | Percentage of context assemblage |
|---|---|---|
| Globular jar | 2684 | 65.1 |
| Bottle | 500.1 | 12.1 |
| Generic jar | 420.3 | 10.2 |
| Conical bowl | 253.4 | 6.1 |
| Semi/hemispherical bowl or dish | 203.2 | 4.9 |
| Generic bowl | 33.2 | 0.8 |
| Tecomate | 30.5 | 0.7 |
| Total | 4124.7 | |

fragments, and many of these were identifiable as globular examples, mostly with out-leaning or out-curving necks (e.g., figures 8.2.A–8.2.D).

Because the category of "hemi/semispherical bowl or dish" is largely composed of a single partial vessel (figure 8.2.E), the ceramics from the Op. LC12 H midden are best described as almost exclusively globular jars. Such a high occurrence of a particular vessel type seems to indicate the specific, perhaps almost singular use to which the deposit's vessels were put. In various global contexts, Prudence Rice has noted that jars often serve as storage vessels (1999) and/or as water containers (1987:113). Elsewhere, Rice (1987:209–10) has stated that storage jars and cooking vessels overlap considerably. In Oaxaca, jars are often identified as cooking vessels,

TABLE A2.10. Combined paste vessel types from the LC12 E-F16–F9 midden

| Vessel type | Weight (g) | Percentage of context assemblage |
|---|---|---|
| Globular jar | 4293.6 | 53.1 |
| Conical bowl | 1158.5 | 14.3 |
| Bottle | 935.8 | 11.6 |
| Semi/hemispherical bowl or dish | 767.2 | 9.5 |
| Generic jar | 489.7 | 6.1 |
| Generic bowl | 172 | 2.1 |
| Collared jar | 125.2 | 1.5 |
| Tecomate | 70.7 | 0.9 |
| Grater bowl | 30.6 | 0.4 |
| Cylindrical vessel | 20.6 | 0.3 |
| Bule | 20 | 0.2 |
| Total | 8083.9 | |

by virtue of the carbonized food remains they sometimes contain (Cira Martínez López, personal communication, 2014). Jars at La Consentida include a variety of out-curving, out-leaning, and in-leaning neck types. Such variation suggests that Tlacuache phase jars may have served a variety of purposes. At least some of these jars bore carbonized food on the interior, suggesting that they were used for cooking. The radiocarbon date from one fragment, which was collected in stratum LC12 H-F4-s2, is calibrated to 1880–1625 cal BC (see table 1.1).

## Op. LC09 B Unit 1H, 1J Midden

In the northern-most extension of the Op. LC09 B area, excavations uncovered a deep layer of ashy sediment (LC09 B-F17) that contained a modest number of well-preserved and often decorated ceramic vessel fragments and other artifacts (e.g., figures 7.20.A and 8.7.B). Although LC09 B-F17 produced fewer diagnostic ceramics (791.1g in total) than did other deposits discussed in this appendix, it is nonetheless interesting as an example of the vessel assemblage from a context containing ceramics of the earliest style present at La Consentida and as a deposit at the extreme western end of Platform 1 (see figure 3.4). Analysis of this deposit demonstrates an emphasis on bowls, which comprised 47 percent of medium wares and 48 percent of coarse wares (tables A2.14–A2.16). It is likely due to this relatively high frequency of bowls that the deposit produced such a high percentage of decorated sherds (see discussion below). The context also had low percentages of jars,

**TABLE A2.11.** Medium brown ware vessel types from the LC12 H-F4 midden context

| Vessel type | Weight (g)x | Percentage of context assemblage |
|---|---|---|
| Globular jar | 5407 | 68.2 |
| Generic jar | 1852.8 | 23.4 |
| Semi/hemispherical bowl or dish | 448.8 | 5.7 |
| Bottle | 159.9 | 2.0 |
| Grater bowl | 30.1 | 0.4 |
| Generic bowl | 9.6 | 0.1 |
| Tecomate | 7.4 | 0.1 |
| Conical bowl | 6.8 | 0.1 |
| Total | 7922.4 | |

**TABLE A2.12.** Coarse brown ware vessel types from the LC12 H-F4 midden context

| Vessel type | Weight (g) | Percentage of context assemblage |
|---|---|---|
| Globular jar | 3048.4 | 86.3 |
| Generic jar | 416.1 | 11.8 |
| Semi/hemispherical bowl or dish | 63.4 | 1.8 |
| Grater bowl | 5.1 | 0.1 |
| Total | 3533 | |

**TABLE A2.13.** Combined paste vessel types from the LC12 H-F4 midden context

| Vessel type | Weight (g) | Percentage of context assemblage |
|---|---|---|
| Globular jar | 8455.4 | 73.8 |
| Generic jar | 2268.9 | 19.8 |
| Semi/hemispherical bowl or dish | 512.2 | 4.5 |
| Bottle | 159.9 | 1.4 |
| Grater bowl | 35.2 | 0.3 |
| Generic bowl | 9.6 | 0.1 |
| Tecomate | 7.4 | 0.1 |
| Conical bowl | 6.8 | 0.1 |
| Total | 11455.4 | |

which comprised 34 percent of medium wares and 46 percent of coarse wares. It is worth noting that the midden's small sample may bias these results. On the basis of the high frequency of bowls (some of which are well finished and decorated), the relative lack of storage or cooking jars, and iconographic artifacts such as a mask fragment (figure 7.20.A), it is likely that this midden results from feasting or other public events.

### Op. LC12 C Domestic Context

The Op. LC12 C domestic area was defined as the Structure 1 building (a possible house or domestic outbuilding) and surrounding occupational surfaces (LC12 C-F9 through C-F2; tables A2.17 through A2.19). While this approach likely combines artifacts from deposits that range somewhat in their chronology, it permits a more holistic analysis of the vessel types associated with probable households atop the northern end of the Substructure 2 mound than would an analysis restricted to materials recovered within Structure 1 itself (see chapters 4 and 5). Jars tend to dominate the sample, particularly among the coarse wares. Serving wares, such as bottles and bowls, are also present.

### Op. LC12 G Domestic Context

The Op. LC12 G domestic area was identified as the Structure 2 house remains and surrounding occupational area (LC12 G-F13 through G-F2) with probable domestic refuse (see chapters 4 and 5). The context is located just south of the Op. LC12 C domestic area on Substructure 2, and just upslope of the Op. LC12 H midden (see figure 3.4). The Op. LC12 G domestic area produced 6725.5 g of diagnostic vessel fragments, when worked sherd discs are included (see tables A2.20–A2.22, A2.27). The context is notable for its high relative quantity of jars (88 percent) and low relative quantities of bowls (10 percent) and bottles (1 percent) when paste types are combined.

### Sitewide Variation Among Vessel Types

For the purpose of studying sitewide distributions of vessel categories, I combined diagnostic fragments of each vessel type regardless of paste. Among vessel types identified at the site, the high percentage of jars in the collection as a whole is noteworthy. Describing jar variation by context, one can assess the relative percentages of jars among deposits in more detail. The resulting data (table A2.23) indicate the heavy emphasis on jars in all the contexts analyzed, but particularly in the LC12

**Table A2.14.** Medium brown ware vessel types from the LC09 B-F17 midden deposit

| Vessel type | Weight (g) | Percentage of context assemblage |
|---|---|---|
| Conical bowl | 202.7 | 33.6 |
| Generic jar | 162.6 | 26.9 |
| Bottle | 113.5 | 18.8 |
| Semi/hemispherical bowl or dish | 75.8 | 12.6 |
| Globular jar | 28.9 | 4.8 |
| Collared jar | 12.8 | 2.1 |
| Grater bowl | 7.4 | 1.2 |
| Total | 603.7 | |

**Table A2.15.** Coarse brown ware vessel types from the LC09 B-F17 midden deposit

| Vessel type | Weight (g) | Percentage of context assemblage |
|---|---|---|
| Globular jar | 85.7 | 45.7 |
| Semi/hemispherical bowl or dish | 70.7 | 37.7 |
| Conical bowl | 20.1 | 10.7 |
| Tecomate | 10.9 | 5.8 |
| Total | 187.4 | |

**Table A2.16.** Combined paste vessel types from the LC09 B-F17 midden deposit

| Vessel type | Weight (g) | Percentage of context assemblage |
|---|---|---|
| Conical bowl | 222.8 | 28.2 |
| Generic jar | 162.6 | 20.6 |
| Semi/hemispherical bowl or dish | 146.5 | 18.5 |
| Globular jar | 114.6 | 14.5 |
| Bottle | 113.5 | 14.3 |
| Collared jar | 12.8 | 1.6 |
| Tecomate | 10.9 | 1.4 |
| Grater bowl | 7.4 | 0.9 |
| Total | 791.1 | |

**TABLE A2.17.** Medium brown ware vessel types from domestic area near Structure 1

| Vessel type | Weight (g) | Percentage of context assemblage |
|---|---|---|
| Globular jar | 984.3 | 50.2 |
| Conical bowl | 272.1 | 13.9 |
| Bottle | 179.2 | 9.1 |
| Semi/hemispherical bowl or dish | 154.9 | 7.9 |
| Generic bowl | 107.8 | 5.5 |
| Grater bowl | 102.1 | 5.2 |
| Generic jar | 81 | 4.1 |
| Collared jar | 45 | 2.3 |
| Tecomate | 35.3 | 1.8 |
| Total | 1961.7 | |

**TABLE A2.18.** Coarse brown ware vessel types from domestic area near Structure 1

| Vessel type | Weight (g) | Percentage of context assemblage |
|---|---|---|
| Globular jar | 1263.4 | 75.0 |
| Semi/hemispherical bowl or dish | 177.5 | 10.5 |
| Generic jar | 70.5 | 4.2 |
| Conical bowl | 66.5 | 3.9 |
| Bottle | 44.7 | 2.7 |
| Generic bowl | 30.7 | 1.8 |
| Tecomate | 17.6 | 1.0 |
| Grater bowl | 14.7 | 0.9 |
| Total | 1685.6 | |

H-F4 midden and, to a lesser extent, the Structure 2 domestic area. There is less relative emphasis on jars in the LC09 B-F17, LC12 D-F11 through F8, and LC12 E-F16–F9 midden contexts. As aforementioned ethnographic information suggests, jars are likely associated with storage, transport, food preparation, or cooking practices rather than with serving. This interpretation is supported by their relative lack of decoration (see chapter 8 and the discussion of decoration below). The relatively lower emphasis on jars in the LC12 D-F11 through F8, LC12 E-F16–F9, and LC09B-F17 middens, and the relatively greater emphasis on serving wares and decorated vessels in those contexts, supports their identification as probable feasting deposits.

TABLE A2.19. Combined paste vessel types from domestic area near Structure 1

| Vessel type | Weight (g) | Percentage of context assemblage |
|---|---|---|
| Globular jar | 2247.7 | 61.6 |
| Conical bowl | 338.6 | 9.3 |
| Semi/hemispherical bowl or dish | 332.4 | 9.1 |
| Bottle | 223.9 | 6.1 |
| Generic jar | 151.5 | 4.2 |
| Generic bowl | 138.5 | 3.8 |
| Grater bowl | 116.8 | 3.2 |
| Tecomate | 52.9 | 1.5 |
| Collared jar | 45 | 1.2 |
| Total | 3647.3 | |

TABLE A2.20. Medium brown ware vessel types from domestic area near Structure 2

| Vessel type | Weight (g) | Percentage of context assemblage |
|---|---|---|
| Globular jar | 2521.1 | 72.9 |
| Generic jar | 316.2 | 9.1 |
| Generic bowl | 176.5 | 5.1 |
| Conical bowl | 147.7 | 4.3 |
| Semi/hemispherical bowl or dish | 134.9 | 3.9 |
| Grater bowl | 80.1 | 2.3 |
| Bottle | 50.7 | 1.5 |
| Collared jar | 30 | 0.9 |
| Total | 3457.2 | |

TABLE A2.21. Coarse brown ware vessel types from domestic area near Structure 2

| Vessel type | Weight (g) | Percentage of context assemblage |
|---|---|---|
| Globular jar | 2614.2 | 85.0 |
| Generic jar | 291.9 | 9.5 |
| Semi/hemispherical bowl or dish | 66.5 | 2.2 |
| Grater bowl | 32.9 | 1.1 |
| Bottle | 26.7 | 0.9 |
| Generic bowl | 26.3 | 0.9 |
| Conical bowl | 16.3 | 0.5 |
| Total | 3074.8 | |

**TABLE A2.22.** Combined paste vessel types from domestic area near Structure 2

| Vessel type | Weight (g) | Percentage of context assemblage |
|---|---|---|
| Globular jar | 5135.3 | 78.6 |
| Generic jar | 608.1 | 9.3 |
| Generic bowl | 202.8 | 3.1 |
| Semi/hemispherical bowl or dish | 201.4 | 3.1 |
| Conical bowl | 164 | 2.5 |
| Grater bowl | 113 | 1.7 |
| Bottle | 77.4 | 1.2 |
| Collared jar | 30 | 0.5 |
| Total | 6532 | |

**TABLE A2.23.** Jars by context

| Context | Total diagnostics (g) | Jars (g) | Percentage of jars relative to diagnostics |
|---|---|---|---|
| Op D middens | 2153.9 | 1087.0 | 50.5 |
| LC12 E-F16 through F9 midden | 8117.5 | 4908.5 | 60.5 |
| LC12 H-F4 midden | 11519.2 | 10724.3 | 93.1 |
| Op C (Structure 1) domestic area | 3647.3 | 2444.2 | 67.0 |
| Op G (Structure 2) domestic area | 6725.5 | 5773.4 | 85.8 |
| LC09 B-F17 midden | 791.1 | 290.0 | 36.7 |
| Averages | 5492.4 | 4204.6 | 65.6 |

Bowls are the second most common basic vessel category in the Tlacuache phase assemblage. Their patterns of within-site distribution (table A2.24) demonstrate that the LC09B-F17, LC12 D-F11 through F8, and LC12 E-F16–F9 middens and the Op. LC12 C (Structure 1) domestic area produced a notably higher percentage of bowls than did the Op. LC12 G (Structure 2) domestic area or the LC12 H-F4 midden. The Op. LC12 E midden produced the largest number of bowl fragments among analyzed contexts, and the LC09B-F17 midden produced by far the highest percentage of bowl remains relative to all diagnostic ceramics. Middens with the highest percentages of bowl fragments were likely associated with events that included the use of serving vessels, as would be the case with public feasts. The lack of bowls and the emphasis on jars in the LC12 H-F4 midden supports its identification as a food production midden. The notable differences between the two

TABLE A2.24. Bowls by context

| Context | Total diagnostics (g) | Bowls (g) | Percentage of bowls relative to diagnostics |
|---|---|---|---|
| Op D middens | 2153.9 | 571.4 | 26.5 |
| LC12 E-F16 through F9 midden | 8117.5 | 2148.9 | 26.5 |
| LC12 H-F4 midden | 11519.2 | 563.8 | 4.9 |
| Op C (Structure 1) domestic area | 3647.3 | 926.3 | 25.4 |
| Op G (Structure 2) domestic area | 6725.5 | 681.2 | 10.1 |
| LC09 B-F17 midden | 791.1 | 376.7 | 47.1 |
| Averages | 5492.4 | 878.1 | 23.4 |

domestic areas highlight possible economic differences among households, or perhaps some chronological variation not yet well understood or otherwise accounted for. A more detailed discussion of the use of specific bowls is possible with the analysis of grater bowl recovery context, as presented below.

Grater bowls among the Tlacuache phase ceramics are a relatively infrequent but ubiquitous vessel type at La Consentida. As demonstrated by the data summarized below (table A2.25), grater bowl fragments occur in all contexts discussed in this appendix. The identification of grater bowls is aided by their diagnostic interior incisions, which may artificially inflate their numbers relative to other vessel types due to the fact that rim and base fragments are often not necessary for their identification. Conversely, the small size of these artifacts means that the grams of grater bowls recorded may underestimate the numbers and/or importance of the actual vessels. Regardless, the comparison of relative quantities of grater bowls among contexts suggests some possibly significant patterns regarding craft production. Notably, the two contexts producing the highest percentages of grater bowls are both identified as domestic areas. The Op. LC12 C (Structure 1) domestic area produced the most grater bowl fragments per unit of diagnostic sherds, with nearly double the percentage calculated for the Op. LC12 G (Structure 2) context. The relative scarcity of grater bowls in deposits interpreted as feasting middens (LC09B-F17, LC12 D-F11 through F8, and LC12 E-F16–F9) suggests that, despite their intricate patterns of interior geometric incisions, these artifacts were infrequently used as fancy serving vessels. As discussed below with regard to worked sherd discs, differences among domestic contexts in the percentages of grater bowl fragments recovered relative to other vessel types may suggest a degree of economic specialization among households. This interpretation is further supported by the inference that grater bowls, perhaps more than any other single vessel type in the Tlacuache

TABLE A2.25. Grater bowls by context

| Context | Total diagnostics (g) | Grater bowls (g) | Percentage of grater bowls relative to diagnostics |
|---|---|---|---|
| Op D middens | 2153.9 | 17.2 | 0.8 |
| LC12 E-F16 through F9 midden | 8117.5 | 30.6 | 0.4 |
| LC12 H-F4 midden | 11519.2 | 35.2 | 0.3 |
| Op C (Structure 1) domestic area | 3647.3 | 116.8 | 3.2 |
| Op G (Structure 2) domestic area | 6725.5 | 113.0 | 1.7 |
| LC09 B-F17 midden | 791.1 | 7.4 | 0.9 |
| Averages | 5492.4 | 53.4 | 1.2 |

phase assemblage, may have been used for some specific activity such as food or pigment processing (for discussion of grater bowls in other parts of Mesoamerica, see Flannery and Marcus 1994:fig. 12.74, 12.101; Martínez López et al. 2000:165–66). These possible economic differences among households could also account for discrepancies in relative frequencies of worked sherd discs, as discussed below.

Among basic vessel categories of the Tlacuache phase, bottles are the third most common type, behind jars and bowls. Their greater relative frequency in the LC09B-F17, LC12 D-F11 through F8, and LC12 E-F16–F9 middens, in comparison with the Structure 1 and Structure 2 domestic areas and especially with the LC12 H-F4 midden, is noteworthy (table A2.26). In general, the discard context of bottles appears to be associated with remains of feasting deposits rather than storage/cooking middens (i.e., LC12 H-F4) or domestic areas. Within domestic contexts, the relative frequencies of bottles once again suggest some differences between the Op. LC12 C (Structure 1) and LC12 G (Structure 2) household areas. Rosemary Joyce (2003:249–50) has identified bottles as among the vessel types used for serving at public functions in ancient Mesoamerica, along with "finely-finished and decorated serving bowls." Their presence in some of La Consentida's middens further suggests the probable feasting events associated with these deposits, while their occurrence at the Op. LC12 C domestic area may indicate a locus of their production, decoration, storage, or a staging area for public feasts.

One category of relatively rare ceramic artifact that is not a vessel type per se, but which may indicate a specific crafting activity or set of related activities, is that of worked sherd discs. These perhaps served as grinders, weights, or lids, or for some unknown purpose. As discussed in chapter 6, similar sherd disc tools occur in the Valley of Oaxaca's Tierras Largas phase (Ramírez Urrea 1993:fig. 72). It is

**Table A2.26.** Bottles by context

| Context | Total diagnostics (g) | Bottles (g) | Percentage of bottles relative to diagnostics |
|---|---|---|---|
| Op D middens | 2153.9 | 327.9 | 15.2 |
| LC12 E-F16 through F9 midden | 8117.5 | 935.8 | 11.5 |
| LC12 H-F4 midden | 11519.2 | 159.9 | 1.4 |
| Op C (Structure 1) domestic area | 3647.3 | 223.9 | 6.1 |
| Op G (Structure 2) domestic area | 6725.5 | 77.4 | 1.2 |
| LC09 B-F17 midden | 791.1 | 113.5 | 14.2 |
| Averages | 5492.4 | 306.4 | 8.3 |

noteworthy that 3 percent (193.5 g) of all diagnostic ceramics from the Op. LC12 G (Structure 2) domestic context comprises worked sherd discs. While this amount may seem small, when all contexts are compared with regard to the presence of sherd disc tools, the Op. LC12 G domestic area produced 64 percent of all examples for contexts discussed in this appendix, regardless of the amount of sediment excavated or the amount of diagnostic pottery recovered (see table A2.27). The Op. LC12 G domestic area had a relative frequency of worked sherd discs nearly five times higher than that of any other context discussed in this appendix and may have been involved in different crafting activities than other areas (table A2.27).

## VESSEL DECORATION

The argument that some middens, such as that in Op. LC12 E, resulted from feasting is in part based on the presence of decorated ceramic vessels. Similarly, Clark and Blake (1994) have argued that decorated pottery of the Barra phase was used in public feasting events. While decorated vessel fragments are generally rare at La Consentida, their recovery in different quantities—and in conjunction with other evidence of feasting such as serving wares (i.e., bowls and bottles) and faunal remains—suggests the use of decorated ceramics in public events such as feasts. When the percentages of decorated ceramics in different middens at La Consentida are compared, it is clear that the Op. LC12 H midden contained a far lower relative quantity (1.9 percent) of decorated ceramics than did other deposits (table A2.28). The Op. LC12 E midden and Op. LC12 D sheet midden deposits (located at the interfaces of the LC12 D-F11 through F8 fill layers) contained relatively high quantities of decorated ceramics (9.5 and 8.5 percent, respectively). The Op. LC09 B midden (LC09 B-F17) produced by far the highest relative quantity of decorated ceramics (53.5 percent), though the

**TABLE A2.27.** Sherd disc tools by context

| Context | Total diagnostics (g) | Sherd disc tools (g) | Percentage of sherd disc tools relative to diagnostics |
|---|---|---|---|
| Op D middens | 2153.9 | 11.9 | 0.6 |
| LC12 E-F16 through F9 midden | 8117.5 | 33.6 | 0.4 |
| LC12 H-F4 midden | 11519.2 | 63.8 | 0.6 |
| Op C (Structure 1) domestic area | 3647.3 | 0.0 | 0.0 |
| Op G (Structure 2) domestic area | 6725.5 | 193.5 | 2.9 |
| LC09 B-F17 midden | 791.1 | 0.0 | 0.0 |
| Averages | 5492.4 | 50.5 | 0.8 |

small sample of diagnostic pottery recovered from this context may partly explain this circumstance. Many scholars (e.g., Clark and Cheetham 2002:294; Hayden 1990; G. Lowe 1975; Rosenswig 2007) view feasting as part of a suite of communal events promoting initial social complexity. The feasts during which La Consentida's decorated ceramics were likely employed may also have been venues for the use of artifacts implying public performance, including musical instruments and ceramic masks.

## COMPARING ANALYZED CONTEXTS

As discussed above, comparison of the middens excavated at La Consentida suggests that they were produced by different activities. Comparisons of domestic areas indicate that they were relatively similar to one another, but nonetheless demonstrate some differences when carefully analyzed. In this section, I discuss similarities and differences among various contexts and also briefly relate the results of Chi-square and Fisher's exact test nonparametric statistics to test my interpretations.[2] Two contexts that contained surprisingly similar artifact assemblages are the Op. LC12 D sheet middens and the Op. LC12 E midden. The Op. LC12 E midden produced a slightly lower percentage of medium brown ware bowls (42 percent, counting cylindrical vessels) than did the Op. LC12 D middens (44 percent) (tables A2.5 and A2.8). The Op. LC12 E midden also contained a lower percentage of jars (75 percent) among coarse wares than did the Op. LC12 D middens (92 percent) (tables A2.6 and A2.9). In general, however, the Op. LC12 D and LC12 E middens contained similar ratios of vessel types, though the latter produced a lower percentage of tecomates, a lower percentage of bottles, and a higher percentage of jars overall (tables A2.7 and A2.10). Both the Op. LC12 D and LC12 E middens contained nicely slipped and decorated sherds, though better-preserved examples occurred

**TABLE A2.28.** Relative frequencies of decorated ceramics by midden context

| Context | Total diagnostics (g) | Decorated ceramics (g) | Percentage of decorated ceramics relative to diagnostics |
|---|---|---|---|
| Op D middens | 2153.9 | 184 | 8.5 |
| Op E midden | 8117.5 | 773 | 9.5 |
| Op H midden | 11519.2 | 218 | 1.9 |
| Op LC09 B 1H, 1J midden | 791.1 | 428 | 53.5 |
| Averages | 5645.4 | 400.8 | 18.4 |

in the more extensively excavated LC12 E area (see figures 8.8 and 8.9.B). As demonstrated in table A2.28, both these contexts produced relatively high percentages of decorated ceramics compared to a probable food preparation midden (LC12 H-F4), though neither context produced the extremely high relative frequency of decorated vessel fragments that the LC09 B-F17 midden context did.

The similar assemblages of the Op. LC12 D and LC12 E middens suggest that the deposits result from similar activities, which likely differed from those producing the Op LC12 H midden (see discussion above). The high percentage of jars and a near lack of bowls (particularly of conical bowls, which bear fancy surface treatment and decoration more frequently than do hemi/semispherical examples) suggest different uses for the ceramics in the LC12 H midden than those in the other two areas. Based on the ethnographic evidence regarding jar use that I discussed above, as well as the ashy nature of the midden matrix itself, it is probable that the Op. LC12 H midden resulted from food preparation (see Skibo and Feinman 1999). The rapid deposition of numerous large jars in just a single or very few discard events may suggest that the refuse resulted from an infrequent event such as cooking for a public feast.

Due to its domestic nature, the Op. LC12 C context is most comparable to the LC12 G domestic area. It is worth noting that the vessel assemblage in the Op. LC12 C area is diverse and that the ratios are similar to those of the Op. LC12 D and LC12 E middens. In comparison with the Op. LC12 C domestic context, Op. LC12 G produced more jars (83 percent among medium brown wares and 95 percent among coarse wares; tables A2.19–A2.21). This context contained a high percentage of jars second only to that in the LC12 H-F4 midden, which is located nearby, and which may result from activities on the same southern end of Substructure 2 (see figure 3.4). This emphasis on jars in the Op. LC12 G domestic area also means that percentages of other vessel types such as hemi/semispherical bowls, conical bowls, grater bowls, and bottles are generally lower than they are for the Op. LC12

C domestic context. Based on the presence of the probable domestic structures in these areas and the relative proximity of the two operations to one another (see figure 3.4), one might expect their vessel assemblages to be more similar. The fact that they are not may underscore subtle economic differences at the household level. Discrepancies in the quantities of worked sherd discs and grater bowls, in particular, may serve as evidence of these differences.

When the contexts described in this appendix are compared with one another, one of the most notable patterns is the high relative frequency of jars in the Op. LC12 H midden and the LC12 G domestic area. If all analyzed contexts are compared using Chi-square nonparametric statistics (see Hepp 2015:fig. A2.24), one finds that they vary in terms of their ceramic assemblages in a statistically significant way ($p < 0.0001$). With the exception of the LC12 H cooking midden, the other middens have high relative frequencies of serving vessels such as bowls and bottles. When middens are compared with one another using Chi-square analysis (see Hepp 2015:fig. A2.25), the differences are again clear and are statistically significant ($p < 0.0001$). Ceramics in the Op. LC12 H cooking midden are nearly exclusively jars. Middens that likely resulted from public feasting contain more serving wares, as is especially clear when analyzing the Op. LC09 B midden. These results are also consistent with the comparison of vessel decoration (table A2.28), which suggests that some middens contained more serving wares than did others. When the Op. LC12 C (Structure 1) and LC12 G (Structure 2) domestic areas are compared, the low number of bottles makes Chi-square analysis unreliable. Instead, a Fisher's exact test (see Hepp 2015:fig. A2.26) demonstrates that the two contexts are significantly different ($p = 0.0037$).

## DISCUSSION

The patterns of vessel type variation discussed in this appendix demonstrate similarities and differences among excavated contexts at La Consentida. In general, the LC09B-F17, LC12 D-F11 through F8, and LC12 E-F16–F9 middens contain the serving wares (i.e., bowls and bottles) and a relative lack of jars consistent with their identification as feasting deposits. Their relative quantities of vessel decoration support this interpretation (table A2.28). The faunal remains from the Op. LC12 E midden (especially of fish and shellfish) serve as evidence of a large event or a few events incorporating food consumption. Cross-fitting ceramic remains from various depths within the LC12 E midden suggest that the deposit resulted from a few large events. The LC12 H midden, where the ceramics consisted almost entirely of globular jars, appears to have resulted from the rapid deposition of specific-use ceramics. The very ashy nature of the Op LC12 H midden matrix (see appendix 1) suggests that it was a cooking deposit, perhaps resulting from the preparation of a feast such as those that produced the other middens discussed here. On the basis of this evidence, and on the decorated ceramics found in certain deposits, I argue that public feasting events were a key component of Early Formative period social interaction at La Consentida.

The nature of possible economic differences among domestic contexts in Ops. LC12 C and LC12 G is difficult to identify on the basis of the relatively modest quantities of diagnostic ceramic remains recovered there. Nevertheless, discrepancies in the relative frequencies of jars, bowls—grater bowls specifically—bottles, and worked sherd discs suggest that the households may have fulfilled different roles in the community. Whether those roles result entirely from economic differences, or whether sampling effects or fine-grained chronological variations are partly responsible, is not clear. The differences in relative quantities of bottles and grater bowls may suggest that the Structure 2 house slightly postdates the Structure 1 house, as those two vessel types appear to have varied in their relative frequencies over the course of site occupation.

Other contextual questions could be asked of the diagnostic ceramics at La Consentida. Small primary contexts, such as briefly occupied surfaces below and within fill deposits in Op. LC12 A, are not discussed here, due to the low quantities of ceramics they produced. Coarse-grained chronological data could also be gleaned from a comparison of fill deposits in Op. LC09 A, LC12 F, and the deep LC12 A trench. These results are not presented here because I chose to prioritize the analysis of primary contexts. Ongoing and future studies of curated materials from the 2009 and 2012 excavations at La Consentida, as well as those of artifacts recovered in subsequent research projects at the site, will likely shed light on the general patterns of vessel variation I have identified.

# Notes

## CHAPTER 1: LA CONSENTIDA AS AN EARLY FORMATIVE MESOAMERICAN VILLAGE

1. In this book, dates reported with "BCE" and "CE" are uncalibrated. Calibrated radiocarbon dates, and the time spans based on them, are reported with "cal BC."

2. Some readers may take exception to my use of the term "village," rather than "hamlet," to describe La Consentida. Historically (e.g., MacNeish 1964), archaeologists have sometimes defined a hamlet as a settled community of a few dozen people, while they described villages as comprising perhaps one hundred to a few hundred people. As I discuss in chapter 5, I estimate that La Consentida's population probably fluctuated around an average of eighty. Defining this group as a hamlet or a village, over a difference of perhaps only twenty people, strikes me as arbitrary. As I hope to demonstrate in this book, the social processes taking place at La Consentida were those of the first settled and increasingly agricultural communities of Mesoamerica. Rather than making fine distinctions between such communities over population differences that I doubt were socially meaningful, I will simply refer to them, and to La Consentida, as villages.

3. AMS radiocarbon calibration performed with IntCal 13 curve by OxCal 4.3.2 and rounded to five-year increments. Unless otherwise stated, I report calibrated dates with $2\sigma$ probability (Reimer et al. 2013). Dates appearing in this text differ slightly from those presented previously (e.g., Hepp 2015:5) due to the use of different rounding and calibration conventions (Hepp in press).

4. By conventional definition, the presence of ceramics establishes La Consentida as dating to the Early Formative period, rather than to the Late Archaic period (see R. Joyce 2004b).

5. The calibrated date ranges for the Tierras Largas and Lagunita phases are my own estimates based on published, uncalibrated dates and the IntCal 13 curve by OxCal 4.3.2.

6. Marcus Winter (personal communication, 2013) suggests that Espiridión should be incorporated into the Tierras Largas phase, due to similarities in the ceramics. Since no radiocarbon dates exist for this phase, the ranges I present are hypothetical.

7. This is an amendment to previous descriptions of the site (e.g., Hepp 2015) as being located about 5 kilometers from the modern coastline.

8. Platform 1 is the only major platform at the site. La Consentida's earthen architecture essentially comprises Platform 1 and several substructural mounds.

## CHAPTER 3: METHODS AND MAPPING FOR THE LA CONSENTIDA ARCHAEOLOGICAL PROJECT

1. Sarah Barber (2009) and Arthur Joyce codirected the 2008 regional GPR study. The GPR was performed under Barber's INAH permit and with assistance from University of Central Florida (UCF) students. TAG Research by Sturm, Inc. was responsible for data analysis. A 500 MHz antenna was used, and anomalies were detected at a depth of about 1.7 m (Barber 2009:4–10).

2. Combining these two bone dates was accomplished by using the R_Combine command in OxCal 4.3.2, rather than simply reporting the mean of the two dates. I am grateful to Jon Lohse (personal communication, 2017) for his guidance on this process.

## CHAPTER 6: DIET AND CHANGING CULINARY TASTES

1. For detailed discussion of the methods employed for faunal analysis, as well as of the faunal remains identified for each of the contexts considered, see Pérez Hernández and Hepp (2015).

2. Silvia Pérez Hernández (personal communication, 2015) interpreted some fish bone from La Consentida as having been boiled, based on discoloration of the bone.

3. Note that statistical analyses discussed here (performed using JMPTM Pro 11 software) relate only to the nine sets of human remains for which dental attrition, age, and chronological data were available. A sample size of 9 is too small for robust statistical tests. These results should thus be considered a heuristic device to inform understandings of dietary change.

## CHAPTER 7: SOCIAL ORGANIZATION

1. I acknowledge the potential for nonbinary gendered identities (neither feminine nor masculine) in ancient Oaxaca. In fact, as Stephen (2002) has discussed regarding the *muxe*,

NOTES    277

Oaxaca is especially relevant for such discussions. I have yet to find compelling evidence for such an interpretation among coastal Oaxacan Formative period figurines, however.

2. Raymond Mueller, personal communication, 2014

3. For an example of previous archaeoastronomy research in coastal Oaxaca, see Sánchez Nava and Šprajc (2012).

### CHAPTER 8: NO VILLAGE IS AN ISLAND

1. Comparison based on published reports of Barra phase vessel ratios (Clark and Blake 1994:25) and my own estimated percentages from several Tierras Largas contexts (Flannery and Marcus 1994:tables 10.1, 10.2, and 11.1). Tierras Largas percentages are based on counts of diagnostic sherds, and Tlacuache ratios are based on grams of diagnostic sherds. Tierras Largas percentages do not add up to 100, due to unidentified sherds counted in aforementioned tables.

2. All statistical tests performed using JMP™ Pro 11

3. According to Bartolomé and Barabas (1996:170–72), modern Chatino children are weaned at about two years of age and quickly take on mature social roles within the family. Girls, for example, begin making tortillas by the age of three or four. Ethnographic evidence from elsewhere in Oaxaca suggests that other native groups (such as the Zapotecs) wean early and transition their children to an adult diet and adult economic roles quickly (Nader 1969:356; Parsons 1936:85–86; Sellen 2001; Taylor 1960:192, 195, 328). Such data may conflict with the interpretation that the grater bowls were used to prepare weaning foods, though the La Consentida population would have had different subsistence practices from modern groups.

4. Various type specimens for Ojochi and Bajío phase ceramics. Courtesy of the Peabody Museum of Natural History, Division of Anthropology, Yale University; http://peabody.yale.edu.

5. XRF results provided by the Archaeometry Laboratory at the University of Missouri Research Reactor.

### CHAPTER 9: LA CONSENTIDA

1. Type specimens for Ojochi and Bajío phase ceramic bottle necks. Courtesy of the Peabody Museum of Natural History, Division of Anthropology, Yale University; http://peabody.yale.edu.

### APPENDIX 2: THE TLACUACHE CERAMIC ASSEMBLAGE

1. All statistical tests were produced using JMP™ Pro 11

2. Comparisons of all contexts are likely affected by sample sizes. For example, the Op. LC12 E midden produced nearly four times as many grams of diagnostic ceramics as the LC12 D middens, at 8117.5 g. The statistical tests employed here are therefore performed on percentages of vessel types to control for the sizes of excavated areas. Because rare artifact classes such as bules, tecomates, and worked sherd discs needed to be eliminated in order to comply with the theoretical requirements of Chi-square analyses, percentages per contexts do not equal 100. This is not problematic, since vessel category percentages are treated as real numbers to control for sample sizes. Note that each analysis presented compares the contexts in question with site averages (produced by combining the six contexts discussed in this appendix) to demonstrate where particular deposits deviate from the norm.

# Works Cited

Aaberg, Stephen, and Jay Bonsignore. 1975. "A Consideration of Time and Labor Expenditure in the Construction Process at the Teotihuacán Pyramid of the Sun and the Poverty Point Mound." In *Three Papers on Mesoamerican Archaeology*, edited by John Graham and Robert Heizer, 40–78. Contributions of the University of California Archaeological Research Facility 24. University of California, Berkeley.

Abrams, Elliot. 1989. "Architecture and Energy: An Evolutionary Perspective." *Archaeological Method and Theory* 1 (January): 47–87.

Abrams, Elliot. 1994. *How the Maya Built Their World: Energetics and Ancient Architecture*. University of Texas Press, Austin.

Adams, Jenny L. 1988. "Use-Wear Analyses on Manos and Hide-Processing Stones." *Journal of Field Archaeology* 15(3): 307–15.

Adams, Robert McC. 1966. *The Evolution of Urban Society: Early Mesopotamia and Prehispanic Mexico*. Aldine Publishing Company, Chicago.

Aguilar, José, and Guy David Hepp. 2015. Appendix 5: Analysis of Human Remains. In "La Consentida: Initial Early Formative Period Settlement, Subsistence, and Social Organization on the Pacific Coast of Oaxaca, Mexico," 536–56. Unpublished PhD dissertation. University of Colorado, Boulder.

Ahern, Frances. 2010. *Identidad y estilo entre las alfareras mixtecas y amuzgas de la Costa de Oaxaca y Guerrero, México*. Arqueología Oaxaqueña 3. Instituto Nacional de Antropología e Historia, Oaxaca, Mexico.

Altman, Patricia B., and Caroline D. West. 1992. *Threads of Identity: Maya Costume of the 1960s in Highland Guatemala*. Fowler Museum of Cultural History, Los Angeles.

Anawalt, Patricia Rieff. 1998. "They Came to Trade Exquisite Things: Ancient West Mexican-Ecuadorian Contacts." In *Ancient West Mexico: Art and Archaeology of the Unknown Past*, edited by Richard F. Townsend, 232–49. Art Institute of Chicago, Chicago.

Arnold, Jeanne E. 1993. "Labor and the Rise of Complex Hunter-Gatherers." *Journal of Anthropological Archaeology* 12(1): 75–119.

Arnold, Jeanne E. 1996. "The Archaeology of Complex Hunter-Gatherers." *Journal of Archaeological Method and Theory* 3(1): 77–126.

Arnold, Jeanne E., and Julienne Bernard. 2005. "Negotiating the Coasts: Status and the Evolution of Boat Technology in California." *World Archaeology* 37(1): 109–31.

Arnold, Philip J. III. 1999. "Tecomates, Residential Mobility, and Early Formative Occupation in Coastal Lowland Mesoamerica." In *Pottery and People: A Dynamic Interaction*, edited by James M. Skibo and Gary M. Feinman, 157–70. University of Utah Press, Salt Lake City.

Arnold, Philip J. III. 2000. "Sociopolitical Complexity and the Gulf Olmecs: A View from the Tuxtla Mountains, Veracruz, Mexico." In *Olmec Art and Archaeology in Mesoamerica*, edited by John E. Clark and Mary E. Pye, 117–35. National Gallery of Art, Washington, DC.

Arnold, Philip J. III. 2003. "Early Formative Pottery from the Tuxtla Mountains and Implications for Gulf Olmec Origins." *Latin American Antiquity* 14(1): 29–46.

Arnold, Philip J. III. 2009. "Settlement and Subsistence among the Early Formative Gulf Olmec." *Journal of Anthropological Archaeology* 28(4): 397–411.

Ashmore, Wendy. 2002. "'Decisions and Dispositions': Socializing Spatial Archaeology." *American Anthropologist* 104(4): 1172–83.

Ashmore, Wendy. 2004. "Classic Maya Landscapes and Settlement." In *Mesoamerican Archaeology*, edited by Julia A. Hendon and Rosemary A. Joyce, 169–91. Blackwell, Malden, MA.

Bachand, Bruce R. 2013. "Las fases formativas de Chiapa de Corzo: Nuevo evidencia e interpretaciones." *Estudios de Cultura Maya* XLII: 11–52.

Baillie, Harold Barry Andrew. 2012. *Late Classic Río Viejo Mound 1 Construction and Occupation, Oaxaca, Mexico*. Unpublished MA thesis, University of Colorado, Boulder.

Banning, E. B. 1998. "The Neolithic Period: Triumphs of Architecture, Agriculture, and Art." *Near Eastern Archaeology* 61(4): 188–237.

Banning, E. B. 2000. *The Archaeologist's Laboratory: The Analysis of Archaeological Data*. Kluwer Academic / Plenum Publishers, New York.

Banning, E. B. 2003. "Housing Neolithic Farmers." *Near Eastern Archaeology* 66(1/2): 4–21.

Banning, E. B. 2011. "So Fair a House: Göbekli Tepe and the Identification of Temples in the Pre-Pottery Neolithic of the Near East." *Current Anthropology* 52(5): 619–60.

Bar-Yosef, Ofer, and Anna Belfer-Cohen. 1989. "The Origins of Sedentism and Farming Communities in the Levant." *Journal of World Prehistory* 3(4): 447–98.

Barber, Sarah B. 2005. "Heterogeneity, Identity, and Complexity: Negotiating Status and Authority in Terminal Formative Coastal Oaxaca." Unpublished PhD dissertation. Boulder: University of Colorado.

Barber, Sarah B. 2009. "Estudio geofísico del Río Verde Inferior: Informe final entregado al consejo de arqueología, Centro INAH Oaxaca." University of Central Florida, Orlando.

Barber, Sarah B., and Guy David Hepp. 2012. "Ancient Aerophones of Coastal Oaxaca, Mexico: The Archaeological and Social Context of Music." In *Studien zur Musickarchäologie VIII: Sound from the Past: The Interpretation of Musical Artifacts in an Archaeological Context*, edited by R. Eichmann, J. Fang, and L-C. Koch, 259–70. VML, Rahden/Westf.

Barber, Sarah B., and Arthur A. Joyce. 2007. "Polity Produced and Community Consumed: Negotiating Political Centralization through Ritual in the Lower Río Verde Valley, Oaxaca." In *Ritual Economy: Archaeological and Ethnological Perspectives*, edited by E. Christian Wells and Karla L. Davis-Salazar, 221–44. University Press of Colorado, Boulder.

Barber, Sarah B., Arthur A. Joyce, Arion T. Mayes, and Michelle Butler. 2013. "Formative Period Burial Practices and Cemeteries." In *Polity and Ecology in Formative Period Coastal Oaxaca*, edited by Arthur A. Joyce, 97–133. University Press of Colorado, Boulder.

Barber, Sarah B., and Mireya Olvera Sánchez. 2012. "A Divine Wind: The Arts of Death and Music in Terminal Formative Oaxaca." *Ancient Mesoamerica* 23(1): 9–24.

Bartolomé, Miguel A., and Alicia M. Barabas. 1996. *Tierra de la palabra: Historia y etnografía de los chatinos de Oaxaca*. 2nd ed. Instituto Oaxaqueño de las Culturas, Instituto Nacional de Antropología e Historia, Mexico City, Oaxaca City.

Baxter, Jane Eva. 2008. "The Archaeology of Childhood." *Annual Review of Anthropology* 37 (October): 159–75.

Bazelmans, Jos. 2002. "Moralities of Dress and the Dress of the Dead in Early Medieval Europe." In *Thinking through the Body: Archaeologies of Corporeality*, edited by Yannis Hamilakis, Mark Pluciennik, and Sarah Tarlow, 71–84. Kluwer Academic/Plenum, New York.

Beck, Daniel D., and Charles H. Lowe. 1991. "Ecology of the Beaded Lizard, Heloderma horridum, in a Tropical Dry Forest in Jalisco, México." *Journal of Herpetology* 25(4): 395–406.

Berlinski, David. 1976. *On Systems Analysis*. MIT Press, Cambridge.

Berry, Kimberly A., and Patricia A. McAnany. 2007. "Reckoning with the Wetlands and Their Role in Ancient Maya Society." In *The Political Economy of Ancient*

*Mesoamerica: Transformations during the Formative and Classic Periods*, edited by Vernon L. Scarborough and John E. Clark, 149–62. University of New Mexico Press, Albuquerque.

Binford, Lewis R. 1968. "Post-Pleistocene Adaptations." In *New Perspectives in Archaeology*, edited by Lewis R. Binford and Sally R. Binford, 313–41. Aldine, Chicago.

Binford, Lewis R. 1971. "Mortuary Practices: Their Study and Their Potential." *Memoirs of the Society for American Archaeology* 25: 6–29.

Binford, Lewis R. 1983. *In Pursuit of the Past*. Thames and Hudson, London.

Blake, Michael. 2006. "Dating the Initial Spread of Zea Mays." In *Histories of Maize: Multidisciplinary Approaches to the Prehistory, Linguistics, Biogeography, Domestication, and Evolution of Maize*, edited by John E. Staller, Robert H. Tykot, and Bruce F. Benz, 55–72. Academic Press, New York.

Blake, Michael, Brian S. Chisholm, John E. Clark, Barbara Voorhies, and Michael W. Love. 1992. "Prehistoric Subsistence in the Soconusco Region." *Current Anthropology* 33(1): 83–94.

Blake, Michael, and John E. Clark. 1989. "The Emergence of Hereditary Inequality: The Case of Pacific Coastal Chiapas, Mexico." Paper presented the Circum-Pacific Prehistory Conference, Seattle, Washington.

Blake, Michael, and John. E. Clark. 1999. "The Emergence of Hereditary Inequality: The Case of Pacific Coastal Chiapas, Mexico." In *Pacific Latin America in Prehistory: The Evolution of Archaic and Formative Cultures*, edited by Michael Blake, 55–73. Washington State University Press, Pullman.

Blanton, Richard E. 1998. "Beyond Centralization: Steps toward a Theory of Egalitarian Behavior in Archaic States." In *Archaic States*, edited by Gary M. Feinman and Joyce Marcus. School of American Research Press, Santa Fe.

Blanton, Richard E., Jill Appel, Laura Finsten, Stephen A. Kowalewski, Gary M. Feinman, and Eva Fisch. 1979. "Regional Evolution in the Valley of Oaxaca, Mexico." *Journal of Field Archaeology* 6(4): 369–90.

Blanton, Richard E., Gary M. Feinman, Stephen A. Kowalewski, and Linda M. Nicholas. 1999. *Ancient Oaxaca*. Cambridge University Press, Cambridge.

Blanton, Richard E., Gary M. Feinman, Stephen A. Kowalewski, and Peter N. Peregrine. 1996. "A Dual-Processual Theory for the Evolution of Mesoamerican Civilization." *Current Anthropology* 37(1): 1–14.

Blomster, Jeffrey P. 1998. "Context, Cult, and Early Formative Period Public Ritual in the Mixteca Alta: Analysis of a Hollow-Baby Figurine from Etlatongo, Oaxaca." *Ancient Mesoamerica* 9(2): 309–26.

Blomster, Jeffrey P. 2004. *Etlatongo: Social Complexity, Interaction, and Village Life in the Mixteca Alta of Oaxaca, Mexico*. Wadsworth / Thomson Learning, Belmont.

Blomster, Jeffrey P. 2009. "Identity, Gender, and Power: Representational Juxtapositions in Early Formative Figurines from Oaxaca, Mexico." In *Mesoamerican Figurines: Small-Scale Indices of Large-Scale Social Phenomena*, edited by Christina T. Halperin, Katherine A. Faust, Rhonda Taube, and Aurore Giguet, 119–48. University Press of Florida, Gainesville.

Blomster, Jeffrey P., and Michael D. Glascock. 2010. "Procurement and Consumption of Obsidian in the Early Formative Mixteca Alta: A View from the Nochixtlán Valley, Oaxaca, Mexico." In *Crossing the Straits: Prehispanic Obsidian Source Exploitation in the North Pacific Rim*, edited by Yaroslav V. Kuzmin and Michael D. Glascock, 183–200. Archaeopress, Oxford.

Borejsza, Aleksander, Charles D. Frederick, Luis Morett Alatorre, and Arthur A. Joyce. 2014. "Alluvial Stratigraphy and the Search for Preceramic Open-Air Sites in Highland Mesoamerica." *Latin American Antiquity* 25(3): 278–99.

Borstein, Joshua A. 2002. "Tripping over Colossal Heads: Settlement Patterns and Population Development in the Upland Olmec Heartland." Unpublished PhD dissertation. Pennsylvania State University, State College.

Bourdieu, Pierre. 1977. *Outline of a Theory of Practice*. Ed. Jack Goody. *Cambridge Studies in Social Anthropology*. Vol. 16. Cambridge University Press.

Bove, Frederick J. 1989. "Dedicated to the Costeños: Introduction and New Insights." In *New Frontiers in the Archaeology of the Pacific Coast of Southern Mesoamerica*, edited by Frederick J. Bove and Lynette Heller, 1–13. Anthropological Research Papers No. 39, Arizona State University, Tempe.

Boyd, Brian. 2006. "On 'Sedentism' in the Later Epipalaeolithic (Natufian) Levant." *World Archaeology* 38(2): 164–78.

Boyd, M., T. Varney, C. Surette, and J. Surette. 2008. "Reassessing the Northern Limit of Maize Consumption in North America: Stable Isotope, Plant Microfossil, and Trace Element Content of Carbonized Food Residue." *Journal of Archaeological Science* 35(9): 2545–56.

Bradley, John E. 1994. "Trondadora Vieja: An Archaic and Early Formative Site in the Arenal Region." In *Archaeology, Volcanism, and Remote Sensing in the Arenal Region, Costa Rica*, edited by Payson D. Sheets and Brian R. McKee, 73–86. University of Texas Press, Austin.

Bradley, Richard. 1993. *Altering the Earth: The Origins of Monuments in Britain and Continental Europe*. Society of Antiquaries of Scotland, Edinburgh.

Bradley, Richard. 1998. *The Significance of Monuments: On the Shaping of Human Experience in Neolithic and Bronze Age Europe*. Routledge, New York.

Bradley, Richard. 2000. *An Archaeology of Natural Places*. Routledge, New York.

Bradley, Richard. 2005. *Ritual and Domestic Life in Prehistoric Europe*. Routledge, New York.

Brady, James E., and Wendy Ashmore. 1999. "Mountains, Caves, Water: Ideational Landscapes of the Ancient Maya." In *Archaeologies of Landscape: Contemporary Perspectives*, edited by Wendy Ashmore and A. Bernard Knapp, 124–48. Blackwell, Malden, MA.

Bronson, Bennet. 1966. "Roots and the Subsistence of the Ancient Maya." *Southwestern Journal of Anthropology* 22(3): 251–79.

Brumfiel, Elizabeth M. 2003. "It's A Material World: History, Artifacts, and Anthropology." *Annual Review of Anthropology* 32(1): 205–23.

Brumfiel, Elizabeth M. 2006. "Cloth, Gender, Continuity, and Change: Fabricating Unity in Anthropology." *American Anthropologist* 108(4): 862–77.

Brush, Charles F. 1965. "Pox Pottery: Earliest Identified Mexican Ceramic." *Science* 149(3680): 194–95.

Brush, Charles F. 1969. A Contribution to the Archaeology of Coastal Guerrero, Mexico. Unpublished PhD dissertation. New York: Columbia University.

Brzezinski, Jeffrey S., Arthur A. Joyce, and Sarah B. Barber. 2017. "Constituting Animacy and Community in a Terminal Formative Bundled Offering from the Coast of Oaxaca, Mexico." *Cambridge Archaeological Journal* 27(3): 511–31.

Byrd, Brian F. 1994. "Public and Private, Domestic and Corporate: The Emergence of the Southwest Asian Village." *American Antiquity* 59(4): 639–66.

Caldwell, Joseph R. 1958. *Trend and Tradition in the Prehistory of the Eastern United States*. Vol. 88. American Anthropological Association, Menasha.

Carballo, David M. 2009. "Household and Status in Formative Central Mexico: Domestic Structures, Assemblages, and practices at La Laguna, Tlaxcala." *Latin American Antiquity* 20(3): 473–502.

Carballo, David M., ed. 2013. *Cooperation and Collective Action: Archaeological Perspectives*. University Press of Colorado, Boulder.

Carmack, Robert M., Janine Gasco, Marilyn A. Masson, and Michael E. Smith. 2016. "Late Postclassic Mesoamerica." In *The Legacy of Mesoamerica*, edited by Robert M. Carmack, Janine Gasco, and Gary H. Gossen, 78–119. 2nd ed. Routledge, New York.

Carr, Christopher. 1995. "Mortuary Practices: Their Social, Philosophical-Religious, Circumstantial, and Physical Determinants." *Journal of Archaeological Method and Theory* 2(2): 105–200.

Carrasco, Michael D., and Joshua D. Englehardt. 2015. "Diphrastic Kennings on the Cascajal Block and the Emergence of Mesoamerican Writing." *Cambridge Archaeological Journal* 25(3): 635–56.

Caso, Alfonso, Ignacio Bernal, and Jorge R. Acosta. 1967. *La Cerámica de Monte Albán*. Vol. 12. Instituto Nacional de Antropología e Historia, Mexico City.

Catlin, George. 1953. "A Choktaw Ball Game." In *Primitive Heritage: An Anthropological Anthology*, edited by Margaret Mead and Nicholas Calas, 289–95. Random House, New York.

Cervantes Pérez, José Manuel, Tito Cuauhtémoc Mijangos García, and Augustín E. Andrade Cuautle. 2017. "Collective Memory in San Sebastián Etla, Oaxaca: Bioarchaeological approaches to an Early Formative period (1400–1200 BCE) mortuary space." *Journal of Archaeological Science: Reports* 13 (June): 737–43.

Chandler-Ezell, Karol, Deborah M. Pearsall, and James A. Zeidler. 2006. "Root and Tuber Phytoliths and Starch Grains Document Manioc (Manihot Esculenta), Arrowroot (Maranta Arundinacea), and Lleren (Calathea sp.) at the Real Alto Site, Ecuador." *Economic Botany* 60(2): 103–20.

Cheetham, David. 2010. "Cultural Imperatives in Clay: Early Olmec Carved Pottery from San Lorenzo and Cantón Corralito." *Ancient Mesoamerica* 21(1): 165–85.

Childe, V. Gordon. 1936. *Man Makes Himself.* Watts, London.

Childe, V. Gordon. 1950. "The Urban Revolution." *Town Planning Review* 21(1): 3–17.

Chisholm, Brian, and Michael Blake. 2006. "Diet in Prehistoric Soconusco." In *Histories of Maize: Multidisciplinary Approaches to the Prehistory, Linguistics, Biogeography, and Evolution of Maize*, edited by John E. Staller, Robert H. Tykot, and Bruce F. Benz, 161–72. Academic Press, New York.

Choe, Chong Pil, and Martin T. Bale. 2002. "Current Perspectives on Settlement, Subsistence, and Cultivation in PreHistoric Korea." *Arctic Anthropology* 39(1/2): 95–121.

Çilingiroğlu, Çiler. 2005. "The Concept of the 'Neolithic Package': Considering Its Meaning and Applicability." *Documenta Praehistorica* 32 (December): 1–13.

Clark, John E. 1987. "Politics, Prismatic Blades, and Mesoamerican Civilization." In *The Organization of Core Technology*, edited by Jay K. Johnson and Carol A. Morrow, 259–84. Westview Press, Boulder.

Clark, John E. 1991. "The Beginnings of Mesoamerica: Apologia for the Soconusco Early Formative." In *The Formation of Complex Society in Southeastern Mesoamerica*, edited by William R. Fowler Jr., 13–26. CRC Press, Boca Raton.

Clark, John E. 1994. "The Development of Early Formative Rank Societies in the Soconusco, Chiapas, Mexico." Unpublished PhD dissertation. Ann Arbor: University of Michigan.

Clark, John E. 1997. "The Arts of Government in Early Mesoamerica." *Annual Review of Anthropology* 26(1): 211–34.

Clark, John E. 2004a. "Mesoamerica Goes Public: Early Ceremonial Centers, Leaders, and Communities." In *Mesoamerican Archaeology: Theory and Practice*, edited by Julia A. Hendon and Rosemary A. Joyce, 43–72. Blackwell, Malden, MA.

Clark, John E. 2004b. "The Birth of Mesoamerican Metaphysics: Sedentism, Engagement, and Moral Superiority." In *Rethinking Materiality: The Engagement of Mind with the Material World*, edited by C. DeMararis, C. Gosden, and C. Renfrew, 205–24. Cambridge University Press, Cambridge.

Clark, John E. 2007. "Mesoamerica's First State." In *The Political Economy of Ancient Mesoamerica*, edited by Vernon L. Scarborough and John E. Clark, 11–46. University of New Mexico Press, Albuquerque.

Clark, John E., and Michael Blake. 1994. "The Power of Prestige: Competitive Generosity and the Emergence of Rank Society in Lowland Mesoamerica." In *Factional Competition and Political Development in the New World*, edited by Elizabeth M. Brumfiel and John W. Fox, 17–30. Cambridge University Press, Cambridge.

Clark, John E., and David Cheetham. 2002. "Mesoamerica's Tribal Foundations." In *The Archaeology of Tribal Societies*, edited by William A. Parkinson, 278–339. International Monographs in Prehistory, Ann Arbor.

Clark, John E., and Dennis C. Gosser. 1995. "Reinventing Mesoamerica's First Pottery." In *The Emergence of Pottery: Technology and Innovation in Ancient Societies*, edited by William Barnett and John W. Hoopes, 209–21. Smithsonian Institution Press, Washington, DC.

Clark, John E., and Thomas A. Jr. Lee. 2007. "The Changing Role of Obsidian Exchange in Central Chiapas." In *Archaeology, Art, and Ethnogenesis in Mesoamerican Prehistory: Papers in Honor of Gareth W. Lowe*, edited by Lynneth S. Lowe and Mary E. Pye, 109–59. Brigham Young University, New World Archaeological Foundation. Provo, UT.

Clark, John E., and Mary E. Pye. 2006. "Los orígenes dello privilegio en el Soconusco, 1650 a.C.: Dos décadas de investigación." In *XIX Simposio de Investigaciones Arqueológicas en Guatemala*, edited by J. P. Laporte, Barbara Arroyo, and H. E. Mejía, 9–20. Instituto de Antropología e Historia, Guatemala City.

Clark, John E., Mary E. Pye, and Dennis C. Gosser. 2007. "Thermolithics and Corn Dependency in Mesoamerica." In *Archaeology, Art, and Ethnogenesis in Mesoamerican Prehistory: Papers in Honor of Gareth W. Lowe*, edited by Lynneth S. Lowe and Mary E. Pye, 23–42. Brigham Young University, New World Archaeological Foundation. Provo, UT.

Clark, John E., and Tamara Salcedo Romero. 1989. "Ocós Obsidian Distribution in Chiapas, Mexico." In *New Frontiers in the Archaeology of the Pacific Coast of Southern Mesoamerica*, edited by Frederick J. Bove and Lynette Heller, 15–24. Arizona State University Anthropological Research Papers, no. 39, Tempe.

Cobean, Robert H. 2002. *A World of Obsidian*. Serie Arqueología de México, Instituto Nacional de Antropología e Historia / University of Pittsburgh, Mexico City / Pittsburgh.

Coe, Michael D. 1961. *La Victoria, an Early Site on the Pacific Coast of Guatemala*. Vol. 53. Harvard University, Cambridge.

Coe, Michael D. 1968. *America's First Civilization*. American Heritage, New York.

Coe, Michael D. 1981. "Gift of the River: Ecology of the San Lorenzo Olmec." In *The Olmec and Their Neighbors: Essays in Memory of Matthew W. Stirling*, edited by Elizabeth P. Benson, 15–39. Dumbarton Oaks, Washington, DC.

Coe, Michael D. 1989. "The Olmec Heartland: Evolution of Ideology." In *Regional Perspectives on the Olmec*, edited by Robert J. Sharer and David C. Grove, 68–82. Cambridge University Press, Cambridge.

Coe, Michael D., and Richard A. Diehl. 1980a. *In the Land of the Olmec: Volume 1. The Archaeology of San Lorenzo Tenochtitlán*. University of Texas Press, Austin.

Coe, Michael D., and Richard A. Diehl. 1980b. *In the Land of the Olmec, Volume 2: The People of the River*. University of Texas Press, Austin.

Coe, Michael D., and Kent V. Flannery. 1967. *Early Cultures and Human Ecology in South Coastal Guatemala*. Smithsonian Contributions to Anthropology. Vol. 3. Washington DC.

Cohen, Mark N. 1985. "Prehistoric Hunter-Gatherers: The Meaning of Social Complexity." In *Prehistoric Hunter-Gatherers*, edited by T. Douglas Price and James A. Brown, 99–119. Academic Press, Orlando.

Cohen, Mark N., and George J. Armelagos, eds. 2013. *Paleopathology at the Origins of Agriculture*. 2nd ed. University Press of Florida, Gainesville.

Cohen, Yehudi A. 1974. *Man in Adaptation: The Cultural Present*. 2nd ed. Aldine, Chicago.

Conklin, Beth A., and Lynn M. Morgan. 1996. "Babies, Bodies, and the Production of Personhood in North America and a Native Amazonian Society." *Ethos* 24(4): 657–94.

Cook, Della C. 1981. "Mortality, Age-Structure and Status in the Interpretation of Stress Indicators in Prehistoric Skeletons: A Dental Example from the Lower Illinois Valley." In *The Archaeology of Death*, edited by Robert Chapman, Ian Kinnes, and Klavs Randsborg, 133–44. Cambridge University Press, Cambridge.

Cooney, Gabriel. 2007. "Parallel Worlds or Multi-Stranded Identities?" In *Going Over: The Mesolithic-Neolithic Transition in North-West Europe*, edited by Alasdair Whittle and Vicki Cummings, 543–66. 144th ed. Oxford University Press, New York.

Crumley, Carole L. 1995. "Heterarchy and the Analysis of Complex Societies." Ed. Robert M. Ehrenreich, Carole L. Crumley, and Janet E. Levy. *Archaeological Papers of the American Anthropological Association* 6(1): 1–5.

Crumley, Carole L. 2004. "Contextual Constraints on State Structure." In *Alternativity in Cultural History: Heterarchy and Homoarchy as Evolutionary Trajectories, Third International Conference "Hierarchy and Power in the History of Civilizations"* edited by Dmitri M. Bondarenko and Alexandre A. Nemirovskiy, 3–22. Center for Civilizational and Regional Studies of the RAS.

Cummings, Linda Scott. 2017. *Bone Collagen Extraction, XAD Purification, and AMS Radiocarbon Age Determination of Samples from La Consentida, Oaxaca, Mexico*. PaleoResearch Institute, Inc., Golden, CO.

Cummings, Vicki, and Alasdair Whittle. 2004. *Places of Special Virtue: Megaliths in the Neolithic Landscape of Wales*. Oxbow, Oxford.

Cyphers, Ann, and Judith Zurita-Noguera. 2012. "Early Olmec Wetland Mounds: Investing Energy to Produce Energy." In *Early New World Monumentality*, edited by Richard L. Burger and Robert M. Rosenswig, 138–73. University Press of Florida, Gainesville.

Cyphers, Ann, Judith Zurita Noguera, and Marci Lane Rodríguez. 2013. *Retos y Riesgos en la Vida Olmeca*. Universidad Nacional Autónoma de México, Instituto de Investigaciones Antropológicas.

Cyphers Guillén, Ann. 1993. "Women, Rituals, and Social Dynamics at Ancient Chalcatzingo." *Latin American Antiquity* 4(3): 209–24.

Danforth, Marie Elaine. 1999. "Nutrition and Politics in Prehistory." *Annual Review of Anthropology* 28(1): 1–25.

Davis, Dale D. 1975. "Patterns of Early Formative Subsistence in Southern Mesoamerica, 1500–1100 B.C." *Man* 10(1): 41–59.

DeBoer, Warren R. 1975. "The Archaeological Evidence for Manioc Cultivation: A Cautionary Note." *American Antiquity* 40(4): 419–33.

de Certeau, Michel. 1984. *The Practice of Everyday Life*. University of California Press, Berkeley.

Diehl, Richard A., and Michael D. Coe. 1996. "Olmec Archaeology." In *Olmec World: Ritual and Rulership*, edited by Elizabeth P. Benson, 11–25. Harry Abrams, New York.

Drennan, Robert D. 1983. "Exchange and Sociopolitical Development in the Tehuacán Valley." In *Trade and Exchange in Early Mesoamerica*, edited by Kenneth G. Hirth, 147–56. University of New Mexico Press, Albuquerque.

Drennan, Robert D. 2003a. "Appendix: Radiocarbon Dates from the Oaxaca Region." In *The Cloud People: Divergent Evolution of the Zapotec and Mixtec Civilizations*, edited by Kent V. Flannery and Joyce Marcus, 363–70. Percheron Press, New York.

Drennan, Robert D. 2003b. "Topic 13: Ritual and Ceremonial Development at the Early Village Level." In *The Cloud People: Divergent Evolution of the Zapotec and Mixtec Civilizations*, edited by Kent V. Flannery and Joyce Marcus, 46–50. Percheron Press, New York.

Drennan, Robert D. 2009. "Religion and Social Evolution in Formative Mesoamerica." In *The Early Mesoamerican Village: Updated Edition*, edited by Kent V. Flannery, 345–68. Left Coast Press, Walnut Creek, CA.

Dunham, Gary H. 1999. "Marking Territory, Making Territory: Burial Mounds in Interior Virginia." In *Material Symbols: Culture and Economy in Prehistory*, edited by John E. Robb, 112–34. Occasional Paper no. 26, Center for Archaeological Investigations. Southern Illinois University, Carbondale.

Erasmus, Charles J. 1965. "Monument Building: Some Field Experiments." *Southwestern Journal of Anthropology* 21(4): 277–301.

Fargher, Lane F., Richard E. Blanton, and Verenice F. Heredia Espinoza. 2010. "Egalitarian Ideology and Political Power in Prehispanic Central Mexico: The Case of Tlaxcallan." *Latin American Antiquity* 21(3): 227–51.

Farías Sánchez, José Antonio. 2006. *Cultivo de moluscos*. Alfaomega Grupo Editor, Mexico City.

Faust, Katherine A., and Christina T. Halperin. 2009. "Approaching Mesoamerican Figurines." In *Mesoamerican Figurines: Small-Scale Indices of Large-Scale Social Phenomena*, edited by Christina T. Halperin, Katherine A. Faust, Rhonda Taube, and Aurore Giguet, 1–22. University Press of Florida, Gainesville.

Feinman, Gary M. 1991. "Demography, Surplus, and Inequality: Early Political Formations in Highland Mesoamerica." In *Chiefdoms: Power, Economy, and Ideology*, edited by Timothy Earle, 229–62. Cambridge University Press, Cambridge.

Feinman, Gary M. 1995. "The Emergence of Inequality: A Focus on Strategies and Processes." In *Foundations of Social Inequality*, edited by T. Douglas Price and Gary M. Feinman, 255–79. Plenum Press, New York.

Feinman, Gary M., and Joyce Marcus, eds. 1998. *Archaic States*. School of American Research Press, Santa Fe.

Feinman, Gary M., and Linda M. Nicholas. 1987. "Labor, Surplus, and Production: A Regional Analysis of Formative Oaxacan Socio-Economic Change." In *Coasts, Plains, and Deserts: Essays in Honor of Reynold J. Ruppé*, edited by Sylvia W. Gaines, 27–50. Arizona State University Anthropological Research Papers, no. 38, Tempe.

Feinman, Gary M., and Linda M. Nicholas. 1989. "The Role of Risk in Formative Period Agriculture: A Reconsideration." *American Anthropologist* 91(1): 198–203.

Feinman, Gary M., and Linda M. Nicholas. 1990. "Settlement and Land Use in Ancient Oaxaca." In *Debating Oaxaca Archaeology*, edited by Joyce Marcus, 71–113. University of Michigan Anthropological Papers, Museum of Anthropology, no. 84, Ann Arbor.

Feinman, Gary M., and Linda M. Nicholas. 1992. "Human-Land Relations from an Archaeological Perspective: The Case of Ancient Oaxaca." In *Understanding Economic Process*, edited by Sutti Ortiz and Susan Lees, 155–78. Monographs in Economic Anthropology, no. 10. University Press of America, New York.

Feinman, Gary M., Linda M. Nicholas, and Helen R Haines. 2002. "Houses on a Hill: Classic Period Life at El Palmillo, Oaxaca, Mexico." *Latin American Antiquity* 13(3): 251–77.

Fernández, Deepika C. 2004. "Subsistence in the Lower Río Verde, Oaxaca, Mexico: A Zooarchaeological Analysis." Unpublished MA thesis. University of Calgary, Calgary.

Flanagan, James G. 1989. "Hierarchy in Simple 'Egalitarian' Societies." *Annual Review of Anthropology* 18(1): 245–66.

Flannery, Kent V. 1968a. "The Olmec and the Valley of Oaxaca: A Model for Interregional Interaction in Formative Times." In *Dumbarton Oaks Conference on the Olmec, October 28th and 29th, 1967*, edited by Elizabeth P. Benson, 79–117. Dumbarton Oaks Research Library and Collection, Washington DC.

Flannery, Kent V. 1968b. "Archaeological Systems Theory and Early Mesoamerica." In *Anthropological Archaeology in the Americas*, edited by Betty J. Meggars, 67–87. Anthropological Society of Washington, DC, Washington, DC.

Flannery, Kent V. 1972a. "The Cultural Evolution of Civilizations." *Annual Review of Ecology and Systematics* 3(1): 399–426.

Flannery, Kent V. 1972b. "The Origins of the Village as a Settlement Type in Mesoamerica and the Near East: A Comparative Study." In *Man, Settlement, and Urbanism*, edited by Peter J. Ucko, Ruth Tringham, and G. W. Dimbleby, 23–53. Duckworth, London.

Flannery, Kent V. 1973. "The Origins of Agriculture." *Annual Review of Anthropology* 2(1): 271–310.

Flannery, Kent V., ed. 1986. *Guilá Naquitz: Archaic Foraging and Early Agriculture in Oaxaca, Mexico*. Studies in Archaeology. Academic Press, Orlando.

Flannery, Kent V. 2002. "The Origins of the Village Revisited: From Nuclear to Extended Households." *American Antiquity* 67(3): 417–33.

Flannery, Kent V., ed. 2009a. *The Early Mesoamerican Village: Updated Edition*. Left Coast Press, Walnut Creek, CA.

Flannery, Kent V. 2009b. "Contextual Analysis of Ritual Paraphernalia from Formative Oaxaca." In *The Early Mesoamerican Village: Updated Edition*, edited by Kent V. Flannery, 333–45. Left Coast Press, Walnut Creek, CA.

Flannery, Kent V., and Joyce Marcus. 1994. *Early Formative Pottery of the Valley of Oaxaca, Mexico*. University of Michigan Press, Ann Arbor.

Flannery, Kent V., and Joyce Marcus, eds. 2003. *The Cloud People: Divergent Evolution of the Zapotec and Mixtec Civilizations*. Percheron Press, New York.

Flannery, Kent V., and Joyce Marucs. 2012. *The Creation of Inequality: How Our Prehistoric Ancestors Set the Stage for Monarchy, Slavery, and Empire*. Harvard University Press, Cambridge, MA.

Ford, James A. 1969. *A Comparison of Formative Cultures in the Americas*. Vol. 11. Smithsonian Contributions to Anthropology. Smithsonian Institution Press, Washington, DC.

Foster, George M. 1944. "Nagualism in Mexico and Guatemala." *Acta Americana* II (May): 85–103.

Fried, Morton H. 1967. *The Evolution of Political Society: An Essay in Political Anthropology*. Random House, New York.

Gabriel, Kathryn. 1996. *Gambler Way: Indian Gaming in Mythology, History and Archaeology in North America.* Johnson Books, Boulder.

Gamble, Lynn H. 2002. "Archaeological Evidence for the Origin of the Plank Canoe in North America." *American Antiquity* 67(2): 301–15.

Garcea, Elena A A. 2006. "Semi-Permanent Foragers in Semi-Arid Environments of North Africa." *World Archaeology* 38(2): 197–219.

García Cook, Ángel, and Beatriz Leonor Merino Carrión. 2005. "El Inicio de la Producción Alfarera en el México Antiguo." In *La producción alfarera en el México Antiguo I*, edited by Beatriz Leonor Merino Carrión and Angel García Cook, 73–119. Instituto Nacional de Antropología e Historia, Mexico City.

Gendron, François, David C. Smith, and Aïcha Gendron-Badou. 2002. "Discovery of Jadeite-Jade in Guatemala Confirmed by Non-Destructive Raman Microscopy." *Journal of Archaeological Science* 29(8): 837–51.

Gibson, Jon L. 2000. *Ancient Mounds of Poverty Point: Place of Rings.* University Press of Florida, Gainesville.

Giddens, Anthony. 1979. *Central Problems in Social Theory.* University of California Press, Berkeley.

Gillespie, Susan D. 1987. *Excavaciones en Charco Redondo 1986.* Informe al Consejo de Arqueología y al Centro INAH Oaxaca del Instituto Nacional de Antropología e Historia, Mexico City, Oaxaca City.

Gillespie, Susan D. 2001. "Personhood, Agency, and Mortuary Ritual: A Case Study from the Ancient Maya." *Journal of Anthropological Archaeology* 20(1): 73–112.

Glascock, Michael D. 2011. *X-ray Fluorescence Analysis of Obsidian Artifacts from La Consentida, Oaxaca, Mexico.* Report produced at University of Missouri Archaeometry Laboratory, Columbia.

Goman, Michelle, Arthur A. Joyce, and Raymond G. Mueller. 2005. Stratigraphic Evidence for Anthropogenically Induced Environmental Change from Oaxaca, Mexico. *Quaternary Research* 63(3): 250–60.

Goman, Michelle, Arthur A. Joyce, and Raymond G. Mueller. 2013. "Paleoecological Evidence for Early Agriculture and Forest Clearance in Coastal Oaxaca." In *Polity and Ecology in Formative Period Coastal Oaxaca*, edited by Arthur A. Joyce, 43–64. University Press of Colorado, Boulder.

Goodman, Alan H. 1991. "Stress, Adaptation, and Enamel Developmental Defects." In *Human Paleopathology: Current Synthesis and Future Options*, edited by Donald J. Ortner and Arthur C. Aufderheide, 280–87. Smithsonian Institution Press, Washington, DC.

Green, Dee F., and Gareth W. Lowe. 1967. *Altamira and Padre Piedra, Early Preclassic Sites in Chiapas, Mexico.* Brigham Young University, Provo, UT.

Grove, David C. 1984. *Chacatzingo: Excavations on the Olmec Frontier*. Thames and Hudson, London.

Grove, David C. 1988. *Archaeological Investigations on the Pacific Coast of Oaxaca*. Report submitted to the National Geographic Society, Illinois University, Urbana-Champaign.

Grove, David C. 1989. "Olmec: What's in a Name?" In *Regional Perspectives on the Olmec*, edited by Robert J. Sharer and David C. Grove, 8–16. Cambridge University Press, Cambridge.

Guernsey, Julia. 2010. "Rulers, Gods, and Potbellies: A Consideration of Sculptural Forms and Themes from the Preclassic Pacific Coast and Piedmont of Mesoamerica." In *The Place of Stone Monuments: Context, Use, and Meaning in Mesoamerica's Preclassic Transition*, edited by Julia Guernsey, John E. Clark, and Barbara Arroyo, 207–30. Dumbarton Oaks, Washington, DC.

Guernsey, Julia. 2012. *Sculpture and Social Dynamics in Preclassic Mesoamerica*. Cambridge University Press, New York.

Gutiérrez, Gerardo, and Mary E. Pye. 2010. "Iconography of the Nahual: Human-Animal Transformations in Preclassic Guerrero and Morelos." In *The Place of Stone Monuments: Context, Use, and Meaning in Mesoamerica's Preclassic Transition*, edited by Julia Guernsey, John E. Clark, and Barbara Arroyo, 27–95. Dumbarton Oaks, Washington DC.

Hard, Robert J., Raymond P. Mauldin, and Gerry R. Raymond. 1996. "Mano Size, Stable Carbon Isotope Ratios, and Macrobotanical Remains as Multiple Lines of Evidence of Maize Dependence in the American Southwest." *Journal of Archaeological Method and Theory* 3(4): 253–318.

Hayden, Brian. 1990. "Nimrods, Piscators, Pluckers, and Planters: The Emergence of Food Production." *Journal of Anthropological Archaeology* 9(1): 31–69.

Hayden, Brian. 1995. "Pathways to Power: Principles for Creating Socioeconomic Inequalities." In *Foundations of Social Inequality*, edited by T. Douglas Price and Gary M. Feinman, 15–86. Plenum Press, New York.

Hayden, Brian. 1998. "Practical and Prestige Technologies: The Evolution of Material Systems." *Journal of Archaeological Method and Theory* 5(1): 1–55.

Hayden, Brian. 2009. "The Proof Is in the Pudding: Feasting and the Origins of Domestication." *Current Anthropology* 50(5): 597–601.

Hayden, Brian. 2011. "Big Man, Big Heart? The Political Role of Aggrandizers in Egalitarian and Transegalitarian Societies." In *For the Greater Good of All: Perspectives on Individualism, Society, and Leadership*, edited by Donelson R. Forsyth and Crystal L. Hoyt, 101–18. Palgrave Macmillan.

Hedgepeth, Jessica D. 2009. *The Domestic Economy of Early Postclassic Río Viejo, Oaxaca, Mexico: Daily Practices and Worldviews of a Commoner Community*. Unpublished MA thesis, University of Colorado, Boulder.

Hendon, Julia A. 1991. "Status and Power in Classic Maya Society: An Archeological Study." *American Anthropologist* 93(4): 894–918.

Hendon, Julia A. 2000. "Having and Holding: Storage, Memory, Knowledge, and Social Relations." *American Anthropologist* 102(1): 42–53.

Hepp, Guy David. 2007. "Formative Period Ceramic Figurines from the Lower Río Verde Valley, Coastal Oaxaca, Mexico." Unpublished MA thesis. Tallahassee: Florida State University.

Hepp, Guy David. 2011a. "El Proyecto La Consentida 2009." In *El Proyecto Río Verde: Informe técnico de la temporada de 2009*, edited by Sarah B. Barber and Arthur A. Joyce, 146–84. Report submitted to the Instituto Nacional de Antropología e Historia, Mexico City.

Hepp, Guy David. 2011b. "Análisis de los artefactos de La Consentida." In *El Proyecto Río Verde: Informe Técnico de la Temporada de 2009*, edited by Sarah B. Barber and Arthur A. Joyce, 289–314. Informe al Consejo de Arqueología y al Centro INAH Oaxaca del Instituto Nacional de Antropología e Historia, Mexico City, Oaxaca City.

Hepp, Guy David. 2011c. "The Material Culture of Incipient Social Complexity in Coastal Oaxaca: The Ceramics of La Consentida." Paper presented at the 76th Annual Meeting of the Society for American Archaeology, Sacramento, CA.

Hepp, Guy David. 2014. "La Consentida: Among Mesoamerica's Earliest for Ceramics and Earthen Architecture." Paper presented at the 79th Annual Meeting of the Society for American Archaeology, Austin, TX.

Hepp, Guy David. 2015. "La Consentida: Initial Early Formative Period Settlement, Subsistence, and Social Organization on the Pacific Coast of Oaxaca, Mexico." Unpublished PhD dissertation. University of Colorado, Boulder.

Hepp, Guy David. In press. "Radiocarbon Evidence for Initial Early Formative Period Occupation in Coastal Oaxaca, Mexico." Latin American Antiquity.

Hepp, Guy David, Sarah B. Barber, and Arthur A. Joyce. 2014. "Communing with Nature, the Ancestors, and the Neighbors: Ancient Ceramic Musical Instruments from Coastal Oaxaca, Mexico." *World Archaeology* 46(3): 380–99.

Hepp, Guy David, and Arthur A. Joyce. 2013. "From Flesh to Clay: Formative Period Ceramic Figurines from Oaxaca's Lower Río Verde Valley." In *Polity and Ecology in Formative Period Coastal Oaxaca*, edited by Arthur A. Joyce, 256–99. University Press of Colorado, Boulder.

Hepp, Guy David, and Ivy A. Rieger. 2014. "Aspects of Dress and Ornamentation in Coastal Oaxaca's Formative Period." In *Wearing Culture: Dress and Regalia in Early Mesoamerica and Central America*, edited by Heather Orr and Matthew Looper, 115–43. University Press of Colorado, Boulder.

Hepp, Guy David, Paul A. Sandberg, and José Aguilar. 2017. "Death on the Early Formative Oaxaca Coast: The Human Remains of La Consentida." *Journal of Archaeological Science: Reports* 13 (June): 703–11.

Heyden, Doris. 1991. "Dryness before the Rains: Toxcatl and Tezcatlipoca." In *To Change Place: Aztec Ceremonial Landscapes*, edited by Davíd Carrasco, 188–204. University Press of Colorado, Niwot.

Higelin Ponce de León, Ricardo, and Guy David Hepp. 2017a. "Talking with the Dead from Southern Mexico: Tracing Bioarchaeological Foundations and New Perspectives in Oaxaca." *Journal of Archaeological Science: Reports* 13 (June): 697–702.

Higelin Ponce de León, Ricardo, and Guy David Hepp. 2017b. Special Section on "Talking with the Dead from Southern Mexico: Tracing Bioarchaeological Foundations and New Perspectives in Oaxaca." *Journal of Archaeological Science: Reports* 13.

Hill, Warren D., and John E. Clark. 2001. "Sports, Gambling, and Government: America's First Social Compact?" *American Anthropologist* 103(2): 331–45.

Hirth, Kenneth G. 1978. "Interregional Trade and the Formation of Prehistoric Gateway Communities." *American Antiquity* 43(1): 35–45.

Hirth, Kenneth, Ann M. Cyphers, Robert H. Cobean, Jason De León, and Michael D. Glascock. 2013. "Early Olmec Obsidian Trade and Economic Organization at San Lorenzo." *Journal of Archaeological Science* 40(6): 2784–98.

Hodder, Ian. 2012. *Entangled: An Archaeology of the Relationships between Humans and Things*. Wiley-Blackwell, Malden, MA.

Hodges, Denise C. 1987. "Health and Agricultural Intensification in the Prehistoric Valley of Oaxaca, Mexico." *American Journal of Physical Anthropology* 73(3): 323–32.

Hoopes, John W. 1994. "Ceramic Analysis and Culture History in the Arenal Region." In *Archaeology, Volcanism, and Remote Sensing in the Arenal Region, Costa Rica*, edited by Payson Sheets and Brian R. McKee, 158–210. University of Texas Press, Austin.

Hopkins, Nicholas A. 1984. "Otomanguean Linguistic Prehistory." In *Essays in Otomanguean Culture History*, edited by J. Kathryn Josserand, Marcus Winter, and Nicholas A. Hopkins, 25–64. Vanderbilt University, Nashville.

Howes, David. 2003. *Sensual Relations: Engaging the Senses in Culture and Social Theory*. University of Michigan Press, Ann Arbor.

Howes, David. 2006. "Scent, Sound and Synaesthesia: Intersensoriality and Material Culture Theory." In *Handbook of Material Culture*, edited by Christopher Tilley, Webb Keane, Susanne Küchler, Mike Rowlands, and Patricia Spyer, 161–72. Sage, London.

Inomata, Takeshi, Daniela Triadan, Kazuo Aoyama, Victor Castillo, and Hitoshi Yonenobu. 2013. "Early Ceremonial Constructions at Ceibal, Guatemala, and the Origins of Lowland Maya Civilization." *Science* 340(26): 467–71.

Isendahl, Christian. 2011. "The Domestication and Early Spread of Manioc (MANIHOT ESCULENTA CRANTZ): A Brief Synthesis." *Latin American Antiquity* 22(4): 452–68.

Jackson, Tomas L, and Michael W. Love. 1991. "Blade Running: Middle Preclassic Obsidian Exchange and the Introduction of Prismatic Blades at La Blanca, Guatemala." *Ancient Mesoamerica* 2(1): 47–59.

Jennings, Sarah. 2010. "Mold-Made Figurines of the Lower Río Verde Valley, Oaxaca, Mexico: Insights into Popular Ideology in the Classic and Early Postclassic." Unpublished MA thesis. University of Colorado, Boulder.

Jones, John G., and Barbara Voorhies. 2004. "Human and Plant Interactions." In *Coastal Collectors in the Holocene: The Chantuto People of Southwest Mexico*, edited by Barbara Voorhies, 300–43. University Press of Florida, Gainesville.

Jones, Terry L., and Kathryn A. Klar. 2005. "Diffusionism Reconsidered: Linguistic and Archaeological Evidence for Prehistoric Polynesian Contact with Southern California." *American Antiquity* 70(3): 457–84.

Josserand, J. Kathryn, Marcus Winter, and Nicholas A. Hopkins, eds. 1984. *Essays in Otomanguean Culture History*. Department of Anthropology, Vanderbilt University, Nashville.

Joyce, Arthur A. 1991a. "Formative Period Occupation in the Lower Río Verde Valley, Oaxaca, Mexico: Interregional Interaction and Social Change." Unpublished PhD dissertation. Rutgers the State University of New Jersey, New Brunswick.

Joyce, Arthur A. 1991b. "Formative Period Social Change in the Lower Río Verde Valley, Oaxaca, Mexico." *Latin American Antiquity* 2(2): 126–50.

Joyce, Arthur A. 2000. "The Founding of Monte Albán: Sacred Propositions and Social Practices." In *Agency in Archaeology*, edited by Marcia-Anne Dobres and John Robb, 71–91. Routledge Press, London.

Joyce, Arthur A. 2005. "La Arqueología del Bajo Río Verde." *Acervos: Boletín de los Archivos y Bibliotecas de Oaxaca* 7(29): 16–36.

Joyce, Arthur A. 2006. "The Inhabitation of Río Viejo's Acropolis." In *Space and Spatial Analysis in Archaeology*, edited by Elizabeth C. Robertson, Jeffrey D. Seibert, Deepika C. Fernandez, and Marc U. Zender, 83–96. Calgary Press, New Mexico.

Joyce, Arthur A. 2010. *Mixtecs, Zapotecs, and Chatinos: Ancient Peoples of Southern Mexico*. Wiley-Blackwell, Malden, MA.

Joyce, Arthur A., ed. 2013a. *Polity and Ecology in Formative Period Coastal Oaxaca*. University Press of Colorado, Boulder.

Joyce, Arthur A. 2013b. "Polity and Ecology in Formative Period Coastal Oaxaca: An Introduction." In *Polity and Ecology in Formative Period Coastal Oaxaca*, edited by Arthur A. Joyce, 1–41. University Press of Colorado, Boulder.

Joyce, Arthur A., Sarah B. Barber, Guy David Hepp, Matt Sponheimer, Michelle Butler, Sarah Taylor, Michelle Goman, Aleksander Borejsza, Raymond G. Mueller, and Paul A. Sandberg. 2017. *Landscape and Dietary Change in Formative Period Coastal Oaxaca*. Paper presented at the Society for American Archaeology Annual Conference, Vancouver, BC.

Joyce, Arthur A., J. Michael Elam, Michael D. Glascock, Hector Neff, and Marcus Winter. 1995. "Exchange Implications of Obsidian Source Analysis from the Lower Río Verde Valley, Oaxaca, Mexico." *Latin American Antiquity* 6(1): 3–15.

Joyce, Arthur A., and Michelle Goman. 2012. "Bridging the Theoretical Divide in Holocene Landscape Studies: Social and Ecological Approaches to Ancient Oaxacan Landscapes." *Quaternary Science Reviews* 55 (November): 1–22.

Joyce, Arthur A., Marc N. Levine, and Sarah B. Barber. 2013. "Place-Making and Power in the Terminal Formative: Excavations on Río Viejo's Acropolis." In *Polity and Ecology in Formative Period Coastal Oaxaca*, edited by Arthur A. Joyce, 135–63. University Press of Colorado, Boulder.

Joyce, Arthur A., and Raymond G. Mueller. 1992. "The Social Impact of Anthropogenic Landscape Modification in the Río Verde Drainage Basin, Oaxaca, Mexico." *Geoarchaeology* 7(6): 503–26.

Joyce, Arthur A., Maxine Oland, and Peter Kroefges. 2009a. "Recorrido Regional de Superficie." In *El Proyecto Río Verde*, edited by Arthur A. Joyce and Marc N. Levine, 322–53. Informe al Consejo de Arqueología y al Centro INAH Oaxaca del Instituto Nacional de Antropología e Historia, Mexico City, Oaxaca City.

Joyce, Arthur A., Maxine Oland, and Peter Kroefges. 2009b. "Apéndice E: Registro Público de Monumentos y Zonas Arqueológicos." In *El Proyecto Río Verde: Informe final*, edited by Arthur A. Joyce and Marc N. Levine, 488–557. Informe al Consejo de Arqueología y al Centro INAH Oaxaca del Instituto Nacional de Antropología e Historia, Mexico City, Oaxaca City.

Joyce, Arthur A., Marcus Winter, and Raymond G. Mueller. 1998. *Arqueologíia de la costa de Oaxaca: Asentamientos del Periodo Formativo en el Valle del Río Verde Inferior*. Centro Oaxaca-INAH, Mexico City.

Joyce, Arthur A., Andrew G. Workinger, Byron Ellsworth Hamann, Peter Kroefges, Maxine Oland, and Stacie M. King. 2004. "Lord 8 Deer 'Jaguar Claw' and the Land of the Sky: The Archaeology and History of Tututepec." *Latin American Antiquity* 15(3): 273–97.

Joyce, Rosemary A. 1999. "Social Dimensions of Pre-Classic Burials." In *Social Patterns in Pre-Classic Mesoamerica*, edited by David C. Grove and Rosemary A. Joyce, 15–47. Dumbarton Oaks, Washington, DC.

Joyce, Rosemary A. 2000a. "Girling the Girl and Boying the Boy: The Production of Adulthood in Ancient Mesoamerica." *World Archaeology* 31(3): 473–83.

Joyce, Rosemary A. 2000b. *Gender and Power in Prehispanic Mesoamerica*. University of Texas Press, Austin.

Joyce, Rosemary A. 2002. "Beauty, Sexuality, Body Ornamentation, and Gender in Ancient Mesoamerica." In *In Pursuit of Gender: Worldwide Archaeological Approaches*, edited by Sarah M. Nelson and Myriam Rosen-Ayalon, 81–92. Altamira Press, New York.

Joyce, Rosemary A. 2003. "Making Something of Herself: Embodiment in Life and Death at Playa de los Muertos, Honduras." *Cambridge Archaeological Journal* 13(2): 248–61.

Joyce, Rosemary A. 2004a. "Unintended Consequences? Monumentality as a Novel Experience in Formative Mesoamerica." *Journal of Archaeological Method and Theory* 11(1): 5–29.

Joyce, Rosemary A. 2004b. "Mesoamerica: A Working Model for Archaeology." In *Mesoamerican Archaeology: Theory and Practice*, edited by Julia A. Hendon and Rosemary A. Joyce, 1–42. Mesoamerican Archaeology. Blackwell, Malden, MA.

Joyce, Rosemary A. 2005. "Archaeology of the Body." *Annual Review of Anthropology* 34 (October): 139–58.

Joyce, Rosemary A. 2009. "Making a World of Their Own: Mesoamerican Figurines and Mesoamerican Figurine Analysis." In *Mesoamerican Figurines: Small-Scale Indices of Large-Scale Social Phenomena*, edited by Christina T. Halperin, Katherine A. Faust, Rhonda Taube, and Aurore Giguet, 407–25. University Press of Florida, Gainesville.

Joyce, Rosemary A., and John S. Henderson. 2001. "Beginnings of Village Life in Eastern Mesoamerica." *Latin American Antiquity* 12(1): 5–23.

Joyce, Rosemary A., and John S. Henderson. 2007. "From Feasting to Cuisine: Implications of Archaeological Research in an Early Honduran Village." *American Anthropologist* 109(4): 642–53.

Kailola, Patricia J., and William A. Bussing. 1995. "Ariidae: Bagres Marinos." In *Guía FAO para identificación de especies para los fines de la pesca: Pacífico Centro-Oriental*, edited by W. Fischer, F. Krupp, W. Schneider, C. Sommer, K. E. Carpenter, and V. Niem, 860–86. FAO, Rome.

Kamp, Kathryn A. 2001. "Prehistoric Children Working and Playing: A Southwestern Case Study in Learning." *Journal of Anthropological Research* 57(4): 427–50.

Kaplan, Lucille N. 1956. "Tonal and Nagual in Coastal Oaxaca, Mexico." *Journal of American Folklore* 69(274): 363–68.

Katzenberg, M. Anne. 2000. "Stable Isotope Analysis: A Tool for Studying Past Diet, Demography, and Life History." In *Biological Anthropology of the Human Skeleton*, edited by M. Anne Katzenberg and Shelley R. Saunders, 305–27. Wiley-Liss, New York.

Kelly, Isabel. 1974. "Stirrup Pots from Colima: Some Implications." In *The Archaeology of West Mexico*, edited by Betty Bell, 206–11. Sociedad de Estudios Avanzados del Occidente de México, Ajijic, Jalisco, Mexico.

Kelly, Isabel. 1980. *Ceramic Sequence in Colima: Capacha, an Early Phase*. Anthropological Papers of the University of Arizona, No. 37. University of Arizona Press, Tucson.

Kelly, Lynne. 2015. *Knowledge and Power in Prehistoric Societies: Orality, Memory, and the Transmission of Culture*. Cambridge University Press, New York.

Kelly, Robert L. 1992. "Mobility/Sedentism: Concepts, Archaeological Measures, and Effects." *Annual Review of Anthropology* 21(1): 43–66.

Kennett, Douglas J., Dolores R. Piperno, John G. Jones, Hector Neff, Barbara Voorhies, Megan K. Walsh, and Brendan J. Culleton. 2010. "Pre-Pottery Farmers on the Pacific Coast of Southern Mexico." *Journal of Archaeological Science* 37(12): 3401–11.

Kent, Susan. 1992. "Variability in the Archaeological Record: An Ethnoarchaeological Model for Distinguishing Mobility Patterns." *American Antiquity* 57(4): 635–60.

Killion, Thomas W. 2013. "Nonagricultural Cultivation and Social Complexity: The Olmec, Their Ancestors, and Mexico's Southern Gulf Coast Lowlands." *Current Anthropology* 54(5): 569–606.

King, Stacie Marie. 2003. "Social Practices and Social Organization in Ancient Coastal Oaxacan Households." Unpublished PhD dissertation. University of California, Berkeley.

King, Stacie Marie, and Gonzalo Sánchez Santiago. 2011. "Soundscapes of the Everyday in Ancient Oaxaca, Mexico." *Archaeologies: Journal of the World Archaeological Congress* 7(2): 387–422.

Kipfer, Barbara Ann. 2007. *The Archaeologist's Fieldwork Companion*. Blackwell, Malden, MA.

Kirchhoff, Paul. 1943. "Mesoamérica: Sus límites geográficos, composición étnica y caracteres culturales." *Acta Americana* 1(1): 92–107.

Klein, Kathryn. 1997. *The Unbroken Thread: Conserving the Textile Traditions of Oaxaca*. Getty Conservation Institute, Los Angeles.

Kowalewski, Stephen A. 1990. "The Evolution of Complexity in the Valley of Oaxaca." *Annual Review of Anthropology* 19(1): 39–58.

Kuijt, Ian. 2009. "What Do We Really Know about Food Storage, Surplus, and Feasting in Preagricultural Communities?" *Current Anthropology* 50(5): 641–44.

Lallo, John W., George J. Armelagos, and Robert P. Mensforth. 1977. "The Role of Diet, Disease, and Physiology in the Origin of Porotic Hyperostosis." *Human Biology* 49 (September): 471–83.

Larsen, Clark Spencer. 1987. "Bioarchaeological Interpretations of Subsistence Economy and Behavior from Human Remains." *Advances in Archaeological Method and Theory* 10: 339–445.

Larsen, Clark Spencer. 1995. "Biological Changes in Human Populations with Agriculture." *Annual Review of Anthropology* 24(1): 185–213.

Lesure, Richard G. 1997a. "Figurines and Social Identities in Early Sedentary Societies of Coastal Chiapas, Mexico, 1550–800 b.c." In *Women in Prehistory: North America and Mesoamerica*, edited by Cheryl Claassen and Rosemary A. Joyce, 227–48. University of Pennsylvania Press, Philadelphia.

Lesure, Richard G. 1997b. "Early Formative Platforms at Paso de la Amada, Chiapas, Mexico." *Latin American Antiquity* 8(3): 217–36.

Lesure, Richard G. 1999a. "Figurines as Representations and Products at Paso de la Amada, Mexico." *Cambridge Archaeological Journal* 9(2): 209–20.

Lesure, Richard G. 1999b. "Platform Architecture and Activity Patterns in an Early Mesoamerican Village in Chiapas, Mexico." *Journal of Field Archaeology* 26(4): 391–406.

Lesure, Richard G. 2004. "Shared Art Styles and Long-Distance Contact in Early Mesoamerica." In *Mesoamerican Archaeology*, edited by Julia A. Hendon and Rosemary A. Joyce, 73–96. Blackwell, Malden, MA.

Lesure, Richard G., ed. 2009. *Settlement and Subsistence in Early Formative Soconusco: El Varal and the Problem of Inter-Site Assemblage Variation*. Cotsen Institute of Archaeology Press, Los Angeles.

Lesure, Richard G., ed. 2011a. *Early Mesoamerican Social Transformations*. University of California Press, Berkeley.

Lesure, Richard G. 2011b. "Early Social Transformations in the Soconusco: An Introduction." In *Early Mesoamerican Social Transformations: Archaic and Formative Lifeways in the Soconusco Region*, edited by Richard G. Lesure, 1–24. University of California Press, Berkeley.

Lesure, Richard G. 2011c. *Interpreting Ancient Figurines: Context, Comparison, and Prehistoric Art*. Cambridge University Press, New York.

Lesure, Richard G., and Michael Blake. 2002. "Interpretive Challenges in the Study of Early Complexity: Economy, Ritual, and Architecture at Paso de la Amada, Mexico." *Journal of Anthropological Archaeology* 21(1): 1–24.

Lesure, Richard G., and Thomas A. Wake. 2011. "Archaic to Formative in Soconusco: The Adaptive and Organizational Transformation." In *Early Mesoamerican Social Transformations: Archaic and Formative Lifeways in the Soconusco Region*, edited by Richard G. Lesure, 67–93. University of California Press, Berkeley.

Levine, Marc N. 2007. "Linking Household and Polity at Late Postclassic Period Yucu Dzaa (Tututepec), a Mixtec Capital on the Coast of Oaxaca, Mexico." Unpublished PhD dissertation. University of Colorado, Boulder.

Levine, Marc N. 2011. "Negotiating Political Economy at Late Postclassic Tututepec (Yucu Dzaa), Oaxaca, Mexico." *American Anthropologist* 113(1): 22–39.

Lewenstein, Suzanne M., and Jeff Walker. 1984. "The Obsidian Chip / Manioc Grating Hypothesis and the Mesoamerican Preclassic." *Journal of New World Archaeology* 6(2): 25–38.

Llano, Carina, and Andrew Ugan. 2014. "Alternative Interpretations of Intermediate and Positive d13C Isotope Signals in Prehistoric Human Remains from Southern Mendoza, Argentina: The Role of CAM Species Consumption." *Current Anthropology* 55(6): 822–31.

Lock, Gracie, Michelle Goman, Arthur A. Joyce, Victor Salazar Chavez, and Guy David Hepp. 2014. "Salinas: Expanding Our Understanding of Prehistoric Land Use in the Coastal Zone of the Lower Río Verde Valley, Oaxaca, Mexico." Paper presented at the Association of American Geographers Annual Meeting, Tampa FL.

Lopiparo, Jeanne. 2006. "Crafting Children: Materiality, Social Memory, and the Reproduction of Terminal Classic House Societies in the Ulúa Valley, Honduras." In *The Social Experience of Childhood in Ancient Mesoamerica*, edited by Traci Ardren and Scott R. Hutson, 133–68. University Press of Colorado, Boulder.

Love, Michael W. 1999. "Ideology, Material Culture, and Daily Practice in Pre-Classic Mesoamerica: A Pacific Coastal Perspective." In *Social Patterns in Pre-Classic Mesoamerica: A Symposium at Dumbarton Oaks 9 and 10 October 1993*, edited by David C. Grove and Rosemary A. Joyce, 127–53. Dumbarton Oaks, Washington, DC.

Love, Michael W. 2002. "Early Complex Society in Pacific Guatemala: Settlements and Chronology of the Río Naranjo, Guatemala." Brigham Young University, Papers of the New World Archaeological Foundation 66. Provo, UT.

Love, Michael W. 2007. "Recent Research in the Southern Highlands and Pacific Coast of Mesoamerica." *Journal of Archaeological Research* 15(4): 275–328.

Love, Michael W., and Julia Guernsey. 2011. "La Blanca and the Soconusco Middle Formative." In *Early Mesoamerican Social Transformations: Archaic and Formative Lifeways in the Soconusco Region*, edited by Richard G. Lesure, 170–88. University of California Press, Berkeley.

Lowe, Gareth W. 1967. "Discussion." In *Altamira and Padre Piedra, Early Preclassic Sites in Chiapas, Mexico*, edited by Dee Green and Gareth W. Lowe, 53–79. Brigham Young University, Provo, UT.

Lowe, Gareth W. 1975. "The Early Preclassic Barra Phase of Altamira, Chiapas." Brigham Young University, Papers of the New World Archaeological Foundation 38. Provo, UT.

Lowe, Gareth W. 1977. "The Mixe-Zoque as Competing Neighbors of the Early Lowland Maya." In *The Origins of Maya Civilization*, edited by Richard E. W. Adams. University of New Mexico Press, Albuquerque.

Lowe, Gareth W. 2007. "Early Formative Chiapas: The Beginnings of Civilization in the Central Depression of Chiapas." In *Archaeology, Art, and Ethnogenesis in Mesoamerican Prehistory: Papers in Honor of Gareth W. Lowe*, edited by Lynneth S. Lowe and Mary E. Pye, 63–108. Brigham Young University, New World Archaeological Foundation. Provo, UT.

Lowe, John W. G., and Robert J. Barth. 1980. "Systems in Archaeology: A Comment on Salmon." *American Antiquity* 45(3): 568–75.

MacNeish, Richard S. 1964. "Ancient Mesoamerican Civilization." *Science* 143(3606): 531–37.

MacNeish, Richard S. 1969. "Speculation about How and Why Food Production and Village Life Developed in the Tehuacán Valley, Mexico." *Archaeology* 24(4): 307–15.

MacNeish, Richard S. 1972. "The Evolution of Community Patterns in the Tehuacán Valley of Mexico and Speculations about the Cultural Process." In *Man, Settlement, and Urbanism*, edited by Peter J. Ucko, Ruth Tringham, and G. W. Dimbleby, 67–93. Duckworth, London.

MacNeish, Richard S. 1992. *The Origins of Agriculture and Settled Village Life*. University of Oklahoma Press, Norman.

MacNeish, Richard S., and Mary W. Eubanks. 2000. "Comparative Analysis of the Río Balsas and Tehuacán Models for the Origin of Maize." *Latin American Antiquity* 11(1): 3–20.

MacNeish, Richard S., and Antoinette Nelken-Terner. 1983. "The Preceramic of Mesoamerica." *Journal of Field Archaeology* 10(1): 71–84.

MacNeish, Richard S., Antoinette Nelken-Terner, and Irmgard W. Johnson. 1967. *The Prehistory of the Tehuacán Valley, Volume Two: Non-Ceramic Artifacts*. University of Texas Press, Austin.

MacNeish, Richard S., Frederick A. Peterson, and Kent V. Flannery. 1970. *The Prehistory of the Tehuacán Valley*, Vol. 3: *Ceramics*. University of Texas Press, Austin.

Marceniuk, Alexandre P., and Naércio A. Menezes. 2007. Systematics of the Family Ariidae (Ostariophysi, Siluriformes), with a Redefinition of the Genera. Zootaxa 1416. Magnolia Press, Auckland.

Marcus, Joyce. 1989. "Zapotec Chiefdoms and the Nature of Formative Religions." In *Regional Perspectives on the Olmec*, edited by Robert J. Sharer and David C. Grove, 148–97. Cambridge University Press, Cambridge.

Marcus, Joyce. 1998. *Women's Ritual in Formative Oaxaca: Figurine Making, Divination, Death and the Ancestors*. University of Michigan Press, Ann Arbor.

Marcus, Joyce. 1999. "Men's and Women's Ritual in Formative Oaxaca." In *Social Patterns in Pre-Classic Mesoamerica: A Symposium at Dumbarton Oaks 9 and 10 October 1993*, edited by David C. Grove and Rosemary A. Joyce, 67–94. Dumbarton Oaks, Washington, DC.

Marcus, Joyce. 2008. "The Archaeological Evidence for Social Evolution." *Annual Review of Anthropology* 37 (October): 251–66.

Marcus, Joyce. 2009. "Rethinking Figurines." In *Mesoamerican Figurines: Small-Scale Indices of Large-Scale Social Phenomena*, edited by Christina T. Halperin, Katherine A. Faust, Rhonda Taube, and Aurore Giguet, 25–50. University Press of Florida, Gainesville.

Marcus, Joyce, and Kent V. Flannery. 1996. *Zapotec Civilization: How Urban Society Evolved in Mexico's Oaxaca Valley*. Thames and Hudson, London.

Márquez Morfín, Lourdes, Robert McCaa, Rebecca Storey, and Andres Del Angel. 2002. "Health and Nutrition in Pre-Hispanic Mesoamerica." In *The Backbone of History: Health and Nutrition in the Western Hemisphere*, edited by Richard H. Steckel and Jerome C. Rose, 307–38. Cambridge University Press, New York.

Marroig, Gabriel, and James A. Cheverud. 2005. "Size as a Line of Least Evolutionary Resistance: Diet and Adaptive Morphological Radiation in New World Monkeys." *Evolution* 59(5): 1128–42.

Marshall, Yvonne. 2006. "Introduction: Adopting a Sedentary Lifeway." *World Archaeology* 38(2): 153–63.

Martin, Debra L., Alan H. Goodman, and George J. Armelagos. 1985. "Skeletal Pathologies as Indicators of Quality and Quantity of Diet." In *The Analysis of Prehistoric Diets*, edited by Robert I. Gilbert and James H. Mielke, 229–74. Academic Press, Orlando.

Martínez López, Cira, Robert Markens, Marcus Winter, and Michael D. Lind. 2000. *Cerámica de la fase Xoo (Epoca Monte Albán IIIB–IV) del Valle de Oaxaca*. Instituto Nacional de Antropología e Historia, Oaxaca.

Mayes, Arion T. 2016. "Spiro Mounds, Oklahoma: Dental Evidence for Subsistence Strategies." *International Journal of Osteoarchaeology* 26(5): 749–58.

Mayes, Arion T., and Arthur A. Joyce. 2017. "The Bioarchaeology of the Cerro de la Cruz Cemetery, Oaxaca, Mexico." *Journal of Archaeological Science: Reports* 13: 712–18.

Mays, Simon A. 1995. "The Relationship between Harris Lines and Other Aspects of Skeletal Development in Adults and Juveniles." *Journal of Archaeological Science* 22(4): 511–20.

McClung de Tapia, Emily, and Judith Zurita-Noguera. 2000. "Las primeras sociedades sedentarias." In *El México antiguo, sus áreas culturales, los orígenes y el Horizonte Preclásico*, edited by Linda Manzanilla and Leonardo López Luján, 1: 255–95. 2nd ed. Instituto Nacional de Antropología e Historia, Mexico City.

McDonald, Mary M. 1991. "Technological Organization and Sedentism in the Epipalaeolithic of Dakhleh Oasis, Egypt." *African Archaeological Review* 9(1): 81–109.

McGovern, Patrick E. 2009. *Uncorking the Past: The Quest for Wine, Beer, and Other Alcoholic Beverages*. University of California Press, Berkeley.

McGuire, Randall H. 1983. "Breaking down Cultural Complexity: Inequality and Heterogeneity." *Advances in Archaeological Method and Theory* 6: 91–142.

McIntosh, Susan K., ed. 1999. *Beyond Chiefdoms: Pathways to Complexity in Africa*. Cambridge University Press, Cambridge.

Meillassoux, Claude. 1972. "From Reproduction to Production: A Marxist Approach to Economic Anthropology." *Economy and Society* 1(1): 93–105.

Mitchell, Mark D. 2008. "Making Places: Burned Rock Middens, Feasting, and Changing Land Use in the Upper Arkansas River Basin." In *Archaeological Landscapes on the High Plains*, edited by Laura L. Scheiber and Bonnie J. Clark, 41–70. University Press of Colorado, Boulder.

Morell-Hart, Shanti, Rosemary A. Joyce, and John S. Henderson. 2014. "Multi-Proxy Analysis of Plant Use at Formative Period Los Naranjos, Honduras." *Latin American Antiquity* 25(1): 65–81.

Morgan, Louis Henry. 1877. *Ancient Society; or, Researches in the Lines of Human Progress from Savagery, through Barbarism to Civilization*. H. Holt and Company, New York.

Morris, Ian. 1991. "The Archaeology of Ancestors: The Saxe/Goldstein Hypothesis Revisited." *Cambridge Archaeological Journal* 1(2): 147–69.

Mountjoy, Joseph B. 1994. "Capacha: Una cultura enigmática del Occidente de México." *Arqueología Mexicana* 2(9): 39–42.

Mountjoy, Joseph B. 1998. "The Evolution of Complex Societies in West Mexico: A Comparative Perspective." In *Ancient West Mexico: Art and Archaeology of the Unknown Past*, edited by Richard F. Townsend, 250–65. Art Institute of Chicago, Chicago.

Mountjoy, Joseph B. 2006. "Excavation of Two Middle Formative Cemeteries in the Mascota Valley of Jalisco, México." Electronic document, Foundation for the Advancement of Mesoamerican Studies, Inc. (FAMSI), http://www.famsi.org/reports/03009/, accessed August 30, 2015.

Mountjoy, Joseph B. 2012. *El Pantano y otros sitios del Formativo Medio en el Valle de Mascota, Jalisco*. Secretaría de Cultura, Gobierno de Jalisco, Guadalajara.

Mueller, Raymond G. 1991. "Appendix 2: Technical Report on the Geomorphological Research of the Río Verde Formative Project." In *Formative Period Occupation in the Lower Río Verde Valley, Oaxaca, Mexico: Interregional Interaction and Social Change*, edited by Arthur A. Joyce, 788–839. Unpublished PhD dissertation. Rutgers, State University of New Jersey, New Brunswick.

Mueller, Raymond G., Arthur A. Joyce, Aleksander Borejsza, and Michelle Goman. 2013. "Anthropogenic Landscape Change and the Human Ecology of the Lower Río Verde Valley." In *Polity and Ecology in Formative Period Coastal Oaxaca*, edited by Arthur A. Joyce, 65–96. University Press of Colorado, Boulder.

Mueller, Raymond G., Arthur A. Joyce, Jason Cesta, Matthew Severs, and Michelle Goman. 2014. "Further Research on Floodplain Evolution Associated with the Site of Río Viejo, Lower Río Verde, Oaxaca, Mexico." Paper presented at the 79th Annual Meeting of the Society for American Archaeology, Austin, TX.

Nader, Laura. 1969. "The Zapotec of Oaxaca." In *Handbook of Middle American Indians (Ethnology, I)*, edited by Evon Z. Vogt, 7: 329–59. University of Texas Press, Austin.

Nicholas, Linda M. 1989. "Land Use in Prehispanic Oaxaca." In *Monte Albán's Hinterland, Part II: The Prehispanic Settlement Patterns in Tlacolula, Etla, and Ocotlán, the Valley of Oaxaca, Mexico*, edited by Stephen A. Kowalewski, Gary M. Feinman, Laura M. Finsten, Richard E. Blanton, and Linda M. Nicholas, 449–505. Memoir 23, Museum of Anthropology, University of Michigan, Ann Arbor.

Niederberger, Christine. 1976. *Zohapilco: Cinco milenios de ocupación humana en un sitio lacustre de la Cuenca de México*. Colección Científica 30. INAH, Mexico City.

Niederberger, Christine. 1979. "Early Sedentary Economy in the Basin of Mexico." *Science* 203(4376): 131–42.

Niederberger, Christine. 1987. *Paleopaysages et archéologie pré-urbaine du Bassin de Mexico (Mexique)*. Vol. II. Etudes Mesoamericaines. Centre d'Etudes Mexicaines et Centramericaines, Mexico City.

Niederberger, Christine. 2000. "Ranked Societies, Iconographic Complexity, and Economic Wealth in the Basin of Mexico toward 1200 BC." In *Olmec Art and Archaeology in Mesoamerica*, edited by John E. Clark and Mary E. Pye, 169–92. National Gallery of Art, Washington, DC.

Oliveros Morales, José Arturo. 1974. "Nuevas Exploraciones en El Opeño, Michoacán." In *The Archaeology of West Mexico*, edited by Betty Bell, 182–201. Sociedad de Estudios Avanzados del Occidente de México, A.C., Ajijic, Jalisco, Mexico.

Oliveros Morales, José Arturo, and Magdalena de los Ríos. 1993. "La cronología en El Opeño, Michoacan: Nuevos fechamientos por radiocarbono." *Arqueología* 9(10): 45–48.

Olsen, Bjørnar. 2010. *In Defense of Things: Archaeology and the Ontology of Objects*. AltaMira, Lanham, MD.

Olsen, Bjørnar, Michael Shanks, Timothy Webmoor, and Christopher Witmore. 2012. *Archaeology: The Discipline of Things*. University of California Press, Berkeley.

Olsen, Kenneth M., and Barbara A. Schaal. 1999. "Evidence on the Origin of Cassava: Pylogeography of Manihot esculenta." *Proceedings of the National Academy of Sciences* 96(10): 5586–91.

Ortiz-Martínez, Teresita, and Victor Rico-Gray. 2007. "Spider Monkeys (Ateles Geoffroyi Vellerosus) in a Tropical Deciduous Forest in Tehuantepec, Oaxaca, Mexico." *Southwestern Naturalist* 52(3): 393–99.

Parsons, Elsie Clews. 1936. *Mitla, Town of the Souls, and Other Zapoteco-Speaking Pueblos of Oaxaca, Mexico*. University of Chicago Press, Chicago.

Parsons, Jeffrey. 1971. *Prehistoric Settlement in the Texcoco Region*. Memoirs No. 3 Museum of Anthropology, University of Michigan, Ann Arbor.

Parsons, Jeffrey. 1974. "The Development of a Prehistoric Complex Society: A Regional Perspective from the Valley of Mexico." *Journal of Field Archaeology* 1(1/2): 81–108.

Pauketat, Timothy R., and Susan M. Alt. 2003. "Mounds, Memory, and Contested Mississippian History." In *Archaeologies of Memory*, edited by Ruth M. Van Dyke and Susan E. Alcock. Blackwell, Malden, MA.

Pauketat, Timothy R., and Thomas E. Emerson. 2007. "Alternative Civilizations: Heterarchies, Corporate Polities, and Orthodoxies." In Alternativity in Cultural History: Heterarchy and Homoarchy as Evolutionary Trajectories, Third International Conference "Hierarchy and Power in the History of Civilizations" June 18–21 2004, Moscow. Selected Papers, edited by Dmitri M. Bondarenko and Alexandre A. Nemirovskiy, 107–17. Center for Civilizational and Regional Studies of the RAS, Moscow.

Paynter, Robert. 1989. "The Archaeology of Equality and Inequality." *Annual Review of Anthropology* 18(1): 369–99.

Paynter, Robert, and Randall H. McGuire. 1991. "The Archaeology of Inequality: Material Culture, Domination, and Resistance." In *The Archaeology of Inequality*, edited by Randall H. McGuire and Robert Paynter, 1–27. Blackwell, Oxford.

Pearson, Mike Parker. 1999. *The Archaeology of Death and Burial*. Texas A&M University Press, College Station.

Pearson, Richard. 2006. "Jomon Hot Spot: Increasing Sedentism in South-Western Japan in the Incipient Jomon (14,000–9250 cal. BC) and Earliest Jomon (9250–5300 cal. BC) Periods." *World Archaeology* 38(2): 239–58.

Pérez Hernández, Silvia, and Guy David Hepp. 2015. "Appendix 3: Results of Faunal Analysis." In *La Consentida: Initial Early Formative Period Settlement, Subsistence, and Social Organization on the Pacific Coast of Oaxaca, Mexico*, 486–502. Unpublished PhD dissertation. University of Colorado, Boulder.

Peters, Joris, and Klaus Schmidt. 2004. "Animals in the Symbolic World of Pre-Pottery Neolithic Göbekli Tepe, South-eastern Turkey: A Preliminary Assessment." *Anthropozoologica* 39(1): 179–218.

Piña Chan, Román. 1958. *Tlatilco*. Vol. 2. Serie Investigaciones. Instituto Nacional de Antropología e Historia. Mexico City.

Piperno, Dolores R. 2011. "The Origins of Plant Cultivation and Domestication in the New World Tropics: Patterns, Process, and New Developments." *Current Anthropology* 52(S4): S453–70.

Piperno, Dolores R., Anthony J. Ranere, Irene Holst, José Iriarte, and Ruth Dickau. 2009. "Starch Grain and Phytolith Evidence for Early Ninth Millennium B.P. Maize from the Central Balsas River Valley, Mexico." *PNAS* 106(13): 5019–24.

Pires-Ferreira, Jane Wheeler. 1978. "Obsidian Exchange Networks: Inferences and Speculations on the Development of Social Organization in Formative Mesoamerica." In *Cultural Continuity in Mesoamerica*, edited by David L. Browman, 49–78. Mouton Publishers, The Hague.

Pires-Ferreira, Jane Wheeler. 2009. "Obsidian Exchange in Formative Mesoamerica." In *The Early Mesoamerican Village: Updated Edition*, edited by Kent V. Flannery, 292–306. Left Coast Press, Walnut Creek, CA.

Pohl, Mary E. D., Kevin O. Pope, John G. Jones, John S. Jacob, Dolores R. Piperno, Susan D. DeFrance, David L. Lentz, John A. Gifford, Marie Elaine Danforth, and J. Kathryn Josserand. 1996. "Early Agriculture in the Maya Lowlands." *Latin American Antiquity* 7(4): 355–72.

Pool, Christopher A. 2007. *Olmec Archaeology and Early Mesoamerica*. Cambridge University Press, New York.

Pool, Christopher A. 2013. "Coastal Oaxaca and Formative Developments in Mesoamerica." In *Polity and Ecology in Formative Period Coastal Oaxaca*, edited by Arthur A. Joyce, 301–28. University Press of Colorado, Boulder.

Pool, Christopher A., Ponciano Ortiz Ceballos, María del Carmen Rodríguez Martínez, and Michael L. Loughlin. 2010. "The Early Horizon at Tres Zapotes: Implications for Olmec Interaction." *Ancient Mesoamerica* 21(1): 95–105.

Pope, Kevin O., Mary E. D. Pohl, John G. Jones, David L. Lentz, Christopher von Nagy, Francisco J. Vega, and Irvy R. Quitmyer. 2001. "Origin and Environmental Setting of Ancient Agriculture in the Lowlands of Mesoamerica." *Science* 292(5520): 1370–73.

Powis, Terry G., Ann Cyphers, Nilesh W. Gaikwad, Louis Grivetti, and Kong Cheong. 2011. "Cacao Use and the San Lorenzo Olmec." *Proceedings of the National Academy of Sciences* 108(21): 8595–8600.

Powis, Terry G., W. Jeffrey Hurst, María del Carmen Rodríguez Martínez, Ponciano Ortiz C., Michael Blake, David Cheetham, Michael D. Coe, and John G. Hodgson. 2007. "Oldest Chocolate in the New World." *Antiquity* 314(December): 302–5.

Powis, Terry G., W. Jeffrey Hurst, María del Carmen Rodríguez Martínez, Ponciano Ortiz C., Michael Blake, David Cheetham, Michael D. Coe, and John G. Hodgson. 2008. "The Origins of Cacao Use in Mesoamerica." *Mexicon* 30(2): 35–38.

Price, T. Douglas, James H. Burton, Paul D. Fullagar, Lori E. Wright, Jane E. Buikstra, and Vera Tiesler. 2008. "Strontium Isotopes and the Study of Human Mobility in Ancient Mesoamerica." *Latin American Antiquity* 19(2): 167–80.

Price, T. Douglas, and Gary M. Feinman, eds. 1995. *Foundations of Social Inequality*. Fundamental Issues in Archaeology. Plenum Press, New York.

Rakita, Gordon F., Jane E. Buikstra, Lane A. Beck, and Sloan R. Williams, eds. 2005. *Interacting with the Dead: Perspectives on Mortuary Archaeology for the New Millennium*. University Press of Florida, Gainesville.

Ramírez Urrea, Susana. 1993. "Hacienda Blanca: Una aldea a traves del tiempo, en el Valle de Etla, Oaxaca." Unpublished licenciatura thesis, Universidad Autónoma de Guadalajara, Guadalajara.

Ranere, Anthony J., Dolores R. Piperno, Irene Holst, Ruth Dickau, and José Iriarte. 2009. "The Cultural and Chronological Context of Early Holocene Maize and Squash Domestication in the Central Balsas River Valley, Mexico." *PNAS* 106(13): 5014–18.

Rathje, William. 1971. "The Origin and Development of Lowland Classic Maya Civilization." *American Antiquity* 36(3): 275–85.

Reilly, F. Kent III. 1994. "Enclosed Ritual Spaces and the Watery Underworld in Formative Period Architecture: New Observations on the Function of La Venta Complex A." In *Seventh Palenque Round Table, 1989*, edited by Merle G. Robertson and Virginia M. Fields, 125–35. Pre-Columbian Art Research Institute, San Francisco.

Reilly, F. Kent III. 1995. "Art, Ritual, and Rulership in the Olmec World." In *The Olmec World: Ritual and Rulership*, edited by Jill Guthrie and Elizabeth P. Benson, 27–45. Art Museum, Princeton University, Princeton.

Reimer, Paula J., Edouard Bard, Alex Bayliss, J. Warren Beck, Paul G. Blackwell, Christopher Bronk Ramsey, Caitlin E. Buck, Hai Cheng, R. Lawrence Edwards, Michael Friedrich, Pieter M. Grootes, Thomas P. Guilderson, Haflidi Haflidason, Irka Hajdas, Christine Hatté, Timothy J. Heaton, Dirk L. Hoffmann, Alan G. Hogg, Konrad A. Hughen, K. Felix Kaiser, Bernd Kromer, Sturt W. Manning, Mu Niu, Ron W. Reimer, David A. Richards, E. Marian Scott, John R. Southon, Richard A. Staff, Christian S. M. Turney, and Johannes van der Plicht. 2013. "IntCal13 and Marine13 Radiocarbon Age Calibration Curves 0–50,000 Years cal BP." *Radiocarbon* 55(4): 1869–87.

Reyes González, Liliana Carla, and Marcus Winter. 2010. "The Early Formative Period in the Southern Isthmus: Excavations at Barrio Tepalcate, Ixtepec, Oaxaca." *Ancient Mesoamerica* 21(1): 151–63.

Rice, Prudence M. 1987. *Pottery Analysis: A Sourcebook*. Edinburgh University Press, Edinburgh.

Rice, Prudence M. 1999. "On the Origins of Pottery." *Journal of Archaeological Method and Theory* 6(1): 1–54.

Roberts, Charlotte, and Keith Manchester. 1995. *The Archaeology of Disease*. 2nd ed. Sutton, Stroud, England.

Rodríguez Martínez, María del Carmen, and Ponciano Ortiz C. 1997. "Olmec Ritual and Sacred Geography at Manatí." In *Olmec to Aztec: Settlement Patterns in the Ancient Gulf Lowlands*, edited by Barbara L. Stark and Philip J. Arnold III, 68–95. University of Arizona Press, Tucson.

Rojas, Alfonso Villa. 1947. "Kinship and Nagualism in a Tzeltal Community, Southeastern Mexico." *American Anthropologist* 49(4): 578–87.

Rosenswig, Robert M. 2006. "Sedentism and Food Production in Early Complex Societies of the Soconusco, Mexico." *World Archaeology* 38(2): 330–55.

Rosenswig, Robert M. 2007. "Beyond Identifying Elites: Feasting as a Means to Understand Early Middle Formative Society on the Pacific Coast of Mexico." *Journal of Anthropological Research* 26(1): 1–27.

Rosenswig, Robert M. 2010. *The Beginnings of Mesoamerican Civilization*. Cambridge University Press, Cambridge.

Rosenswig, Robert M. 2011. "An Early Mesoamerican Archipelago of Complexity." In *Early Mesoamerican Social Transformations: Archaic and Formative Lifeways and the Soconusco Region*, edited by Richard G. Lesure, 242–71. University of California Press, Berkeley.

Rosenswig, Robert M. 2015. "A Mosaic of Adaptation: The Archaeological Record for Mesoamerica's Archaic Period." *Journal of Archaeological Research* 23(2): 115–62.

Rosenswig, Robert M., and Marilyn A. Masson. 2002. "Transformation of the Terminal Classic to Postclassic Architectural Landscape at Caye Coco, Belize." *Ancient Mesoamerica* 13(2): 213–35.

Roskams, Steve. 2001. *Excavation*. Cambridge University Press, Cambridge.

De Sahagún, Bernardino. 1974. *Florentine Codex: General History of the Things of New Spain*. School of American Research and the University of Utah, Santa Fe and Salt Lake City.

Salmon, Merrilee H. 1978. "What Can Systems Theory Do for Archaeology?" *American Antiquity* 43(2): 174–83.

Salmon, Merrilee H. 1980. "Reply to Lowe and Barth." *American Antiquity* 45(3): 575–79.

Sánchez Nava, Pedro Francisco, and Ivan A. Šprajc. 2012. *Propiedades astronómicas de la arquitectura y el urbanismo en Mesoamérica: Informe de la temporada 2012*. Report submitted to the Instituto Nacional de Antropología e Historia, Mexico City.

Sánchez Santiago, Gonzalo. 2014. "El Complejo Serpiente-búho en los Silbatos Zapotecos del Clásico." In *Zaachila y au historia prehispánica: Memoria del quincuagésimo aniversario del descubrimiento de las Tumbas 1 y 2*, edited by Ismael G. Vicente Cruz and Gonzalo Sánchez Santiago, 245–60. Secretaria de las Culturas y Artes de Oaxaca y el Ayuntamiento de Villa de Zaachila, Oaxaca.

Sanders, William T. 1956. "The Central Mexican Symbiotic Region: A Study in Prehistoric Settlement Patterns." In *Prehistoric Settlement Patterns in the New World*, edited by Gordon R. Willey, 115–27. Viking Fund Publications in Anthropology 23, New York.

Sanders, William T. 1965. *The Cultural Ecology of the Tehuacán Valley, A Preliminary Report of the Teotihuacan Valley Project*. Pennsylvania State University, University Park.

Sanders, William T. 1968. "Hydraulic Agriculture, Economic Symbiosis and the Evolution of States in Central Mexico." In *Anthropological Archaeology in the Americas*, edited by Betty J. Meggars, 88–107. Anthropological Society of Washington, Washington, DC.

Sanders, William T., and Deborah L. Nichols. 1988. "Ecological Theory and Cultural Evolution in the Valley of Oaxaca [and Comments and Reply]." *Current Anthropology* 29(1): 33–80.

Sanders, William T., Jeffrey Parsons, and Robert S. Santley. 1979. *The Basin of Mexico: Ecological Processes in the Evolution of a Civilization*. Academic Press, New York.

Sanders, William T., and Barbara J Price. 1968. *Mesoamerica: The Evolution of Civilization*. Random House, New York.

Sanders, William T., and David T. Webster. 1978. "Unilinealism, Multilinealism, and the Evolution of Complex Societies." In *Social Archaeology: Beyond Subsistence and Dating*, edited by Charles L. Redman, Mary J. Berman, Edward V. Curtin, William T. Longhorn Jr., Nina M. Versaggi, and Jeffrey C. Wanser, 249–302. Academic Press, New York.

Santley, Robert S., and Philip J. III Arnold. 1996. "Prehispanic Settlement Patterns in the Tuxtla Mountains, Southern Veracruz, Mexico." *Journal of Field Archaeology* 23(2): 225–49.

Santley, Robert S., and Eric K Rose. 1979. "Diet, Nutrition and Population Dynamics in the Basin of Mexico." *World Archaeology* 11(2): 185–207.

Saunders, Nicholas J. 2001. "A Dark Light: Reflections on Obsidian in Mesoamerica." *World Archaeology* 33(2): 220–36.

Saxe, Arthur A. 1971. "Social Dimensions of Mortuary Practices in a Mesolithic Population from Wadi Halfa, Sudan." *Memoirs of the Society for American Archaeology* 25: 39–56.

Schele, Linda, and David A Freidel. 1990. "Sacred Space, Holy Time, and the Maya World." In *A Forest of Kings: The Untold Story of the Ancient Maya*, edited by Linda Schele and David A. Freidel, 64–77. William Morrow, New York.

Schmidt Schoenberg, Paul. 2006. "La Epoca Prehispánica en Guerrero." *Arqueología Mexicana* 14(82): 28–37.

Schwarcz, Henry P., and Margaret J. Schoeninger. 2011. "Stable Isotopes of Carbon and Nitrogen as Tracers for Paleo-Diet Reconstruction." In *Handbook of Environmental Isotope Geochemistry*, edited by Mark Baskaran, 725–42. Springer-Verlag, Berlin.

Sealy, Judith. 2006. "Diet, Mobility, and Settlement Pattern among Holocene Hunter-Gatherers in Southernmost Africa." *Current Anthropology* 47(4): 569–95.

Seeger, Anthony. 1987. *Why Suyá Sing: A Musical Anthropology of an Amazonian People*. Cambridge University Press, New York.

Seinfeld, Daniel M., Christopher von Nagy, and Mary E. D. Pohl. 2009. "Determining Olmec Maize Use through Bulk Stable Carbon Isotope Analysis." *Journal of Archaeological Science* 36(11): 2560–65.

Sellen, Daniel W. 2001. "Comparison of Infant Feeding Patterns Reported for Nonindustrial Populations with Current Recommendations." *Journal of Nutrition* 131(10): 2707–15.

Sewell, William H. Jr. 1992. "A Theory of Structure: Duality, Agency, and Transformation." *American Journal of Sociology* 98(1): 1–29.

Shafer, Harry J. 1975. "Clay Figurines from the Lower Pecos Region, Texas." *American Antiquity* 40(2:1): 148–58.

Shafer, Harry J., and Anna J. Taylor. 1986. "Mimbres Mogollon Pueblo Dynamics and Ceramic Style Change." *Journal of Field Archaeology* 13(1): 43–68.

Sheets, Payson, David L. Lentz, Dolores R. Piperno, John G. Jones, Christine Dixon, George Maloof, and Angela Hood. 2012. "Ancient Manioc Agriculture South of the Ceren Village, El Salvador." *Latin American Antiquity* 23(3): 259–81.

Sheets, Payson, and Michelle Woodward. 2002. "Cultivating Biodiversity: Milpas, Gardens, and the Classic Period Landscape." In *Before the Volcano Erupted: The Ancient Cerén Village in Central America*, edited by Payson Sheets, 184–91. University of Texas Press, Austin.

Sherratt, Andrew. 1990. "The Genesis of Megaliths: Monumentality, Ethnicity, and Social Complexity in Neolithic North-West Europe." *World Archaeology* 22(2): 147–67.

Skibo, James M., and Gary M. Feinman, eds. 1999. *Pottery and People: A Dynamic Interaction*. Foundations of Archaeological Inquiry. University of Utah Press, Salt Lake City.

Skinner, Mark, and Alan H. Goodman. 1992. "Anthropological Uses of Developmental Defects of Enamel." In *Skeletal Biology of Past Peoples: Research Methods*, edited by Shelley R. Saunders and M. Anne Katzenberg, 153–74. Wiley-Liss, New York.

Smalley, John, and Michael Blake. 2003. "Sweet Beginnings: Stalk Sugar and the Domestication of Maize." *Current Anthropology* 44(5): 675–703.

Smith, Ledyard. 1932. "Two Recent Ceramic Finds at Uaxactun." *Contributions to American Archaeology* 2(5): 1–25. No. 436. Carnegie Institution of Washington.

Smyth, Michael P. 1989. "Domestic Storage Behavior in Mesoamerica: An Ethnoarchaeological Approach." *Archaeological Method and Theory* 1 (January): 89–138.

Sousa, Lisa. 1997. "Women and Crime in Colonial Oaxaca: Evidence of Complementary Gender Roles in Mixtec and Zapotec Societies." In *Indian Women of Early Mexico*, edited by Susan Schroeder, Stephanie Wood, and Robert Haskett. University of Oklahoma Press, Norman.

Sousa, Lisa. 1998. "Women in Native Societies and Cultures of Colonial Mexico." Unpublished PhD dissertation, University of California, Los Angeles.

Spencer, Charles R., and Elsa M. Redmond. 2004. "Primary State Formation in Mesoamerica." *Annual Review of Anthropology* 33 (October): 173–99.

Spores, Ronald. 1984. *The Mixtecs in Ancient and Colonial Times*. University of Oklahoma Press, Norman.

Spores, Ronald. 1993. "Tututepec: A Postclassic-Period Mixtec Conquest State." *Ancient Mesoamerica* 4(1): 167–74.

Spores, Ronald. 1997. "Mixteca Cacicas: Status, Wealth, and the Political Accommodation of Native Elite Women in Early Colonial Oaxaca." In *Indian Women of Early Mexico*, edited by Susan Schroeder, Stephanie Wood, and Robert Haskett, 185–98. University of Oklahoma Press, Norman.

Stark, Barbara L., and Barbara Voorhies, eds. 1978. *Prehistoric Coastal Adaptations: The Economy of Maritime Middle America*. Academic Press, New York.

Stephen, Lynn. 1991. *Zapotec Women*. University of Texas Press, Austin.

Stephen, Lynn. 2002. "Sexualities and Genders in Zapotec Oaxaca." *Latin American Perspectives* 29(2): 41–59.

Stern, Stephen J. 1995. *The Secret History of Gender: Women, Men, and Power in Late Colonial Mexico*. University of North Carolina Press, Chapel Hill.

Steward, Julian H. 1955. *Theory of Culture Change*. University of Illinois Press, Urbana.

Stockett, Miranda K. 2005. "On the Importance of Difference: Re-Envisioning Sex and Gender in Ancient Mesoamerica." *World Archaeology* 37(4): 566–78.

Storey, Rebecca, Lourdes Márquez Morfín, and Vernon Smith. 2002. "Social Disruption and the Maya Civilization of Mesoamerica: A Study of Health and Economy of the Last Thousand Years." In *The Backbone of History: Health and Nutrition in the Western Hemisphere*, edited by Richard H. Steckel and Jerome C. Rose, 283–306. Cambridge University Press, New York.

Stuart-Macadam, Patty L. 1989. "Nutritional Deficiency Disease: A Survey of Scurvy, Rickets, and Iron-Deficiency Anemia." In *Reconstruction of Life from the Skeleton*, edited by Memet Y. Iscan and Kenneth A. R. Kennedy, 201–22. Reconstruction of Life from the Skeleton. Liss, New York.

Symonds, Stacey. 2000. "The Ancient Landscape at San Lorenzo Tenochtitlán, Veracruz, Mexico: Settlement and Nature." In *Olmec Art and Archaeology in Mesoamerica*, edited by John E. Clark and Mary E. Pye, 55–74. National Gallery of Art, Washington, DC.

Symonds, Stacey, Ann Cyphers, and Roberto Lunagómez. 2002. *Asentamiento prehispánico en San Lorenzo Tenochtitlán*. Instituto de Investigaciones Antropolgógicas, Universidad Nacional Autónoma de México, Mexico City.

Taube, Karl A. 1986. "The Teotihuacan Cave of Origin: The Iconography and Architecture of Emergence Mythology in Mesoamerica and the American Southwest." *Anthropology and Aesthetics* 12(1): 51–82.

Taube, Karl A. 1992. "The Iconography of Mirrors at Teotihuacan." In *Art, Ideology, and the City of Teotihuacan*, edited by Janet C. Berlo, 169–204. Dumbarton Oaks, Washington, DC.

Taube, Karl A. 1993. *Aztec and Maya Myths*. University of Texas Press, Austin.

Taube, Karl A. 2000. "Lightning Celts and Corn Fetishes: The Formative Olmec and the Development of Maize Symbolism in Mesoamerica and the American Southwest." In *Olmec Art and Archaeology in Mesoamerica*, edited by John E. Clark and Mary E. Pye, 296–337. National Gallery of Art, Washington, DC.

Taube, Karl A. 2010. "Where Earth and Sky Meet: The Sea in Ancient and Contemporary Maya Cosmology." In *Fiery Pool: The Maya and the Mythic Sea*, edited by Daniel Finamore and Stephen D. Houston, 202–19. Peabody Essex Museum, Salem.

Taylor, Robert B. 1960. "Teotitlan del Valle: A Typical Mesoamerican Community." Unpublished PhD dissertation. University of Oregon, Eugene.

Taylor, William B. 1979. *Drinking, Homicide, and Rebellion in Colonial Mexican Villages*. Stanford University Press, Stanford, CA.

Terraciano, Kevin. 1994. "Nudzahui History: Mixtec Writing and Culture in Colonial Oaxaca." Unpublished PhD dissertation, University of California, Los Angeles.

Terraciano, Kevin. 2000. "The Colonial Mixtec Community." *Hispanic American Historical Review* 80(1): 1–42.

Terraciano, Kevin. 2001. *The Mixtecs of Colonial Oaxaca: Nudzahui History, Sixteenth through Eighteenth Centuries*. Stanford University Press, Stanford, CA.

Thomas, J. 1991. *Rethinking the Neolithic*. Cambridge University Press, Cambridge.

Tilley, Christopher. 1994. *A Phenomenology of Landscape*. Berg, Oxford.

Tilley, Christopher. 2007. "The Neolithic Sensory Revolution." In *Going Over: The Mesolithic-Neolithic Transition in North-West Europe*, edited by Alasdair Whittle and Vicki Cummings, 327–45. Oxford University Press, New York.

Tolstoy, Paul. 1975. "Settlement and Population Trends in the Basin of Mexico (Ixtapaluca and Zacatenco Phases)." *Journal of Field Archaeology* 2(4): 331–49.

Tolstoy, Paul. 1989a. "Chiefdoms, States, and Scales of Analysis." *Review of Archaeology* 10(1): 72–78.

Tolstoy, Paul. 1989b. "Coapexco and Tlatilco: Sites with Olmec Materials in the Basin of Mexico." In Regional Perspectives on the Olmec, edited by Robert J. Sharer and David C. Grove, 85–121. Cambridge University Press, Cambridge.

Torrence, Robin. 1983. "Time Budgeting and Hunter-Gatherer Technology." In *Hunter-Gatherer Economy in Prehistory: A Social History*, edited by Geoff Bailey, 11–22. Cambridge University Press, Cambridge.

Tremain, Cara Grace. 2014. "Pre-Columbian 'Jade': Towards an Improved Identification of Green-Colored Stone in Mesoamerica." *Lithic Technology* 39(3): 137–50.

Turner, Victor W. 1967. *The Forest of Symbols: Aspects of Ndembu Ritual*. Cornell University Press, Ithaca.

Turner, Victor W. 1969. *The Ritual Process: Structure and AntiStructure*. Aldine Publishing, Chicago.

Twiss, Katheryn C. 2008. "Transformations in an Early Agricultural Society: Feasting in the Southern Levantine Pre-Pottery Neolithic." *Journal of Anthropological Archaeology* 27(4): 418–42.

Tykot, Robert H. 2006. "Isotope Analyses and the Histories of Maize." In *Histories of Maize: Multidisciplinary Approaches to the Prehistory, Linguistics, Biogeography, Domestication, and Evolution of Maize*, edited by John E. Staller, Robert H. Tykot, and Bruce F. Benz, 131–42. Academic Press, New York.

Tykot, Robert H., and John E. Staller. 2002. "The Importance of Early Maize Agriculture in Coastal Ecuador: New Data from La Emerenciana." *Current Anthropology* 43(4): 666–77.

Urcid, Javier. 2005. *Zapotec Writing: Knowledge, Power, and Memory in Ancient Oaxaca*. Electronic document, Foundation for the Advancement of Mesoamerican Studies, Inc. (FAMSI), http://www.famsi.org/research/williams/, accessed February 24, 2016.

Urcid, Javier, and Arthur A. Joyce. 2001. "Carved Monuments and Calendrical Names: The Rulers of Río Viejo, Oaxaca." *Ancient Mesoamerica* 12(2): 199–216.

Urquhart, Kyle R. 2010. *Analysis of Charco Phase Ceramics from the Site of Corozo, Oaxaca*. Unpublished undergraduate honors thesis. University of Colorado, Boulder.

VanDerwarker, Amber M. 2006. *Farming, Hunting, and Fishing in the Olmec World*. University of Texas Press, Austin.

Van Dyke, Ruth M., and Susan E. Alcock (editors). 2003. *Archaeologies of Memory*. Blackwell, Malden, MA.

Vega-Centeno Sara-Lafosse, Rafael. 2007. "Construction, Labor Organization, and Feasting during the Late Archaic Period in the Central Andes." *Journal of Anthropological Archaeology* 26(2): 150–71.

Voorhies, Barbara. 1976. *The Chantuto People: An Archaic Period Society of the Chiapas Littoral, Mexico*. Brigham Young University, Provo, UT.

Voorhies, Barbara. 1989. "Settlement Patterns in the Western Soconusco: Methods of Site Recovery and Dating Results." In *New Frontiers in the Archaeology of the Pacific Coast of Mesoamerica*, edited by Frederick J. Bove and Lynette Heller, 103–24. Arizona State University, Tempe.

Voorhies, Barbara. 2004. *Coastal Collectors in the Holocene: The Chantuto People of Southwest Mexico*. University Press of Florida, Gainesville.

Voorhies, Barbara, and Douglas J. Kennett. 2011. "A Gender-Based Model for Changes in Subsistence during the Terminal Late Archaic Period on the Coast of Chiapas, Mexico." In *Early Mesoamerican Social Transformations: Archaic and Formative Lifeways in the Soconusco Region*, edited by Richard G. Lesure, 27–46. University of California Press, Berkeley.

Waters, Michael R., Thomas W. Jr. Stafford, Brian Kooyman, and L. V. Hills. 2015. "Late Pleistocene Horse and Camel Hunting at the Southern Margin of the Ice-free Corridor." *PNAS* 112: 4263–67.

Webster, David L. 2011. "Backward Bottlenecks." *Current Anthropology* 52(1): 77–104.

Webster, David L., David Rue, and Alfred Traverse. 2005. "Early Zea Cultivation in Honduras: Implications for the Iltis Hypothesis." *Economic Botany* 59(2): 101–11.

Weisdorf, Jacob L. 2005. "From Foraging to Farming: Explaining the Neolithic Revolution." *Journal of Economic Surveys* 19(4): 561–86.

Whalen, Michael E. 1981. "Excavations at Santo Domingo Tomaltepec: Evolution of a Formative Community in the Valley of Oaxaca, Mexico." In *Prehistory and Human Ecology of the Valley of Oaxaca*, edited by Kent V. Flannery. 12th–13th ed. University of Michigan Press, Ann Arbor.

Whalen, Michael E. 1983. "Reconstructing Early Formative Village Organization in Oaxaca, Mexico." *American Antiquity* 48(1): 17–43.

Whalen, Michael E. 2009. "Zoning within an Early Formative Community in the Valley of Oaxaca." In *The Early Mesoamerican Village: Updated Edition*, edited by Kent V. Flannery, 75–79. Left Coast Press, Walnut Creek, CA.

Whittle, Alasdair. 2003. *The Archaeology of People: Dimensions of Neolithic Life*. Routledge, London.

Williams, David Thomas. 2012. "Typological and Geochemical Analysis of Obsidian Artifacts: A Diachronic Study from the Lower Río Verde Valley, Oaxaca, Mexico." Unpublished MA thesis. University of Colorado, Boulder.

Williams, Eduardo. 2007. "Prehispanic West México: A Mesoamerican Culture Area." Electronic document, Foundation for the Advancement of Mesoamerican Studies, Inc. (FAMSI), http://www.famsi.org/research/williams/, accessed August 30, 2015.

Wing, Elizabeth S. 1978. "Use of Dogs for Food: An Adaptation to the Coastal Environment." In *Prehistoric Coastal Adaptations: The Economy and Ecology of Maritime Middle America*, edited by Barbara L. Stark and Barbara Voorhies, 29–41. Academic Press, New York.

Winter, Marcus. 1989. *Excavaciones en La Consentida, 1988*. Report submitted to the Instituto Nacional de Antropología e Historia, Mexico City.

Winter, Marcus. 1992. *Oaxaca: The Archaeological Record*. P.G.O., Oaxaca, Mexico.

Winter, Marcus. 2002. "Monte Albán: Mortuary Practices as Domestic Ritual and Their Relation to Community Religion." In *Domestic Ritual in Ancient Mesoamerica*, edited by Patricia Plunket. Monograph 46. The Cotsen Institute of Archaeology, University of California, Los Angeles.

Winter, Marcus. 2009. "The Archaeological Household Cluster in the Valley of Oaxaca." In *The Early Mesoamerican Village: Updated Edition*, edited by Kent V. Flannery, 25–31. Left Coast Press, Walnut Creek, CA.

Winter, Marcus, Margarita Gaxiola G., and Gilberto Hernández D. 1984. "Archaeology of the Otomanguean Area." In *Essays in Otomanguean Culture History*, edited by J. Katherine Josserand, Marcus Winter, and Nicholas A. Hopkins, 65–108. Vanderbilt University, Nashville.

Winter, Marcus, and Pablo Mateos. 2010. "Artefactos de Lítica Pulida." In *Proyecto Salvamento Arqueológico Carretera Oaxaca-Istmo: Región Istmo. Parte 1*, edited by Marcus Winter, 181–207. Informe final del Proyecto Salvamento Arqueológico Carretera Oaxaca-Istmo 2004–2007: Tramos Km 177–190 y Km 190–210, INAH Oaxaca.

Winter, Marcus, and Gonzalo Sánchez Santiago. 2014a. "Introducción: Dos Oaxacas." In *Panorama Arqueológico: Dos Oaxacas*, edited by Marcus Winter and Gonzalo Sánchez Santiago, 1–30. Centro INAH Oaxaca, Oaxaca, Mexico.

Winter, Marcus, and Gonzalo Sánchez Santiago, eds. 2014b. *Panorama arqueológico: Dos Oaxacas*. Centro INAH Oaxaca, Oaxaca, Mexico.

Wolf, Eric R. 1959. *Sons of the Shaking Earth: The People of Mexico and Guatemala—Their Land, History, and Culture*. University of Chicago Press, Chicago.

Workinger, Andrew G. 2002. "Coastal/Highland Interaction in Prehispanic Oaxaca, Mexico: The Perspective from San Francisco de Arriba." Unpublished PhD Dissertation. Nashville: Vanderbilt University.

Zárate Morán, Roberto. 1995. "El Corozal, un sitio arqueológico en la Costa del Pacífico de Oaxaca." *Cuadernos del Sur: Ciencias Sociales* 3(10): 9–36.

Zedeño, María Nieves. 2008. "Bundled Worlds: The Roles and Interactions of Complex Objects from the North American Plains." *Journal of Archaeological Method and Theory* 15(4): 362–78.

Zeitlin, Robert N. 1978. "Long Distance Exchange and Growth of a Regional Center on the Southern Isthmus of Tehuantepec, Mexico." In *Prehistoric Coastal Adaptations*, edited by Barbara Stark and Barbara Voorhies, 183–210. Academic Press, New York.

Zeitlin, Robert N. 1979. "Prehistoric Long-Distance Exchange on the Southern Isthmus of Tehuantepec, Mexico." Unpublished PhD dissertation. Yale University, New Haven, CT.

Zeitlin, Robert N. 1982. "Toward a More Comprehensive Model of Interregional Commodity Distribution: Political Variables and Prehistoric Obsidian Procurement in Mesoamerica." *American Antiquity* 47(2): 260–75.

# Index

*Page numbers in italics indicate illustrations*

Aaberg, Stephen, 89
Abrams, Elliot, 89
Agriculture, 3–4, 12, 14, 17–18, 19, 25–27, 28–29, 32–33, 34–35, 36, 50–51, 96, 99, 101, 119, 126–127, 202–203; agricultural practices, 19, 20, 26–28, 29, 32–33, 35, 45, 103, 104–105, 132–133, 198, 201; Class I land, 26, 45; horticulture, 25–26, 27, 28, 29, 31, 33, 51, 64
Agricultural products: beans, 3–4, 26, 29, 31, 33; beer, 41, 134; cacao, 30, 31, 40, 41, 105, 179–180; chiles, 32; guayaba, 32; gruel, 134; legumes, 29; maguey, 32; maize, 3, 4, 7, 13, 18, 19, 21, 22, 24, 26, 27, 28–30, 32, 33, 35, 36, 41, 51, 64, 65, 96, 98, 101, 103, 104–105, 119–123, 124, 126–127, 131, 132, 134, 196, 198, 200–201, 254; malanga, 26, 200; manioc, 26, 31–32, 33, 129, 130, 200; palm fruit, 33; squash, 3, 4, 26, 33, 179; tubers, 18, 29, 56, *59*
Aguilar, José "Pepe," x, 59, 61–62, 160, 166
Ajalpan phase, 184; Early Ajalpan, *9*, 173; Late Ajalpan, 176
Altamira, 32
Amazon Basin, 37–38
Anatolian Plateau, 44, 101–102. *See also* Göbekli Tepe
Ancestors. *See* Religion

Archaic period, 3, 17, 18, 21–22, 23, 26, 27, 30–31, 40, 41, 46, 50, 70, 96, 99, 103, 109, 126–127, 130, 156, 190, 199, 204, 276; Early Archaic, 29, 140; Late Archaic, 3, 7, 21, 32, 37, 86, 98, 201; Middle Archaic, 29
Architecture, 12, 19, 24, 77, 78, 79, 88, 89, 90, 196, *222*, *223*; ceremonial/ritual architecture, 42, 44; domestic architecture, 24, 85, *126*, *128*, 201; labor reconstructions, 19–20, 24–25, 71, 80–81, 87–90; mounded earthen architecture, 5, 7, 16, 20, 22, 23–24, 26, 55, 70–71, 77, 80, 84, 86, 88, 89, *90*, *91*, 98, 102, 103, 194, 196, 201, 202, monumental architecture, 19, 22–23, 27, 41, 85, 101, 102; public architecture, 18, 20, 22–23, 24, 36, 43, 44; stone architecture, *224*
Arnold, Phillip J. III, 20, 21, 22, 23, 32–33, 94, 125–126, 127
Aztecs, 37, 154–155

Bajío phase, 19, 22, 86, 179, 277
Ball game, 36, 37, 146, 202; ballcourt, 20, 24, 37, 42, 43, 44; ballplayer, 38, 146–*147*, 148, 170
Barra phase, 5, 20, 22, 27, 30, 41, 76, 134, 172, 173, 174, 184, 190, 191, 193, 195, 251, 254, 255, 269, 277

317

318  INDEX

Bazelmans, Jos, 32, 47
Belize, 28–29, 31–32
Blake, Michael, 30, 36, 38, 39, 40, 41, 42, 43, 49, 50, 134, 169, 172, 199, 269
Blomster, Jeffrey P., x, 143–144, 186, 187
Bodily adornment/modification, 48, 197; beads, 65, *114*, 149, *150–151*, 153, 161, 162, 197; bracelets, 153, 161; clothing (general), 135, 138, 140, *142*, 144, *145*, 149, 197; ear spools/gauging, 74, 141, 142, 144; hairstyles, 74, 138, 141, 148; headdresses, 74, 144, 148, 149, 169; jewelry, 110, 135, 137, 144, 148–149, *150*, 169, 197; masks, 146, 158, 159–160, 197, 199, 262, 270; mirrors, 38, 47, 149, *150*, *152*, 197; necklaces/pendants, 110, 141, 143, 144; scarification, 142–143
Bonsignore, Jay, 89
Bove, Frederick J., 39
Boyd, Brian, 25
Bradley, Richard, 102
Burial practices, 43, 47. *See also* Cemeteries; Religion
Cabeza de Vaca, 179, 237
Caldwell, Joseph R., 26
Cantón Corralito, 177
Capacha phase, *10*, 171, 179, 180–81, *182*, 184, 190, 192, 195
Carr, Christopher, 47
Cemeteries, 5, 7, 25, 59, 65, 85, 161, 165, 194, 201, 202
Ceramics: Barra phase pottery, 20, 22, 27, 30, 32, 41, 75–76, 94, 172–174, 176, 184, 190–191, 192–193, 195, 197, 251–252, 254–255, 269, 277; bottles, 8, 14, 63, 74, 75–76, 127, 130, 134, 161, 167, 172, 173–174, 176–177, 179–184, *185–187*, 192, 196, *216*, 238, 239, *240–241*, 244, *245*, 250, *251*, 253–255, 256, 257, *258–261*, 262, *263–266*, 268, 269, 270, 271, 272, 273, 277; bowls, 8, 14, 63, 74, 75, 76, 167, 172, 173, 174, 176, 179, 192, 193, 195, 238, 239, 240, 241, 242, 243, 245–247, 250, 253, 255, 256–257, 260, 262, 266–267, 268, 269, 270–272; Capacha phase pottery, 171, 179–181, *182*, 183, 184, 192, 195; effigy vessels, 145, *146*, 159, 180, 181, 196–197, 227, *245*, 250; grater bowls, 63, 153, 164, 174, 176, 177–179, 240–241, 245, 247, 253–255, *258–261*, *263–266*, 267–268, 271–272, 273, 277; jars, 5, 8, 14, 63, 64, 65, 76–77, 153, 162, 164, 167, 172, 173–174, 175–176, 179, 181, *182*, *183*, 192, 193, 195, 231,
*235*, 238, 239–*241*, 243–245, 246, 248–250, 252, 253, 255, 256–266, 268, 270–271, *272–273*; malacates, 84; phytomorphic vessels, 172, 179, 251; rocker stamping, 176; tecomates, 14, 20, 21, 30, 63, 70, 94, 103, 172, 173–174, 176, 179, 191, 192, 238, 240, 244–245, 251–252, 255, 256, 270, 278; pox pottery, *11*; Tierras Largas phase pottery, 76, 137, 173–174, 176, 184, 190, 192–193, 247; Tlacuache phase pottery, 14, 70, 81, 88, 171, 172–184, 190–191, 192–193, 195, 197, 202, 227, 237–273; worked sherd discs, 63, 71, 131, *133*, 179, 181, *186*, 195, 238, *247*, 256, 262, 268–269, *270*, 272, 273, 278
Cerén, 31–32
Cerro de la Cruz, 117–118, 161
Chalcatzingo, 177
Chantuto culture, 21, 30; Chantuto B phase, 86
Cheetham, David, *11*, 19, 20, 98
Cherla phase, *11*
Chiapas, 18, 172, 187
Chicharras phase, *10*, 83
Childe, V. Gordon, 203
Children, 39, 47, 87–88, 138, 153, 157, 161, 165, 253, 277; child burial, 153, 157–158, 164–165, 179, 248. *See also* Cemeteries; Identity
Chilo, 47
Chisholm, Brian, 134
Chumash, 131
Clark, John E., x, *11*, 19, 20, 21–22, 29, 30–31, 36–38, 39–40, 41–42, 44, 47, 49, 96, 103, 172, 190, 200–201, 269
Classic period, 15, 31–32, 50, 79, 83–84, *118*, 127, 130, 148, 154, 214–220, *222*; Early Classic, *16*, 60, 83–84, 88, 130, 214, *222*; Late Classic, 7, *16*
Coapecxo, 87
Coatzacoalcos River, 19
Cob Swamp, 28
Coe, Michael D., 26–27, 203
Conchas phase, 177
Cooney, Gabriel, 101–102
Cotorra 1-A phase, *11*
Cotorra 1-B phase, *11*
Coyame phase, *11*
Cruz A phase, *10*, 75, 144, 187, 190, 222
Cruz B phase, *11*, 130, 187
Cuadros phase, *11*
Cuilapan, x, 53, 61
Cummings, Vicki, 79, 101
Cyphers, Ann, 22

# INDEX

DeBoer, Warren, 31
Dental pathologies, 104, 119, 121, 123, 134, 254;
    dental attrition, 119, 121, 123, *124*, 196, 198,
    276; dental caries, 119, 121, 198; linear enamel
    hypoplasia, 48, 119, 179; mandibular abscesses,
    119, *121*, 198; porotic hyperostosis, 48

Ecuador, 184
El Salvador, 31–32
Espiridión phase, 5, *9*, 173, 276
Estero Rabón, 19
Etla, 26
Exchange/importation/trade, 3, 13, 19, 27, 35,
    36, 41, 42–43, 61, 64, 120, 131, 149, 167, 181,
    184–187, 190–192, 193, 195, 201–202, 203
Exchange goods, 27, 36, 191–193, 202; basalt, 96,
    97, 124, 189; chalcedony, 63, 129; chert, 63,
    129, 130–131, *132*, 161–162; figurine imagery,
    195; greenstone, 38, 149, *150–151*, 171, 189, 191,
    192, 195, 202; quartzite, 129, 237–238; salt,
    24, 192; shell, 32, 43, 58, 62, 66, 71, 72, 74,
    75–76, 77, 79, 80, 86, 109, 110, 114–115, *116*,
    117, 131, 149, *150*, 153, 197, *206*, *213*, *218*, 221, *226*,
    *228–230*, 237–238, 256, 257. *See also* Ceramics;
    Obsidian

Feasting, 4, 28, 29, 30, 33, 36, 40–41, 76, 77, 91,
    131, 134, 135, 166–170, 172, 177, 196, 197, 202,
    221–222, 231, 239, 246, 253–254, 255, 256, 257,
    262, 264, 267, 268, 269, 270, 271, 272–273
Fernández, Deepika, 62, 117–118
Figurines, 13, 15, 25, 38, 46, 52, 59, 62, 63, 71, 74,
    75, 77, 79, 81, 83, 130, 135, 136–144, 148–149,
    152, 158–159, 162, 163, 169–170, 171–172, 184,
    190, *191*, 195, 196–197, 199, 201, *219*, 222–223,
    *229*, 277; anthropomorphs, 25, 47, 48–49,
    71, 75, 136–138, *139*–142, 144, 156, 159, 190,
    199, 201; transformationals, 81, *147*, 148; zoo-
    morphs, 156. *See also* Musical instruments
Fishhooks, 130–131
Flannery, Kent V., 26, 27–28, 44, 45, 153, 154,
    202, 203
Ford, James, 184, 192
Formative period: Early Formative, 3, 4, 7, 8–*9*,
    12, 13, 15, 16, 17, 21, 29, 34, 37, 41, 50, 51, 60,
    65, 78, 79–80, 103, 130, 131, 133, 135, 139, 147,
    166, 175, 176, 179, 184, 187, 190, 191–192, 193,
    194, 199, 200–201, 214, 221, 273, 276; Middle
    Formative, 4, 7, 8, 14, 15, *16*, 18, 22, 26, 27, 33,
    34, 42, 43, 60, 62, 83, 116, 117, 118, 129, 143,
    169, 177, 181; Late Formative, *16*, 19, 20, 34, 83,
    103, 116–117, 118, 143
Fried, Morton, 49

Glascock, Michael, 186–187
Gender, 41, 46, 48, 49, 136, 137, 141, 148, 153,
    276–277. *See also* Identity
Göbekli Tepe, 23, 44, 101, 102
Golfo phase, *11*
Goman, Michelle, 110
Green, Dee, 31, 32
Ground Penetrating Radar (GPR), x, 12, 52, 53,
    *54*, 59, 205, 207, 276
Ground stone, 19, 20–22, 32, 52, 62, 63, 64,
    66, 70, 74, 79, 81, 85–86, 92–101, 104, 124,
    126–127, 134, 190, 196, 201, 207, 221, *232*;
    ground stone tools, 13, 21–22, 63, 64, 66, 70,
    74, 85–86, 90, 92–93, 94–95, 97, 99, 100–101,
    103, 104–105, 120, 123–125, 133, 198, 201, 254;
    manos, 13, 21, 22, 64, 70, 81, 90, 95, 96, 97, 98,
    99, 100, *101*, 103, 123, 124–126, 127, 128, 130,
    134, 189, 190, 196, 198, *224*, 254; metates, 13,
    20, 21, 22, 24, 29, 64, 70, 81, 90, 96, 97–98,
    99–100, 101, 103, 119, 123, 124, 125, 127, 128,
    133–134, 190, 196, 198, *224*, 254; mortars, 21,
    22, 64, 103, 190; pestles, 21, 22, 64, *95*, 96, *97*,
    103, *125*, polisher/hammer stones, *95*, 96, 97,
    99–100, 101, 103, 124, *125*
Grove, David, 177
Guatemala, 18, 26, 172, 187, 189
Guernsey, Julia, x, 138, 143
Gulf Coast, 4, *9–11*, 13, 18, 19, 20, 22, 32, 42, 86,
    103, 104, 134, 179, 186, 189, 191, 195, 196

Hayden, Brian, 36, 49, 203
Henderson, John S., 22, 40, 41
Hirth, Kenneth, 83
Holocene epoch, 17, 37
Honduras, 20, 22, 36, 40, 195
Hubbell, Gordon, 153

Iconography, 13, 15, 40, 42, 48–49, 61, 136–137,
    138, 142–144, 145, 156, 169–170, 171–172,
    180, 184, 189, 191, 201, 262. *See also* Figurines;
    Musical instruments
Identity, 13, 37, 41, 47–48, 81, 101–102, *137*, 142,
    144–145, 148, 153, 155; age, 48, 49, 61, 121, *124*,
    136, 148, 153, 162, *168*, 179, 276, 277; children,

153, 157–158, 164–165, 179, *248*; gender, 41, 46, 48, 49, 136, 137–138, 141, 148, 153; sex, 48, 49, 61, 137, 140, 148, *168*

INAH (Instituto Nacional de Antropología e Historia), ix, x, 53, 58, 61, 238, 276

Isthmus of Tehuantepec, 110, 172, 181, 187, 190

Joyce, Arthur A., x, xi, 8, 43, 46, 88–89, 117, 161, *188*, 276

Joyce, Rosemary A., x, 22, 39, 40–41, 47, 139, 268

Kelly, Isabel, 176, 179, 180, 181, 184, 192

Laboratory methods, 12, 52–53, 61, 194; AMS radiocarbon dating, 4–5, 6, 7–8, 15–16, 53, 65, 72–73, 75, 77, 79, 82, 120, 153, 166, *168*, 176, 191, 194, 205, 208, 275; stable isotope analysis, 53, 61, 63–64, 104–105, 121–123, 131, 133, 201; XAD purification, *6*, 53, 65, 79, 120; X-ray fluorescence (XRF), 13, 53, 61, 63, 184–185, 186, 277

Languages/language families: Mixe-Zoquean, 191, 202; Otomanguean, 191, 195, 202

La Joya, 20, 21, 23

La Victoria, 32

Laguna Zope, 110, 131

Lagunita phase, 5, *9*, 276

Lesure, Richard G., 20, 21, 27, 38, 39, 41, 42, 43, 50, 169, 192, 199

Locona phase, *9*, 42, 47, 86, 134, 172, 191, 192, 193

Lock, Gracie, 192

Lowe, Gareth W., 30–31, 32

Love, Michael W., 41–43, 47

Machalilla culture, 184, 192

MacNeish, Richard S. "Scotty," 27

Mangrove, 75, 80, 110, 131, 133

Marcus, Joyce, 26, 44, 141, 153

Maritime economy, 130–131

Maya, 23, 42, 154

Miocene epoch, 153–154

Mitchell, Mark D., 23, 98

Mitla, 138

Mixtecs, 7, 46; Mixteca Alta region, *10–11*, 35, 75, 130, 144, 187, 190

Mokaya, 23, 30

Monte Albán, 35

Morell-Hart, Shanti, 31, 254

Mountjoy, Joseph B., x, 181, *182*

Mueller, Raymond G., 69, 87, 277

Musical instruments, 13, 15, 62, 63, 78, 135, 136, 137, 138, 148, 156, 157, 159–160, 163, 190, 199; musicians, 159, 170, 197; Ocarinas, 78, 137, 153, 156–157, 162, 163, 166, 197, *218;* whistles,: 137, 156, *157*. *See also* Figurines

Neolithic period, 25, 33, 101, 203

Nichols, Deborah L., 35

Nisa phase, 177

Nochixtlán Valley, 186, 187–188

Nomadism, 24, 103

Obsidian, 8, 13, 31, 38, 53, 61, 63, 69, 70, 71, 76, 79, 83, 129, 130, 161, 171, 184, 187, 192, 195, 202, *210–211*, *232*; El Chayal, 41–42, 63, 186–187, 188; Guadalupe Victoria, 63, 186–187, *188*, *189*; Ixtepeque, 63, 186; obsidian, tools, 31; Otumba, 63, 186, 188, *189*; Pachuca, 63, 83, 130, *188*; Paredón, 63, 186, 187, *189*; Pico de Orizaba, *188*, *189*; prismatic blades, 31, 60, 83, 129, 130

Ocós phase, *10*, 30, 32, 47

Ocotillo phase, *10*, 22, 40, 180, 195

Ojochi phase, *9*, 179, 195, 277

Olmec, 38, 42, 146–147, 170, 180; Colossal heads, 38, 146; "Mother Culture" model, 42; "X Complex," 42

Opeño phase, *10*, 179, 180, 181, 184, 195

Panama, 28–29, 31

Parsons, Elsie Clews, 138

Paso de la Amada, 18, 20, 25, 35, 37, 38, 41, 42, 43; Mound 6, 20, 37, 38; Mound 19, 32; Structures 2–6, 42

Pe phase, 177

Pérez Cervantes, José, x

Pérez Hernández, Silvia, xi, 61–62, 113–114, 276

Piedmont zones, 71, 117

Piña Chan, Roman, 180

Pohl, Mary E. D., 28–29, 31, 98

Postclassic period, 7, 15, 19, 34, 84, 154, 184, 214–215; Early Postclassic, *16*, 120; Late Postclassic, *16*

Poverty Point, 20, 101

Powis, Terry G., 179–180

Pox phase, *11*

Puerto Escondido, Honduran archaeological site, 40

Puerto Escondido, Oaxaca, xi
Puerto Hormiga, 192
Puerto Marquez, *9, 11*
Pulltrouser Swamp, 28
Purrón phase, *9*, 173, 175–176, 190, 192
Pye, Mary E., 41

Red-on-Buff horizon, 171, 172–73, 176, 181, 190, 191, 192–193, 195, 202. *See also* Ceramics
Religion: ancestor remembrance, 39, 199; bloodletting, 38, 153–154; calendrical ritual, 155; cardinal directions, 154, 155, 156; cremation, 152; dance, 159, 163, 170, 197, 198; Huitzilopochtli, 155; mortuary/funerary practices, 5, 24, 35, 36, 38, 42, 43, 44, 47–48, 50, 59–60, 63, 79, 128, 135, 138, 142, 149–155, 160–166, 169–170, 192, 201, 214; Maya cosmology, 154–155; nagualism, 147–148; Quetzalcoatl, 155; ritual caching, 62, 77–78, 105, *110*, 113, *154, 155, 156*, 166, *218*; ritual practitioners, 146, 147–148, 153, 155, 158, 197, 199; shamanism, 148, 199; spirit, 138, 146; summer solstice, 165–166; Tezcatlipoca, 155; Underworld, 154–155, 166; Xipe Totec, 154–155
Río Balsas, 28–29
Río Naranjo, 18, 43
Río Verde, 15, 62, 69, 117, 133, 238; Río Verde Formative Project, 7; Río Verde Region/Valley, 7, *12*, 14, *16*, 24, 34, *55*, 56, 62–63, 65, 66, 67, 70, 71, 80, 84, 86, 116, 117, 118, 120, 130, 143, 144, 148, 166, 169, 171, 237

Salcedo Romero, Tamara, 41–42
Salmon, Merrilee H., 45
Sanders, William T., 27–28, 35
San José Mogote, 19–20, 25, 26, 35, 44, 46, 188
San José phase, *10*, 43–44, 46–47, 137, 153, 156, 177
San José del Progreso, x, 61
San Lorenzo, *10, 11*, 19, 33, 42, 83; San Lorenzo A phase, *11*, 19
San Martín Jilotepeque, 41–42
Sandberg, Paul A., xi, 61, 119
Santo Domingo Tomaltepec, 43
Santley, Robert S., 22
Sculpture, 4 2; Potbelly sculptures, 142–143
Sedentism, 3–4, 13, 14, 17–18, 19–20, 21–25, 26–27, 28, 29, 32, 33, 40, 50–51, 60, 85, 92–93, 97, *100*, 101, 102, *103*, 165, 196, 198, 200, 201, 202

Sheets, Payson, x
Skeletal pathologies, 121, 160, 201
Smyth, Michael P., 23
Social organization, 4, 8, 13, 14, 15, 17, 28, 33, 34, 36, 43, 47, 48–49, 50, 51, 60–61, 88, 101, 135, 149, 150, 153, 157–158, 160, 167, 168, 169, 170, 171, 194, 198, 201–202. *See also* Theory
Soconusco, 4, 5, *9–11*, 13, 18, 20, 21, 22, 24, 27, 28, 29–30, 32, 33, 35, 36, 38, 40, 42, 43, 44, 47, 51, 76, 86, 103, 104, 109, 131, 134, 143, 172, 177, 180, 187, 191, 192, 193, 195, 196, 200, 202; Mazatán, 4, 18, 20, 27, 28, 29, 30, 35, 36, 37, 42, 43, 44, 47
Steward, Julian, 35
Symonds, Stacey, 19

Tajumulco, 41–42
Taube, Karl A., 154
Tehuacán, Region/Valley: *9*, 18, 27, 173, 175–76, 184, 190, 192, 195, 247
Textiles, 150, 153; basketry, 153, 179; mat motif, 153; woven basketry/mat, 153, 179
Theory: agency, 35; aggrandizer model, 35, 36, 38–40, 41, 43, 49; communal domesticity, 46; communitas, 37, 199; core-periphery model, 42; craft specialization, 36, 41, 169, 203; ecological systems theory, 28, 46; economic specialization, 48, 267; general systems theory, 45; heterarchy, 48–49, 197; hierarchy, 7, 15, 36, 42, 43, 44, 48, 49, 148, 169, 170, 197; interaction sphere model, 26, 42, 171, 176, 180, 192, 195; monuments/monumentality, 42; social complexity, 3, 4, *12*, 14, 18, 25, 26–27, 28, 29, 30, 33, 34–36, 37, 38–39, 41, 42–45, 47, 48, 49, 50–51, 59, 135, 146, 169, 170, 172, 200, 202, 203, 270
Tierras Largas phase, 5, *9, 11*, 13, 19–20, 43, 44, 46, 76, 131, 137, 156, 173–174, 176, 179, 184, 190, 192, 193, 198, 247, 268, 276, 277
Tlacuache phase, *11*, 14, 15, 70, 81, 88, 172, 176, 180, 181, 193, 195, 202, 227, 237, 238, 239, 243, 244, 245, 247, 252, 253, 255, 256, 260, 266, 267, 268
Tlatilco, 175, 179, 180, 195, 247
Transegalitarian societies, 36, 41, 49, 167, 197, 202. *See also* Theory
Tulipan phase, *10*, 20, 23, 94
Tututepec, x, 7
Tuxtla Mountains, 20, 22

Valdivia, 184, 192
VanDerwarker, Amber M., 33
Vivero, 47

Walker, Jeff, 129
Warí, 47
Whittle, Alasdair, 101
Wild resources: *ariopsis guatemalensis* (catfish), 105, *107–112*, 116, 117, 118, 133; camote de agua root, 56, *59*, 220; cichlids, 117; crocodiles, 30, 105, 111, *115*, 116, 118, 162, 254; deer, 30, 32, 70, 105, 106, *107–109*, *111*, *113*, 117, 122; fish, 32, 33, 76, 105, 106, 110, *112*, 114, 115, 116, 117, 118, 122, 133, *218*, 272, 276; gar, 30, *108*; *Heloderma horridum*, 110, 113, 154, *155*, 197, *218*; iguanas, *107–108*, *112*, 117; jacks, 117; mojarra, 30; mollusks, 116–117; Mytilid mussels (*tichinda*), 80, 110; Osteichthyes, 105, *107–111*, 114; shellfish, 24, 33, 76, 106, 110, 117, 118, 192, 272; snakes, 30; teosinte, 96; turtles, 30, *107–110*, 153, 154, *218*; ungulates, 33; waterfowl, 33
Wing, Elizabeth S., 33, 116
Woodward, Michelle, 32

Xoo phase, 46, 177–178

Yucatán, 23

Zaachila, 138
Zaragoza, 187–188
Zapotecs, 46, 138, 277
Zeitlin, Robert N., 131, 187
Zohapilco, 174, 176, 179, 180, 189–190, 195, 247
Zurita-Noguera, Judith, 22